Chromosomes: Eukaryotic, Prokaryotic, and Viral

Volume I

Repetitive Human DNA: The Shape of Things to Come. Sequence-Specific DNA-Binding Proteins Involved in Gene Transcription. Physical and Topological Properties of Closed Circular DNA. Structure of the 300A Chromatin Filament. Three-Dimensional Computer Reconstructions of Chromosomes in Human Mitotic Cells. The Kinetochore and its Role in Cell Division. X Inactivation in Mammals, An Update. The Y Chromosome of *Drosophila*.

Volume II

Meiosis. Chromosome Structure and Function During Oogenesis and Early Embryogenesis. Chromatin Organization in Sperm. Structure and Function of Polytene Chromosomes. *Saccharomyces cerevisiae:* Structure and Behavior of Natural and Artificial Chromosomes. DNA Replication in Higher Plants. Chromatin Structure of Plant Genes. Ploidy Manipulations in the Potato. The Chloroplast Genome and Regulation of its Expression.

Volume III

Bacterial Chromatin (A Critical Review of Structure-Function Relationships). The Chromosomal DNA Replication Origin, *oriC*, in Bacteria. Replication and Segregation Control of *Escherichia coli* Chromosomes. Termination of Replication in *Bacillus subtilis, Escherichia coli* and R6K. Polyoma and SV40 Chromosomes. The Genome of Cauliflower Mosaic Virus: Organization and General Characteristics. Reinitiation of DNA Replication in Bacteriophage Lambda. *In Vivo* Fate of Bacteriophage T4 DNA. Double-Stranded DNA Packaged in Bacteriophages: Conformation, Energetics and Packaging Pathway. Bacteriophage P1 DNA Packaging. Bacteriophage P22 DNA Packaging.

Chromosomes:
Eukaryotic,
Prokaryotic,
and Viral

Volume I

Editor

Kenneth W. Adolph, Ph.D.
Associate Professor
Department of Biochemistry
University of Minnesota Medical School
Minneapolis, Minnesota

CRC Press, Inc.
Boca Raton, Florida

Library of Congress Cataloging-in-Publication Data

Chromosomes: eukaryotic, prokaryotic, and viral.

Includes bibliographies and index.
1. Chromosomes. I. Adolph, Kenneth W., 1944-
QH600.C498   1989   574.87'322   88-35340
ISBN 0-8493-4397-6 (v. 1)

Direct all inquiries to CRC Press, Inc., 2000 Corporate Blvd., N.W., Boca Raton, Florida, 33431.

© 1990 by CRC Press, Inc.

International Standard Book Number 0-8493-4397-6 (Volume I)
International Standard Book Number 0-8493-4398-4 (Volume II)
International Standard Book Number 0-8493-4399-2 (Volume III)

Library of Congress Card Number 88-35340
Printed in the United States

PREFACE

All animal cells, plant cells, and viruses capable of reproduction contain genetic material. For each living system, the genetic material exists as a complex of the coding molecules of DNA or RNA with proteins. The protein molecules serve to protect the DNA from degrading enzymes and shearing, and also control expression of the genetic information. Eukaryotic chromosomes are organized by the association of DNA with histones: these small, highly charged proteins coat the DNA to form the bead-like structures of the nucleosomes. Chromatin fibers are produced by the helical coiling of the beads-on-a-string filament of nucleosomes, and folding of the chromatin fiber results in the recognizable morphology of mitotic chromosomes. Nonhistone proteins further contribute to the structure and function of eukaryotic chromosomes, particularly in the regulation of gene expression.

The characteristic structures of mitotic chromosomes are observed during the process of normal cell division. But other processes and activities can influence DNA packaging in eukaryotic cells. The reduction of the diploid to haploid complement of chromosomes in meiosis is accompanied by special DNA packaging in the formation of eggs and sperm. In addition, chromosomes are active in replication and transcription, and this requires changes in the organization of chromosomes.

Prokaryotes, included bacteria and cyanobacteria (blue-green algae), have much smaller genomes than eukaryotes and their chromosomes are correspondingly less complex. The intestinal bacterium *E. coli* has 1000 × less DNA than human cells, lacks the histone proteins, and doesn't undergo mitosis or meiosis. Yet the DNA is contained in a defined nucleoid and is organized as supercoiled loops. This arrangement permits the efficient expression of genes and the segregation of replicated chromosomes to daughter cells. The lower complexity of prokaryotic chromosomes is an advantage for studies of the molecular biology and biochemistry of transcription and replication.

Viral chromosomes have even smaller genomes as a consequence of their parasitic reproduction cycles. For example, animal viruses such as polyoma and SV40 have DNA molecules that are 1000 × shorter than those of *E. coli*. Basic differences in the replication of animal, plant, and bacterial viruses have produced a variety of means of packaging viral genomes. Some animal viruses, included polyoma and SV40, have true minichromosomes with histones and nucleosomes. The heads of bacterial viruses such as T7 and P22 contain tighly folded or wound DNA strands that are free of bound protein. And most plant viruses (cauliflower mosaic virus is a notable exception) have RNA genomes enmeshed in protective coat protein subunits.

Major factors which determine the structures of chromosomes therefore include: the biochemical nature of DNA as a linear molecule composed of A, T, C and G subunits, the size of the genome and extent of DNA coiling and interaction with proteins, the nature of the histone proteins as globular and highly-charged, the activity of the genome in replication and transcription, and the requirement for special DNA packaging in mitosis and meiosis. These and additional topics will be examined in the following chapters, which will hopefully convey the variety and fascination of eukaryotic, prokaryotic, and viral chromosomes.

The chapters are divided into five sections; the first three of these are concerned with eukaryotic chromosomes, while the remaining two sections are devoted to prokaryotic and viral chromosomes. Each section begins with an introduction to give a brief overview of the subject matter and to relate it to the topics in other sections. Reviewing the results of research on different systems is important because, although chromosomes possess vastly different degrees of complexity, all chromosomes share similar features. Genetic information is encoded in the same DNA (or RNA) molecules and the sequence of genes must be protected and compacted by interacting with proteins. However, even though all chromosomes have the underlying unity of being protein-nucleic acid complexes, the special features observed for chromosomes of different sources makes studying chromosomes particularly challenging.

Chromosomes: Eukaryotic, Prokaryotic, and Viral should be a valuable resource for readers with a variety of interests and backgrounds. It is hoped that the information presented will be useful and that a sense is imparted of the excitement of research on chromosomes.

THE EDITOR

Kenneth W. Adolph, Ph.D., is presently a faculty member in the Department of Biochemistry of the University of Minnesota Medical School in Minneapolis. His research concerns two fundamental aspects of chromosomes: the structure of chromosomes determined by analysis of electron micrographs and the roles of nonhistone proteins in chromosome organization. He also maintains an interest in virus assembly. Kenneth W. Adolph has been a faculty member at the University of Minnesota since 1978 and is currently an Associate Professor. Postdoctoral training at Princeton University in the Department of Biochemical Sciences preceded this appointment; metaphase chromosome substructure and nuclear substructure were investigated in the laboratory of U.K. Laemmli. Earlier postdoctoral and graduate research was concerned with the assembly of viruses, protein-nucleic acid complexes much simpler than eukaryotic chromosomes and equally interesting. The structure of an icosahedral plant virus was studied during a postdoctoral year working with D. L. D. Caspar at the Rosenstiel Center, Brandeis University. Prior to this, the editor had a postdoctoral position at the Medical Research Council Laboratory of Molecular Biology in Cambridge, England. Research in the laboratory of Aaron Klug involved the *in vitro* reassembly of another simple, icosahedral plant virus. Kenneth W. Adolph received his Ph.D. from the Department of Biophysics, University of Chicago. His thesis concerned the isolation and characterization of cyanophages, and his advisor was R. Haselkorn. B.S. and M.S. degrees were received from the Department of Physics at the University of Wisconsin, Milwaukee.

CONTRIBUTORS

Kenneth W. Adolph, Ph.D.
Associate Professor
Department of Biochemistry
University of Minnesota Medical School
Minneapolis, Minnesota

Ronald D. Balczon
Department of Cell Biology and Anatomy
University of Alabama
Birmingham, Alabama

William R. Bauer, Ph.D.
Professor
Department of Microbiology
State University of New York
Stony Brook, New York

B. R. Brinkley, Ph.D.
Professor
Department of Cell Biology and Anatomy
University of Alabama
Birmingham, Alabama

Niren Deka, Ph.D.
Department of Chemistry
University of California
Davis, California

Robert Gallo
Program of Molecular Biology
State University of New York
Stony Brook, New York

Stanley M. Gartler, Ph.D.
Professor
Department of Medicine and Genetics
University of Washington
Seattle, Washington

Graham H. Goodwin, Ph.D.
Research Team Leader
Department of Cell and Molecular
  Biology
Institute of Cancer Research
London, England

Wolfgang Hennig, Ph.D.
Professor
Department of Molecular and
  Developmental Genetics
Katholieke Universiteit Nijmegen
Nijmegen, The Netherlands

Elisabeth A. Keitges, Ph.D.
Assistant Director
Department of Cytogenetics
Swedish Hospital Medical Center
Seattle, Washington

Charles K. Knox, Ph.D.
Associate Professor
Department of Physiology
University of Minnesota
Minneapolis, Minnesota

A. Gregory Matera, B.A.
Department of Chemistry
University of California
Davis, California

Geoffrey A. Partington, Ph.D.
Department of Cell and Molecular
  Biology
Institute of Cancer Research
London, England

Neil D. Perkins
Department of Cell and Molecular
  Biology
Institute of Cancer Research
London, England

Carl W. Schmid, Ph.D.
Professor
Department of Chemistry
University of California
Davis, California

Jonathan Widom, Ph.D.
Assistant Professor
Department of Chemistry
University of Illinois
Urbana, Illinois

TABLE OF CONTENTS

Volume 1

SECTION I. EUKARYOTIC CHROMOSOMES: DNA AND PROTEINS

SECTION II. EUKARYOTIC CHROMOSOMES: MITOSIS; X AND Y CHROMOSOMES

Section I. Eukaryotic Chromosomes: DNA and Proteins

INTRODUCTION

The primary component of chromosomes is DNA. The linear sequence of nucleotides along the strands of the DNA double helix defines the genes encoded by the DNA molecule. For this reason, "chromosomes" and "DNA sequence" are thought of almost as synonyms in molecular biology. But consideration of higher levels of DNA structure is also important to understand such processes as DNA replication and gene transcription. The most striking feature of DNA in eukaryotic chromosomes is the length of the molecule. Human cells contain about 1 meter of DNA, which may consist of a single molecule that is continuous through each of the 46 chromosomes of the diploid cell. Even simple eukaryotes contain an extraordinary length of DNA. (Each *Drosophila* cell has 5.6 cm of DNA and each yeast cell has 0.5 cm.) The intricate mode of DNA packaging is another major feature of eukaryotic DNA. Brought about by interactions with histones and other proteins, packaging involves folding of the DNA and structural alterations such as supercoiling. Another characteristic aspect of eukaryotic DNA is that a large proportion of chromosomal DNA is made up of sequences that are not unique and can be highly repetitious.

Repetitive DNA sequences comprise nearly one quarter of the human genome and this must have consequences for DNA organization. Proretroviral sequences or processed RNA pseudogenes, both resulting from insertion of RNA-derived precursors, account for about half of these repetitive sequences. The Alu family of short interspersed repeats is another abundant category of mobile DNA. The presence of these repetitive DNAs could result in gene duplications and deletions, gene mutations, and alterations in transcription and post-transcriptional processing.

Interacting with DNA in eukaryotic chromosomes are proteins, including sequence-specific DNA-binding proteins and the histones. The nature of the DNA-protein interactions has been characterized with techniques of molecular biology, biochemistry, genetics and biophysics. Such biophysical and biochemical studies have helped to clarify the mechanisms that regulate gene expression.

Supercoiling is a macroscopic feature of DNA structure that is involved in chromosome organization. This higher-order twisting of the double helix exists for DNA constrained into domains and for circular DNA. Since DNA loops are the predominant means of packing DNA in eukaryotic chromosomes, supercoiling is likely to be important in determining chromosome morphology. In addition to supercoiling, local DNA structures exist (Z DNA, cruciform structures, etc.) which could contribute to the regulation of DNA function.

Most of the DNA of eukaryotic chromosomes is complexed with histones. The binding of DNA and histones to form nucleosomes and the coiling of the 10-nm filament of nucleosomes into the 30-nm fiber creates a structure unique to eukaryotes. Besides protecting and condensing the DNA, the DNA-histone fibers have been implicated in the control of gene expression. The organization of histones with DNA is therefore related to the activity of DNA in transcription and replication. But determining chromatin fiber structure has been difficult because of the inherent looseness of fiber substructure.

In view of the significance of these topics for understanding eukaryotic chromosomes, Section I includes chapters concerning fundamental aspects of the biochemistry and molecular biology of chromosomal proteins and DNA.

Chapter 1

REPETITIVE HUMAN DNA: THE SHAPE OF THINGS TO COME

Carl W. Schmid, Niren Deka, and A. Gregory Matera

TABLE OF CONTENTS

I. BACKGROUND

Depending on the criteria for sequence similarity, nearly one quarter of the human genome consists of repetitive DNA sequences. Almost all of this DNA can be apportioned into a small number of families. Similar sequence families comprise most of the repetitive DNA in other mammalian genomes. This review primarily focuses on human repetitive sequences for the purpose of illustration.

The functions, if any, of these repetitive DNAs are unknown. As a model system, we therefore examine a group of sequences present in multiple copies having a known function, the human α-like globin genes. The structure and evolution of repetitive genes are shown in some ways to resemble those properties of repetitive sequences which do not code for known proteins.

Approximately half of the repetitive sequence content of human DNA results from the insertion of RNA-derived precursors. These mobile elements resemble either proretroviral sequences or processed RNA pseudogenes. The distinction is that proretroviral sequences might encode factors required for their own transposition and may retain their transcriptional competence upon insertion, whereas processed RNA pseudogenes are likely to be the passive products of retrovirally driven transpositions. Each of these two sequence classes is broadly distributed throughout the human genome, suggesting that their transposition might affect gene structure and expression. For these reasons, this review primarily focuses on the properties of these mobile elements, rather than tandemly organized repetitive sequences, which are briefly reviewed.

The most abundant family of short interspersed repeats in human DNA is the *Alu* family, which is accordingly examined in detail. New members of this family of mobile elements are generated by a very small number of select founder sequences that are evidently active in contemporary populations. However, most human *Alu* repeats were stably incorporated into the genome prior to the divergence of the human and chimpanzee lineages. Subsequent to their insertion, *Alu* repeats acquire mutations at the rate expected for nonselected sequences, without noteworthy sequence alterations such as conversion or excessive transposition.

Although repetitive DNAs, as defined herein, do not have known functions, sequences comprising one quarter of the human genome are certain to cause effects. Effects summarized here include gene duplications and deletions resulting from unequal homologous recombination between interspersed repeats, alterations in the transcription of neighboring genes, changes in both post-transcriptional processing and regulation of mRNA, and mutations in protein coding sequences. The plasticity with which the genome accommodates the insertion of repetitive elements is remarkable. The human genome contains a great deal of repetitive DNA and has an astonishing capacity to retain this genetic load and adapt these sequences to cellular function.

II. OVERVIEW OF HUMAN AND OTHER MAMMALIAN GENOMES

A. MAMMALIAN DNAs CONTAIN GENE DUPLICATES AND OTHER REPEATS

Within the near future it is quite probable that the entire human genome, as well as several other mammalian genomes, will be physically mapped.[1] Optimists also confidently foresee the complete sequence determination and analysis of several eukaryotic DNAs, including, especially, human DNA. The significance of this mass of sequence information is debatable and it remains an open question as to what will be learned from its availability. However, a number of structural features of the mammalian genome are already apparent, if only at a superficial level. One remarkable feature is that much or even all of the mammalian genome, depending on one's criteria, might consist of repetitive sequences exhibiting a spectrum of sequence similarity.[2,3]

For theoretical reasons, gene duplication has long been the favored mechanism for forming new genes; homologous protein sequences established their importance long before the modern era of DNA sequence determination.[4,5] In both mouse and human DNAs, careful sequence scrutiny has revealed the inverted duplication of exons in adjacent introns.[6] Even single-copy sequences are repeated.[6] In many cases, repetitive mammalian DNA sequences are either recognizable genes or were derived from genes. The majority of repetitive sequences, however, do not correspond to known genes, but are instead, because of their repetition, ideal candidates to serve a number of noncoding functions such as regulating gene expression or determining chromosome structure.[7-12] These possibilities have historically motivated most investigations of the nongenic repeats. Despite intensive research, the suggestion that repetitive DNAs serve noncoding functions remains an attractive, but still unproven, hypothesis. Several investigators question whether the hypothesis of function is even justified.[10,13,14] Although the putative functions of DNA repeats remain elusive, because of their abundance, broad genomic dispersion, and mobility, repeats are likely to affect gene structure and expression. This review documents such effects.

Since there are already several very good recent reviews of mammalian repetitive DNAs, we shall, where possible, cite the more comprehensive conclusions of these reviews, rather than reevaluate the primary literature.[9,10,15-18] This review emphasizes new evidence pertaining to the mobility of repeated DNAs and the plasticity of the genome in accommodating inserted sequences. Additionally, in anticipation of future interest in mapping and sequencing mammalian DNAs, an overview of the sequence composition of human DNA is presented.

B. REPETITIVE GENES: MODELS FOR NONCODING REPEATS
1. Structure and Evolution of the Human α-Globin Gene Cluster

Families of repeats coding for protein products should share many features in common with repeats that do not code for known proteins. For this reason, one example of a repetitive gene family is examined in detail. Many of the lessons inferred from this study will be shown later to apply to other noncoding repeats. The human α-globin gene cluster serves as the archetype of a repetitive gene family.

In the 5′ to 3′ direction, the human α-globin gene cluster consists of two embryonic genes (ζ2 and ζ1), two independently inactivated pseudogenes (Ψα2 and Ψα1), two functional adult genes (α2 and α1), and the θ gene of unknown function (Figure 1).[19-23] Phylogenetic comparisons show that the functional adult α2- and α1- globin genes were duplicated after the divergence of the prosimian and anthropoid lineages, but prior to the divergence of the monkey and human lineages, i.e., between 60 and 30 million years ago.[24] Despite their great age, the duplicate α2 and α1 genes are nearly identical sequences.[21,25,82] Similarly, the duplicate α-globin genes in other higher primates, e.g., chimpanzee, are almost identical

FIGURE 1.   Schematic representation of human α-globin genes and *Alu* repeats. Arrows represent interspersed Alu repeats; boxes represent genes. Known pseudogenes are designated by the Greek letter Ψ. ζ2 and Ψζ1 are embryonic α-globin genes, Ψα2 and Ψα1 are anciently inactivated adult α-globin genes, α2 and α1 are functional adult genes, and θ1 is a gene of unknown function.[19-23,29]

sequences, but differ from the duplicate human globin genes.[26] The intraspecies similarity and interspecies difference of gene duplicates is evidence for the existence of a mechanism for correcting the duplicate genes within a species to the identical sequence. This process is commonly called gene conversion or gene correction. The best example of the effectiveness of gene conversion is the intraspecies homogenization of the repetitive ribosomal RNA genes. These genes within a species are nearly identical in sequence, yet they differ from those in another, closely related species.[27,28]

Although gene conversion may erase the differences between gene duplicates, duplication can also create new genes. Random mutations coupled with selection could easily adapt the function of a duplicate protein product to some new specialized task; after all, the gene already codes for a functional protein.[4,5] Recent observations on repetitive sequences inserted into genes, as reviewed in Section II.D.5, provide novel insight into the strength with which new protein products may be selected. Entirely unrelated repetitive sequences can be incorporated into a gene and used to code for an altered protein product. The embryonic ζ- and adult α-globin genes are more conventional examples of specialized functions being served by divergent gene duplicates.[23] The two ζ-globin genes are also an unusual example of gene duplication. The nearly identical sequences of these genes implies the occurrence of an extremely recent gene conversion event. However, in some individuals the ζ1 gene has been silenced by an even more recent single nucleotide mutation resulting in a termination codon.[23]

Selective pressure will not always be sufficient to impose a new function on a randomly mutated gene duplicate. The Ψα1 gene may be viewed as a "failure" in the search for new function. The Ψα1 genes in human, chimpanzee, and monkey share a number of common mutations, any of which would be sufficient to inactivate the gene.[29] Although this gene has been inactive for at least as long as the divergence time of human and monkey, it has not been selectively removed from either primate lineage. A recurring theme is that repetitive DNA sequences may be functionless pseudogenes.[13,14] The example of the Ψα1 gene show that functionless DNA (DNA no longer able to code for either a protein or even a transcriptional product) might enjoy an extremely long residence before it is selectively pruned out of the genome. As an even more striking example of the tolerance of the genome for functionless DNA, the human Ψα2 gene is barely recognizable as an α-globin gene and was probably inactivated at a much earlier time than even the anciently inactivated Ψα1 gene.[22] The continued presence of this fossilized gene suggests that long-term decay, rather than amputation, may be the preferred mechanism for removing inactive sequences. The high degree of sequence divergence of various repetitive DNA families decribed in Section II.D implies that their sequences, like these pseudogenes, may be merely decomposing.

The number of gene duplications in the human α-globin cluster is also remarkable (Figure 1). Unequal recombination between duplicate globin genes is sufficient to account for subsequent duplications; this mechanism is supported by sequence comparisons of primate α-globin gene clusters.[24] Like globin genes, homologous copies of repetitive sequences distributed throughout the genome are also candidate sites for homologous unequal crossing over. Section II.D.5 describes several examples of unequal recombination between interspersed repeats.

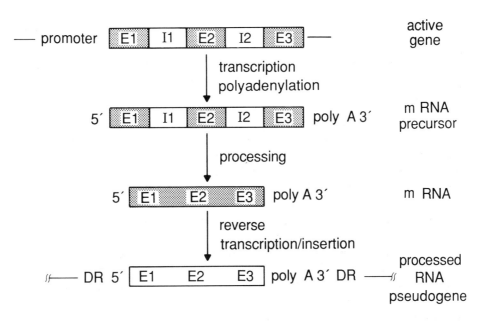

FIGURE 2.   Formation of a processed RNA pseudogene. The letters E and I identify exons and introns, respectively, in an active gene. Transcription of this gene from the 5′ transcriptional start site in step 1 results in a polyadenylated mRNA precursor. The precursor is subsequently processed into a mature mRNA in step 2. The mRNA is converted into a processed RNA pseudogene in step 3. The detailed mechanism of this conversion is largely hypothetical, but must ultimately include both reverse transcription of mRNA into DNA and insertion of this sequence into a genomic entry site "DR", resulting in its duplication and the formation of the flanking direct repeats "DR".[16,30,31]

2. Pseudogenes Formed from Retroposition of Processed RNA

Mouse DNA contains a very different kind of α-globin pseudogene which was evidently formed from an RNA intermediate.[30] This pseudogene transposed to a different chromosome than that of the parent gene, and its introns were correctly processed.[30] Other hallmark features of processed RNA pseudogenes include a 5′ end that corresponds to the precise transcriptional start site, polyadenylation of the 3′ end and short, flanking, direct repeats formed from a duplication of the genomic entry site (Figure 2).[15-17,30,31] The formation of such pseudogenes requires as a minimum the transcription of the parent gene, processing of the RNA, reverse transcription, and ultimately the insertion of a DNA copy of the processed RNA.[15-17,31] "Retroposition" has also been observed for a variety of genes expressed both in different cell types and at different developmental times.[15-17,30] For brevity, the products of this retroposition mechanism are called retroposons.[17]

Ignoring the difficulties in the proposed mechanism for retroposition, there is also the question of how a blood cell transcript becomes stably incorporated in the mouse genome.[31] One possibility is that α-globin happened to be misexpressed in the developing germ line. Although one might postulate this rare event, the existence of many different processed RNA pseudogenes, from mRNAs which normally are expressed only in somatic cells, contradicts the proposal that only genes expressed in the developing germ line give rise to processed RNA pseudogenes.[15,16,30,31]

There is now clear evidence favoring a horizontal transmission of RNA pseudogenes.[32] A defective retrovirus has been observed to package cytoplasmic RNA and to direct the retroposition of that RNA into a recipient host. This same observation additionally provides a source of reverse transcriptase and integrase. Plausibly, the retroposon class of repetitive sequences described below is also horizontally transmitted by a retroviral-mediated process.

Generally, retroposons resulting from processed RNA are pseudogenes. In many instances, these retroposons have point mutations that either inactivate possible protein products or terminate their translation.[16] Although these changes might in part reflect a lack of selection for the pseudogene sequence, it is also likely that some are introduced during the retroposition event; reverse transcription under *in vivo* conditions is relatively error prone.[33] Processing of RNA transcripts from retroposons is also unlikely as the introns of the parent gene are absent and polyadenylation of the retroposon removes at least part of its normal polyadenylation signal. The omission of these steps could interfere with nuclear export and/or translation of retroposon transcripts. Additionally, retroposition implies that the processed RNA pseudogene would abandon required upstream transcriptional control signals, including, in particular, its promoter elements, and could be inserted in an unfavorable context for its normal regulated transcription. Transcripts of processed RNA pseudogenes would thus normally result from the upstream promoter of the transcription unit into which it happened to be inserted. Hence, retroposition is an inefficient mechanism for creating new gene functions.

Soares et al. have observed one exception to the generalization that retroposition would normally give rise to a pseudogene.[34] A semiprocessed rodent preproinsulin gene has several hallmarks of retroposition. Of particular significance, its 5′ flanking sequence includes 0.5 kb of homology to the parent gene, such that the retroposed sequence was apparently transcribed from an aberrant upstream promoter. Plausibly, the inclusion of transcriptional control elements in this retroposon resulted in its retention of functionality.

As reported in Section II.C, a major fraction of the human genome (ca. 10%) consists of repetitive sequences which are derived through retroposition. In analogy to processed RNA pseudogenes, it is likely that these repeats are also inactive with respect to producing a gene product. However, it is also likely that such a large quantity of inserted DNA would provide opportunities to mold new functionalities. Recent observations also document this possibility.

C. CLASSIFICATION AND GENOMIC CONTENT OF REPETITIVE SEQUENCES

1. Many Repetitive Sequences Do Not Code for Known Products

Although much or all of the mammalian genome might be derived by sequence duplication, it is still possible to distinguish between repetitive and single-copy DNA. This distinction depends on the establishment of somewhat arbitrary criteria for sequence similarity.[2,3,7] Based on DNA renaturation under conditions for which sequences that are at least 70% similar can form duplex structure, approximately 20% of human DNA consists of sequences which are present in ten or more copies (Table 1).[35]

Most of the repetitive sequence content of the human genome can be accounted for by a few families of repeated DNAs (Table 1). Although the estimated total amount of identified repetitive DNA in Table 1 exceeds the previously cited estimate that about 20% of human DNA is repetitive, it is also certain that the list of repetitive sequences in Table 1 is incomplete and excludes several small RNA pseudogenes, ribosomal genes, and tRNA genes, to cite a few examples.[16] Allowing for the substantial theoretical and experimental uncertainties in these estimates, it is reasonably certain that a number of repetitive human DNA sequences remain unidentified. However, it also seems certain that the major mass fraction of repetitive human DNAs has been identified (Table 1). Included in Table 1 are all of the different known classes of repeated human DNA sequences, as well as the most abundant representative family in each class. Weiner et al. have recently published a more inclusive list of known repeated DNAs.[16] The following discussion briefly distinguishes between the different types of repeated DNAs in Table 1.

2. Satellite DNAs Are Not Interspersed

Satellite DNA consists of long tandem arrays of a sequence element, which are usually uninterrupted by single-copy DNA.[9,10] In the human, two major families of satellite DNAs

TABLE 1
Classification and Abundance of Several Major Repetitive DNA Families

Family	Organization	Classifications	Copy #	Length	% Genome	Ref.
Satellite I—IV	Tandem repeat	Satellite	—	5 bp subunit and variations	2—5 7	9,36,38
α Satellite	Tandem repeat	Satellite	—	171 bp subunit	~5	151,155
Alu	Interspersed with single copy	Sine/retroposon	0.5×10^6 1.1×10^6	300 bp	6—13	74,152
L1	Interspersed with single copy	Line/retroposon	~50,000 (3' end)	Variable 5' truncations ~300 bp to 7 kb av. = 2,000 bp?	4	18,74,90,153
THE 1	Interspersed with single copy	Proretroviral transposon	10,000	2,300 bp	0.8	102
Poly CA	Tandem repeats interspersed with single copy	Minisatellite	50,000	Variable	0.1	128,153,154
				av. = 50 bp?	18—30 = Total	

Note: The abundance of each repeat sequence family is subject to the following theoretical and experimental uncertainties. No single method is capable of revealing the total amount of any satellite sequence and the identity of satellites I to IV is poorly defined. The length of *Alu* repeats is well established, but estimates of their numerical abundance based on renaturation techniques vary about twofold. The length of *L1* repeats varies over the range of ~300 to 7,000 bp due to their 5' truncation. Consequently, the 3' end is also numerically more abundant that the 5' end. Poly(CA) runs also have variable lengths.

— α satellite and the numbered satellites I, II, III, and IV — have been identified.[36,37] The identification of the four numbered satellites has primarily depended on the uncertainties of the density gradient techniques used in their isolation. Although more modern cloning techniques have established the interrelatedness of at least some of these four satellites, their overall identity and precise genomic mass fraction are far from certain.[38,156] It is probable that some of the numbered satellites are merely reisolated subfractions of each other.[38,156]

3. Other Repetitive DNAs Are Interspersed with Single Copy Sequences

The especially abundant *Alu* and *L1* families are retroposons which fulfill the structural criteria that identify processed RNA pseudogenes (Figure 2). The structures supporting these identifications are reviewed below. Repetitive mammalian retroposons have been further divided into short and long interspersed repeats, "Sines" and "Lines", respectively.[39] Many repetitive Sines, or at least their founder sequences, correspond to sequences which are transcribed by RNA polymerase III, whereas in the case of the most thoroughly studied Line, the L1 founder sequence(s) evidently codes for a protein and would be an RNA polymerase II transcription unit.[16]

As previously discussed, retroposons are probably inactive, at least with respect to coding for a functional transcript. In contrast, proretroviral sequences contain necessary transcriptional control signals in their flanking long-terminal repeats (LTRs) so that, upon retrotransposition, they can retain their transcriptional competence.[40,41] Additionally, proretroviral sequences code for several protein products that are required for their own transposition — unlike retroposon pseudogenes, which are the passive victims of transposition. For these reasons, we distinguish between these two classes of transposons as being either *retroposons* or *proretroviral transposons*. The best-studied examples of proretroviral transposons include *Ty* elements in yeast and copia-like elements in *Drosophila*.[42,43] The human THE-1 family is the most abundant human sequence having the structural organization of a proretroviral transposon (Table 1).

RNA intermediates are required in the formation of both retroposons and retroviral transposons. In contrast, minisatellite sequences consist of tandem arrays of short sequence elements and are interspersed with single-copy sequences (Table 1). An RNA intermediate seems unlikely in minisatellite formation and it is more likely that they arise from such DNA-mediated events as replication slippage or unequal recombination.[15]

Not listed in Table 1 are a variety of repetitive genes for both structural RNAs and protein-coding mRNAs. Additionally, several lower copy retroposons have also not been tabulated.[16] However, the entries in Table 1 do include the highest copy number family, which exemplifies each known class of repetitive DNA. Just as it is unlikely that all families of repetitive DNA have been identified, it is also unlikely that every class of repetitive sequence has been identified. The remainder of this review summarizes pertinent new discoveries for each of these repetitive sequence classes. To simplify the discussion, repetitive gene clusters are not further considered. The reason for this exclusion lies in our belief that the remaining repetitive sequences comprising 20% of the genome might serve other important, albeit unknown, functions. Regardless of whether these sequences have functions, their mobility and ubiquitous representation throughout the genome guarantees that they have important effects.[13,14] These assertions are documented below.

D. REPETITIVE SEQUENCES FORMED THROUGH RNA INTERMEDIATES
1. Most Mammalian Interspersed Repeats are Retroposons

The retroposon and proretroviral classes of interspersed repeats transpose through an RNA intermediate. As described in somewhat more detail in sections II.D.2 and II.D.6, the human *Alu* and *L1* families of sequences exhibit many of the structural features of the retroposons or processed RNA pseudogenes (Figure 2).[15,16] Thus, retroposons are by either mass or number more abundant than the proretroviral transposons (Table 1). This result is curious. In lower eukaryotes such as *Drosophila* and yeast, there is far less interspersed repetitive DNA and the greater proportion of these repeats is composed of proretroviral transposons.[42-44] Also, the best current model (*vide supra*) is that the passive insertion of retroposons requires a variety of *trans*-acting factors, which are probably provided by retroviral sequences. Indeed, mutant retrovirus is sufficient to cause the retroposition of co-packaged RNAs into recipient cells.[32] A rationalization is that the abundance of *Alu* and *L1* repeats may be partially attributable to their patterns of expression and that perhaps they are abundantly transcribed in germ cells.

Because of their relative abundance, the remainder of this review primarily focuses on the properties of retroposons.

2. Human *Alu* Repeats Exemplify Mammalian Sines

The human *Alu* family consists of about 1 million members, most of which share a 282-bp consensus sequence, a precisely defined 5′ end that corresponds to the *in vitro* pol III transcription start site, a polyadenylated 3′ end, and flanking direct repeats, which in several instances are known to be duplicated genomic entry sites.[15] These features identify the *Alu* family as a collection of retroposons (Figure 2). The 1 million member sequences are, with some notable exceptions, rather evenly interspersed throughout the genome.[15,45-47] For example, there are 11 known *Alu*'s within the 30-kb human α-globin cluster, corresponding to about 600,000 *Alu*'s if they occurred with the same frequency throughout the genome (Figure 1). The human *Alu* consensus sequence is a head-to-tail dimer of two monomer units, which are related in sequence to 7S RNA.[15,16] The 7S RNA can be viewed as an *Alu* monomer containing a 150-nt insert. Because 7S RNA is so highly conserved in evolution, the human *Alu* founder sequence(s) must ultimately be 7S RNA derivatives.

Rodent DNAs contain two major Sine families, which have been variously called B1 and B2 families of the Type I and Type II *Alu* equivalent families.[15,16] The B1 family is essentially an *Alu*-like monomer sequence, whereas the B2 sequence is derived from a tRNA related precursor.[16] Rabbit and artiodactyl Sines are also tRNA derivatives. Sines in the prosimian primate, galago, include a potpourri of dimeric *Alu*'s, monomeric *Alu*'s, tRNA derivatives and composites of *Alu* and tRNA derivatives.[48]

3. Function of Sines: Nonexistent or Unknown?

The search for function where none exists would certainly be a sterile exercise. However, so much mammalian DNA consists of Sines that it is at least worth considering the possibilities. Thus, for the purpose of this discussion, we assume the existence of a Sine function. *Alu* repeats could function at the level of their DNA sequences as structural elements in chromosome folding or might be recognized by DNA binding proteins for some other purpose. Little is known about the relationship of a DNA sequence to its chromosome structure, making it difficult to evaluate this possibility. Based on sequence comparisons, Jelinek et al. suggested that *Alu* repeats might serve as origins of DNA replication.[123] Elegant experiments by Johnson and Jelinek show that *Alu*-containing plasmid clones replicate in cell culture.[124] Alternatively, the function of Sines could involve their transcripts.

The very different sequences of mammalian Sines, e.g., some are tRNA-like and others 7S-like, implies that the putative function must be fulfilled at some higher level of structure. One speculation is that they may be involved in processing primary transcripts. Konarska et al. have postulated that *Alu* sequences may be involved in the *trans*-splicing of RNA.[50] Sines are abundantly transcribed by virtue of their ubiquitous interspersion and are largely removed during mRNA maturation. The folded Sine transcripts, either 7S or tRNA derivatives, could serve as processing recognition sites. Another possibility is that Sines could affect mRNA stability. Clemens has postulated that B2 might base-pair with AU-rich sequences on the 3' end of some short-lived messages, thereby protecting them.[51]

A third role for Sines, inferred from their structure, is a translational one. Both 7S RNA and tRNA interact with the ribosome and so too might their pseudogene derivatives. Theoretical predictions suggest that the *Alu*-like moiety of 7S RNA folds into a tRNA tertiary structure,[158] thus providing a plausible mechanism for the induction of translation arrest by the signal recognition particle.[52] We speculate that transcripts of the dissimilar 7S and tRNA derivative Sines might have the same tRNA tertiary structure and thus dissimilar sequences could perform identical functions.

The transcriptional activity of Sines is complex. Symmetric Sine transcripts are invariably represented in hnRNA, presumably because of their ubiquity.[49] Human *Alu*, despite its internal RNA polymerase III promoter, is not detectably transcribed from this promoter in HeLa cells.[53] It has been recently reported that an *Alu* variant, consisting of the left monomer unit, is transcribed in neural tissue.[54] The isolation of this sequence, called here a primate ID, depended initially on the use of a rodent Sine transcript, which needs to be first considered before describing the primate ID transcript.

Rodent B1 sequences are, like their human *Alu* counterpart, transcriptionally quiescent despite their internal RNA polymerase III promoters.[55] A B1 repeat in a transfected minigene construct was observed to be heavily transcribed, suggesting that transcription of the genomic copies must be efficiently repressed.[56] The tRNA-derived B2 retroposons are transcribed in various rodent cell lines and tissues; however, Singh et al. observed a significant increase in B2 transcripts in SV40 transformed cells.[57-60] There are several examples of an increase in the level of a Sine transcript upon cell transformation or a shift to a proliferative state.[16,61] In the case of the ID sequence, the regulation is post-transcriptional.[61] In summary, the highly repetitive mammalian Sine sequences can be transcribed *in vivo*; this potential transcriptional activity is largely repressed and the level of its repression is regulated. The transcriptional inactivity of 7S RNA pseudogenes provides insight into the transcriptional repression of Sines. These 7S RNA pseudogenes, like mammalian Sines, contain RNA polymerase III promoters, which are active *in vitro*, but unlike the functional 7S gene, they lack necessary upstream control elements and are thus inactive *in vivo*.[16]

The rodent ID sequence was originally thought to be a marker for neural specific transcripts, i.e., a brain identifier sequence.[62] Subsequent observations showed this report to be incorrect.[61,63-65] However, for clarity of identification, we continue here to refer to it as the ID sequence. The 82-nt ID sequence is a tRNA-derived retroposon.[16] This same sequence was subsequently used to detect and identify a putative primate ID sequence.[54] The rodent ID hybridizes to a discrete-length 200 nt transcript in monkey brain RNA. Extension of the primer $dT_{12}CC$ by reverse transcriptase results in a 135-nt long cDNA band and a clone corresponding to this cDNA, called the primate ID sequence, is found to be a left monomer of the human *Alu* family with no other notable differences from the consensus. Since the rodent ID sequence is not a tissue-specific transcript identifier and since it is unrelated in sequence to the putative primate ID (one is 7S related and the other is tRNA related), the suggestion that this half *Alu* is an identifier sequence has no merit. It is, however, reportedly more abundant in monkey brain than in liver and spleen RNA. However, it is transcribed in both human HeLa and osteosarcoma cells, disproving its characterization as an identifier.[54] We also question whether the isolated half *Alu* bears any relationship to the monkey transcript which was originally detected by the unrelated rodent ID sequence. Regardless of these criticisms, the observations do imply that a variant of the *Alu* family may be actively transcribed *in vivo*. This finding is important. Regarding this *Alu* monomer, it is noteworthy that almost all human *Alu* sequences have either the dimeric consensus or can be regarded as deletions of dimeric structure.[66] However, the tissue plasminogen activator gene contains three examples of left monomeric *Alu*'s.[46]

In conclusion, the majority of Sines either have no function or their function does not depend on their sequence. Some Sines, but certainly not all, may code for functional transcripts. This last point is pertinent to the conclusion reported in Section II.D.4 that only a very small number of *Alu* sequences code for new family members.

4. *Alu* Sequence Conversion and Transposition Are Infrequent

Globin genes serve as reference points to examine the long-term evolutionary history of primate *Alu* repeats.[24] At each of seven tested positions, human and chimpanzee are found to share orthologous *Alu* repeats (Figure 1) and, similarly, they share orthologous *Alu*'s at two positions tested within the β-globin gene cluster.[67,68] The divergence of these human and chimpanzee *Alu* pairs approximates the value expected for unselected sequences. Koop et al. have made corresponding observations on seven orthologous *Alu* pairs in the human and orangutan β-globin gene clusters.[69] The implications of these combined observations are (1) *Alu*'s have not been especially mobile since the divergence of human and apes, (2) once inserted, *Alu*'s decay as unselected sequence, and (3) none of these *Alu*'s has been converted to a different member sequence since human and chimpanzee diverged.[24,69] The *Alu* positioned 5′ to the Ψα1 gene (Figure 1) is orthologous in higher primates and, again, its divergence suggests no selection.[24] A prosimian primate, galago, has been shown to have an entirely different *Alu* family; the galago α-globin gene cluster does not share any common *Alu*'s with the human gene cluster.[24,70,71] The appearance of these *Alu*'s in the primate α- and β-globin gene clusters is phylogenetically summarized in Figure 3.

The insertion time of human *Alu*'s has been estimated by two other independent sequence comparisons with similar results (Figure 3). The inexact short direct repeats surrounding *Alu* family members must have accumulated at least part of their sequence divergence subsequent to their creation by an *Alu* insertion. The average divergence of the direct repeats has been estimated to range from 5 to 12%.[15,72] The substantial uncertainty of this estimate in part reflects the ambiguity of identifying the direct repeats. However, this broad range also suggests that *Alu*'s were inserted following the human-prosimian divergence and pre-

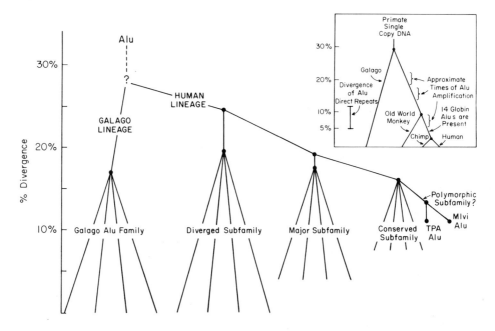

FIGURE 3.   Insertion of *Alu* subfamilies. A minimum of three human *Alu* subfamilies, the "Conserved", "Major", and "Diverged" subfamilies has been detected.[66] The Diverged subfamily is most nearly related to the galago *Alu* family, whereas the Conserved subfamily is most distantly related to the galago *Alu* family. Two *Alu* repeats, "Mlvi" and "TPA", which are polymorphic in the human genome, resemble the Conserved subfamily sequence, but are equally similar in sequence to each other (Figure 4).[46,66,73] The insert compares the divergence of the membership of these subfamilies to the single-copy divergence of various primate DNAs. In agreement with these estimates, a total of 14 *Alu*s in the human α-globin gene cluster and human β-globin gene cluster are known to be inserted prior to the divergence of human and chimpanzee and human and orangutan, respectively, as depicted in the insert.[24,67,69] The divergence of the short direct repeats flanking Alu repeats provides another independent estimate of the insertion time of *Alu* repeats, as depicted by the bar in the insert.[15,72]

ceding the divergence of human and chimpanzee and perhaps human and monkey (Figure 3). A more precise estimate is possible by comparing the divergence of *Alu* member sequences from each other and thus indirectly to their respective founder sequences. A complication in this analysis is the number and identity of the presumed founder sequence(s). Statistical analysis of a parsimony alignment suggests a minimum of three episodic bursts of *Alu* insertion into the human genome (Figure 3).[66] Each episode merely represents the progeny of a single, parsimonious founder, but the actual kinetics could be more complex and occur over a substantial evolutionary period.

Although this episodic model may be oversimplified in some details, one of its major predictions has been confirmed. One *Alu* subfamily contains members having very similar sequences relative to their divergence from the entire family (Figure 3).[66] It is therefore predicted that this "conserved" subfamily contains the most recently inserted *Alu* repeats. Although the transposition of *Alu*'s is somewhat infrequent, two *Alu*'s, which are polymorphic in the human genome, have been identified; each is very closely related (4% and 5% divergence, respectively) in sequence to the putative, newly inserted *Alu* subfamily (Figures 3, 4).[46,66,73] There is also an *Alu* which has inserted into the gorilla β-globin gene cluster subsequent to its divergence from the human lineage.[59] This recently inserted gorilla sequence also closely matches the human conserved subfamily (2.5% divergence) (Figure 4). In this case, we must allow for the additional divergence of the putative gorilla and human *Alu* founder sequences. Thus, *Alu* transpositions have occurred in both lineages in relatively recent evolutionary times following the human and gorilla divergence and is evidently occurring in the contemporary human population. Further, newly inserted members

```
GGCCGGGCGC GGTGGCTCAC GCCTGTAATC CCAGCACTTT GGGAGGCCGA GGCGGGCGGA TCACGAGGTC AGGAGATCGA GACCATCCTG GCTAACACGG  Conserved
T.........  ..........  ..........  ..........  ..........  ..........  ..........  ......C...  ...A.....   ..A.....    Mlvi
..........  ..........  ..........  ..........  ..........  ..........  ..........  ......C...  ...A.....   ..A.....A   TPA
..........  .......A..  ..........  ..........  ..........  ..........  ..........  ..........  .........   ........A   Gorilla

TGAAACCCCG TCTCTACTAA AAAATACAAA-AATTAGCCGG GCGTGGTGGC GGGCGCCTGT AGTCCCAGCT ACTCGGGAGG CTGAGGCAGG AGAATGGCGT  Conserved
..........  .......-..  ...TA....A  ......-..A  ..........  ...CG--...  ..........  .......T..  .........   ........    Mlvi
..........  .......-..C  ....A...-.  .....A....  ..........  ..........  ....TG.   ...T....    ...A.....   ........    TPA
..........  ..........  .......A..  ..........  ..........  ..........  ..........  ......G...  ...-.....   ........    Gorilla

GAACCCGGGA GGCGGAGCTT GCAGTGAGCC GAGATCGCGC CACTGCACTC CAGCCTGGGC GACAGAGCGA GACTCCGTCT C  Conserved
..........  ....--....  ..........  ......C...  ..........  ..........  ..........  .........  .   Mlvi
..........  ..........  ..........  ......C...  ..........  ......A...  ..........  .........  .   TPA
..........  .......-..  ..........  ..........  ..........  ..........  ..........  .........  .   Gorilla
```

FIGURE 4. Comparison of new *Alu*'s to the Conserved subfamily sequence. The consensus of the Conserved subfamily was previously identified and proposed to resemble the consensus sequence of more recently inserted *Alu* repeats.[66] *Alu*'s associated with the Mlvi locus and the *TPA* gene are each polymorphic in the human population and thus identified as newly inserted *Alu*'s.[46,73] Each closely resembles the putative consensus, as is also depicted in Figure 3. The gorilla *Alu* is an insertion into the gorilla β-globin cluster, following the human-gorilla divergence.[59] The similarity between the human subfamily consensus and gorilla *Alu* consensus (7 differences in 280 positions or 2.5% divergence) is strong evidence that the Conserved subfamily consensus resembles the *Alu* founder sequence(s). Allowance must also be made for additional divergence of the human and gorilla *Alu* founders.

must be generated by either a single founder sequence or a small, select group of closely related founder sequences. As described in Section II.D.3, conceivably only *Alu*'s having necessary upstream control elements would be transcribed *in vivo* by RNA polymerase III. These control elements could plausibly identify the transpositionally competent founder sequence(s).

The two known human *Alu* polymorphisms can be compared to the other 150 reported *Alu* sequences as a very rough measure of *Alu* transposition frequency.[46,47,66] In principle, this is an underestimate of the true value since many of the *Alu* reports did not test for polymorphisms, but the data are too sparse to refine the estimate. As a gross upper limit of the polymorphism attributable to *Alu* insertion, 14% of the human *Alu*'s belong to the conserved subfamily, which should contain all of the more recently inserted *Alu*'s in the human lineage. Some conserved subfamily *Alu* members are known to have inserted prior to the divergence of the human and ape lineages. Thus, 14% is certainly an upper limit to the number of *Alu*'s which are polymorphic in the human populations. The two polymorphic *Alu*'s differ from each other and from their presumed founder sequence (represented by the conserved subfamily consensus) by about 4% (Figure 4). Since they have not resided in the genome long enough to accumulate 4% divergence, they were probably substantially different upon their insertion. These sequence differences might be attributable to different founders. Moreover, as noted above, there is very good evidence that *in vivo* reverse transcription and retroposition is susceptible to error.[33]

DNA renaturation results contradict the previous conclusion that human *Alu*'s are relatively immobile. Using the potentially very precise technique of strand titration, Hwu et al. estimate that there are 0.9 million human *Alu*'s, as compared to 0.3 million *Alu*'s in chimpanzee.[74] This putative difference between the number of human and chimpanzee *Alu*'s would be extremely important; however, we take issue with these results. As mentioned, human and chimpanzee share orthologous *Alu*'s at all positions tested within the human α- and β-globin gene clusters. Of course, these studies sample only a small number of *Alu* repeats, but the divergence of the three *Alu* subfamilies (Figure 3) also indicates that most *Alu*'s have resided in the genome far longer than the divergence of the human and chimpanzee lineages. The renaturation studies leading to this putative difference may be technically flawed since very different preparations were used for human and great ape DNA without a satisfactory control for their DNA fragment length. This suggestion of a systematic error is also consistent with Hwu et al.'s additional report that the number of *L1* repeats is also higher in human than in ape DNAs by similar ratios. Although we criticize these results, they are not disproven. This important issue remains to be settled.

5. Effects Attributable to *Alu* and Other Sines
a. Changes in Gene Expression due to Sines

Alu family members are often positioned between genes, as in the case of the human α-globin genes (Figure 1), or within introns, as described in several examples.[45-47] Presumably, the effects of these *Alu*'s on the expression of their neighboring genes would in general be subtle. An interesting exception is an *Alu* positioned in an intron of a decay-accelerating factor gene.[75] Splice sites within this *Alu* provide the opportunity for alternate processing of the transcript, resulting in two different mRNAs, an "*Alu* plus mRNA" containing 120 nt of *Alu* sequence and an "*Alu* minus mRNA". The incorporated *Alu* sequence causes a frame shift in the message, resulting in an entirely different C-terminal protein having 39 amino acids encoded by the *Alu* sequence and an additional 40 frame-shifted amino acids encoded by the 3' end of the message, including its normally untranslated 3' end. The corresponding region of the *Alu* minus mRNA codes for an entirely different 20-amino-acid sequence. Additional evidence suggests that the altered *Alu* plus product may be a secretory protein, whereas the *Alu* minus product is a membrane protein.

This observation is extremely surprising as it is very unlikely that *Alu* family members or their founder sequence(s), which are ancestrally related to 7S RNA, would normally have any protein coding capacity. The transcriptional orientation of the *Alu* in the decay-accelerating factor mRNA is opposite to that of 7S RNA. As an added surprise, Sharma et al. have discovered a complete *Alu*, in the same orientation as 7S RNA, that codes for 32 amino acids on the C terminus of a human B-cell growth factor.[76] This *Alu* introduces a termination codon situated near the center of the *Alu* sequence and thus also forms part of the 3' untranslated region of the mRNA. In each case, the inserted *Alu* mutates the parent gene product, but, regardless of transcriptional orientation, now codes for a presumably functional product. Newly inserted sequences are amazingly plastic. A rodent identifier sequence codes for seven C-terminal amino acids and forms the 3' untranslated region of a rat mRNA.[61] In this and the previous example of the human B-cell growth factor mRNA, the Sine may also provide a polyadenylation signal for the resulting mRNA. A B2 repeat has also been observed to direct mRNA polyadenylation.[77] Thus, in three independent examples a Sine mutates the protein product of a gene, and in three examples a Sine forms all or part of the 3' untranslated region (UTR) of an mRNA. Since the function of the 3' UTR is not entirely known, it is not possible to appreciate fully the significance of these 3' UTR mutations. However, sequence insertions in the 3' UTR are liable to affect the post-transcriptional regulation of mRNA.[78] The mRNA containing the rat identifier sequence is found to accumulate in transformed cells. Insertion of this element in the 3' UTR of a transfected β-globin clone leads to accumulation of the resulting transcript through a change in its post-transcriptional regulation.[61] Although currently the significance of Sine elements contained in noncoding regions of genes cannot be completely assessed, these examples demonstrate their importance.

b. Changes in Gene Structure due to Sines

Sines positioned between genes might also have unrecognized effects on gene expression, but certainly have had recognizable effects on the evolution of gene structure. For example, unequal homologous recombination between interspersed *Alu* repeats caused an internal duplication of the LDL receptor gene, thereby resulting in one form of familial hypercholesterolemia.[79] As discussed in the example of the duplication of primate α-globin genes, unequal recombination between interspersed repeats is expected. Barsh et al. found that the human growth hormone genes were in part also duplicated by unequal recombination between interspersed *Alu* repeats.[80] Rouyer et al. have shown that an *Alu* recombination caused a sex chromosome rearrangement in human, resulting in XX maleness.[81] Interspersed *Alu* repeats have served as one of the end points in a variety of deletions in both the α-globin and β-globin gene clusters.[15,79] *Alu* repeats also precisely mark the end points of the corrected duplication units in the human α2- and α1-globin genes, as well as goat globin genes.[82,83] The role of *Alu*'s in these latter examples is less certain. Possibly, these observations merely reflect the extremely large number of *Alu* repeats present in human DNAs.

6. Unlike Sines, Lines Have Long ORFs

The human *L1* family of repeats, "*L1*H", consists of about 50,000 members, most of which have a common polyadenylated 3' end and variable truncations on their 5' ends.[15,16,18] *L1* repeats are also usually flanked by short direct repeats, thus fulfilling some of the structural requirements of retroposons. Plausibly, the 5' truncations could result from incomplete reverse transcription of the template strand.

Full length primate *L1* repeats are nearly 7 kb in length and two long open-reading frames (ORF) are present in some, but not all, members.[18,84,85] Random mutations in unselected sequences or error prone reverse transcription might explain the inactivation of the ORFs in most members. *L1*H (H designating human) has a 1-kb 5' ORF which is separated by a 33-nt segment from a 4-kb ORF. The 33-nt segment is bounded by stop codons, which are conserved in most of the member sequences. *L1* repeats in mouse, *L1*M, have similar-length ORFs which overlap by five codons.[86] In some regions, such as their 3' UTRs, *L1*M and *L1*H are entirely different sequences; however, in other regions, such as parts of the longer ORF, they are obviously homologous.[87] The predicted protein product for the longer ORF has some limited homology to a retroviral *Pol* gene.[86,88,89] Conceivably, the master *L1* gene(s) codes for its own reverse transcriptase, thereby promoting its own retroposition. However, the activity of this predicted product has not been tested nor, to our knowledge, has the protein been detected *in vivo*.

The simplest conclusion is that *L1* repeats are merely a collection of pseudogenes. Retroposons would normally abandon upstream control elements upon retroposition, resulting in their transcriptional inactivation. Additionally, point mutations would inactivate or terminate the predicted protein product, and since most *L1* repeats are truncated, their corresponding ORFs are deleted in whole or in part. Of course, *L1* repeats are heavily transcribed *in vivo* as they are inserted into a large number of other transcription units.[90] Tandem repeats are present on the 5' ends of some full-length *L1*M repeats.[86,91] Conceivably, these tandem repeats might have a role in promoting L1 transcription and, by participating in the retroposition, would make the newly inserted member transcriptionally competent. This is, however, merely a speculation and the absence of corresponding tandem repeats in full-length *L1*H sequences shows that their role is nonessential.

L1 repeats are present in every mammalian lineage tested, implying the conservation of a required function.[18] We assume the existence of an active *L1* founder sequence(s) which codes for this required product and additionally generates new family members. The difficulty is to isolate this one member from a background of 50,000 pseudogenes. Skowronski et al. have succeeded in detecting full-length *L1* transcripts in human teratocarcinoma cells.[84,85] Sequence heterogeneity of these transcripts shows that more than one member of the family is transcribed. It is not known whether the *L1* genes coding for these RNAs are retroposons or if they are founder elements having intact 5' flanking sequences.

Several studies have noted the relative sequence similarity of *L1* repeats within a species and specific sequence differences between the *L1* repeats of different species.[92-95] These observations have been interpreted as favoring sequence conversion of the *L1* membership to different master sequences in divergent species.[92-94] This interpretation assumes that the *L1* families were formed prior to mammalian radiation and subsequently diverged in each of the different lineages. In the case of *Alu* Sines, this assumption is incorrect (Figure 3); *Alu*'s are young relative to mammalian radiation. The best test for sequence conversion is to compare orthologous sequences, i.e., the corresponding sequences in two individuals, and to determine if one was altered. The comparison of several orthologous *Alu* loci in human versus chimpanzee or orangutan near globin marker genes (*vide supra*) shows that there was no conversion of these *Alu* sequences in either lineage since the divergence of human and chimpanzee. A corresponding experiment has not yet been performed for the case of *L1* repeats. Yet another possibility is that Sines and Lines are merely turning over.[15,95] However, in the case of *Alu*, there is no vestigial evidence for a primordial Sine family.[24]

Again, as in the case of Sines, any mobile sequence comprising 5% of a genome is certain to cause effects. *L1* repeats have provided polyadenylation signals for mRNAs resulting from upstream promoters and, as documented by seven examples, have generated polymorphisms by their retroposition.[18,88,90,96] Previously, we noted that both transcriptional orientations of *Alu*'s can be assimilated into the protein coding region of an mRNA. If *Alu*'s can so readily adopt an ORF which is compatible with coding for a functional protein, it seems even more likely that *L1* repeats, with their long preexisting ORFs, could give rise to novel protein products by forming chimeric mRNAs with surrounding genes, but such examples have yet to be found.

7. A Mechanism for Proretrovirus Transposition

Mammalian proretrovirus typically consists of genes encoding the products required for the retrovirus cycle, including reverse transcriptase, integrase, envelope proteins, and several other specialized functions (Figure 5A).[40,41,97] The coding regions are flanked by LTRs which contain a promoter, an enhancer, and a polyadenylation signal. Proretroviruses are usually flanked by short direct repeats having a length of 5 nt. Since the 5' LTR promotes transcription, whereas the 3' LTR polyadenylates transcription, the resulting retroviral RNA has incomplete left and right LTRs (Figure 5A).[41,97] The LTR element R is repeated on both ends of the viral RNA, whereas the LTR elements U5 and U3 are unique to the 5' and 3' ends of the viral RNA (Figure 5A).[41,97] The complete proretroviral LTRs, "U3RU5", are thus composites of the incomplete viral RNA LTRs and are formed by template switching ("copy choice") during reverse transcription (Figure 5A). Retroviral gene structure, the activities of the corresponding gene products, and the mechanism of retroviral reverse transcription have all been excellently reviewed.[41,97]

Our interest here concerns the probable mobility of mammalian proretroviral transposons and its consequences. The reason for this concern is that proretroviral transposons in yeast and *Drosophila* are known to code for products which direct their own mobility and, upon transposition, affect the expression of neighboring genes. We first document these points. Yeast and *Drosophila* proretroviral transposons code for products having homology to retroviral reverse transcriptases and gag-like proteins, among others.[98-100] RNA from the yeast proretroviral transposon, *Ty*, is associated with viral-like particles which have reverse transcriptase.[101] The virus-like particles are not infectious and can be considered as defective retroviruses. Of particular interest, the transposition of *Ty* elements can be driven by transcribing a *Ty* element from an inducible promoter.[33] Marker mutations and the proper removal of an exogenously added intron in this inducible *Ty* element confirm this transposition to be RNA mediated.[33] In this same example, the mature LTRs are also shown to be composites of the incomplete left, RU5, and right, U3R, LTRs present in the mRNA, as schematically depicted (Figure 5A).

Interest in proretroviral transposons stems from regulatory mutations caused by their insertion.[40,41] Given the transcriptional control signals in the LTRs, it is not surprising that a proretrovirus can have important effects on neighboring genes. In yeast and *Drosophila*, these effects include silencing neighboring gene expression, changing inducible promoters into constitutive promoters, and promoting the expression of neighboring transcription units.[42,43] While this variety of effects has not been compiled for mammalian proretroviral transposons, it is likely that their insertion causes similar regulatory mutations.

8. Mammalian Proretroviral Transposons
a. Their Variety is Partially Attributed to a Propensity to Form Mosaics

The two most repetitive mammalian proretroviral transposons are the mouse *IAP* (intracisternal A particle) family and the human THE 1 family (Table 1).[102-104] *IAP* particles are essentially defective retroviral particles and the *IAP* sequence is recognizably a variant proretrovirus.[103,104] The human THE 1 sequence, as well as its predicted translation products, are unrelated to any known retroviral RNA or protein.[102] Additionally, mouse, human, and nonhuman primates each contain a number of lower copy-number defective retrovirus and solitary LTRs.[105-115,125,146,147]

Horak and colleagues have identified particularly interesting relationships between several proretroviral sequences in mouse.[109,110] These authors isolated a repetitive LTR sequence called LTR-IS by reduced stringency hybridization to a xenotropic LTR called 36.1 (Figure 5B).[109] These two LTRs share a homologous sequence labeled "C". The 36.1 LTR is homologous to the Moloney murine leukemia virus LTR (MoMuLV), which was originally used in its isolation ("A" and "B", Figure 5B). Surprisingly, the LTR-IS and MoMuLV

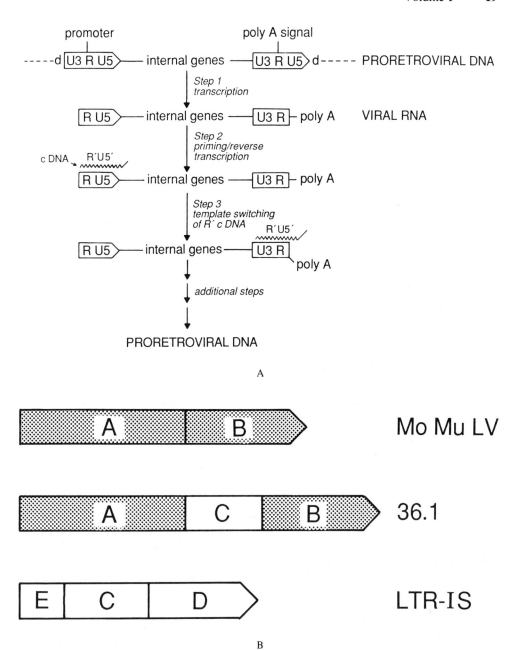

FIGURE 5. (A) Proretroviral transposition involves template switching. Formation of a proretroviral transposon from an existing sequence includes several steps, notably: Step 1, transcription; Step 2, initiation of reverse transcription; Step 3, switching of the reverse transcript primer to a different template and additional steps.[41,97] For our present purposes, we are concerned with the consequence of reverse transcriptase to switch a growing primer to another template.[41] (B) Mosaic relationship of several mouse proretroviral LTRs. These sequences exemplify mosaic structures observed for a number of proretroviral transposons, as discussed in the text. Base sequence analysis shows that MoMuLV (Moloney murine leukemia virus) and LTR-IS are different, although the 36.1 sequence contains elements common to each.[110]

```
ggcatggtctttatCAGCTGCATGAAAACAGATTAATACAGTaaattagtacc          THE 1

||||*||********|||*|||||||**||||||||||||||||**|***|*|||

ggcctgccggggccCAGGTGCATGACCACAGATTAATACAGTtgaggggcacc          gibbon
```

FIGURE 6.   Comparison of human THE 1 and gibbon LTR sequences. Insertion of the gibbon leukemia virus LTR results in the constitutive overexpression of a neighboring interleukin-2 gene.[111] The region depicted is proposed to be part of an enhancer.[111] The human THE 1 sequence is at the 3' end of the LTR ending in the dinucleotide CA, as depicted above in bold-face letters.[102] Lower-case letters depict a breakdown of sequence similarity and the dinucleotide GT depicted in capital, but not bold-face, letters initiates the internal region of the THE 1 sequence.

do not share homologous sequences (Figure 5B). Plausibly, one of these sequences could be formed from the others by template switching during reverse transcription (Figure 5). As further evidence for this interpretation, Propst and Van de Woude have identified another mouse LTR sharing homology to the RU5 region of the LTR-IS.[116] These discoveries highlight the ease with which retroviral RNA recombination can generate novel sequence mosaics; undoubtedly, similar events lead to the recruitment of oncogenes.[41] As one further example of retroviral mosaics, an endogenous human proretrovirus can be represented as a composite of types A, B, C and D retrovirus.[115]

b. IAP Elements

Mouse intracisternal A particles (type A retrovirus, see above) are retroviruses that lack an extracellular form. There are ~1000 repeats per mouse genome, including several shorter structural variants of the parental 7.2-kb element.[144] *IAP* elements are transcribed in various tissues and have been shown to have gene products that are seemingly not part of the retroviral replication pathway. For example, secreted IgE binding factors from a rat-mouse T-cell hybridoma are encoded by the *gag* regions of some of the variant *IAP*s mentioned above.[145] Perhaps the cell has adapted some of these repeats to perform new gene functions.

c. THE 1 Elements

The human THE 1 family is unusual in that, while it structurally resembles a proretroviral transposon, it does not code for any identifiable retroviral products.[102] Plausibly, it results from an RNA-mediated recombination with an unidentified active retrovirus, which caused both its amplification and retroposition. In support of this hypothesis, we have observed that a 5' TTAA 3' recognition sequence is present at the genomic target site of THE 1 insertion.[148] Such a sequence also appears to be the recognition site of the retroviral "integrase" enzyme.[149,150] In addition to the 10,000 THE 1 members, there are an approximately equal number of solitary THE 1 LTRs in human DNA. The large number of these elements in human DNA raises the question of what effects are attributable to its insertion into other transcription units.

The THE 1 LTR does not contain recognizable promoter sequences and it is not observed to promote either *in vivo* or *in vitro* transcription.[117] The THE 1 LTR shares limited similarity with a gibbon leukemia virus LTR (Figure 6). Insertion of this gibbon viral LTR in the 3' end of the interleukin 2 gene resulted in its constitutive expression, prompting the proposal that it contains enhancer elements which include the region of similarity to the THE 1 LTR (Figure 6).[111] Thus, the insertion of mammalian proretroviral transposons exhibits many regulatory effects that are attributable to yeast and *Drosophila* proretroviral transposons.[40] Additionally, the similarity between only one sequence element in the THE 1 and the gibbon viral LTRs resembles Horak's finding that mouse proretroviral transposons exchange LTR elements, giving rise to mosaics.

d. *Transcription and Other Properties of the THE 1 Elements*

A number of discrete-length transcripts in various human tissues and cell lines contain interspersed THE 1 members.[117] Several of these polyadenylated cytoplasmic RNAs are also present in polysomal-size fractions and probably still code for protein products, despite their mutation by THE 1 insertion. THE 1 LTRs provide polyadenylation of the mutated transcripts in two independent cases.[117]

In yet another case, the THE 1 element is inserted into the 3' untranslated region of the resulting mRNA which codes for a Ca^{2+} binding protein.[159] Although the effect of this insertion on the lifetime and translational activity of the resulting mRNA is unknown, it is likely that an additional 2 kb of sequence would affect the level of its expression.

The transposition of THE 1 elements has not been observed; however, these sequences are present on extrachromosomal circular DNAs.[102,118] Circular DNAs are thought to be intermediates in proretroviral transposition and retroposons are also represented on extrachromosomal circular DNA.[119]

Circular DNAs could also result from a simple deletion by homologous recombination. As previously mentioned, deletions involving retroposons such as *Alu* have been observed in several cases. As an example of a THE 1 deletion and its effects, one THE 1 element is part of an immunoglobulin gene duplication.[120] One of these duplicated genes is inactivated by a subsequent deletion which extends into the upstream THE element.

In summary, both retroposon Sines and retroviral THE 1 elements have been deletion end points, have mutated transcripts, and have provided polyadenylation signals for mutated transcripts.

e. *Abrupt Evolutionary Appearance of Novel Proretroviral Transposons*

Although proretroviral transposons may not be highly mobile in contemporary populations, they apparently arise rapidly in an evolutionary sense. The mouse LTR-IS family is not present in rat and some other rodents, but is present in several mouse species, albeit at different loci.[121] Similarly, the human THE 1 element is present in ape and monkey, but is nearly single copy in prosimian primate DNAs, galago, and lemur.[122] Potter and colleagues have developed an ingenius phylogenetic screening approach to detect repetitive DNA which is absent in one species, but present in a closely related species.[125,126] Using this technique, they isolated a proretroviral sequence which is present in ~1000 copies in the white-footed mouse, but absent in house mouse.[125] Similarly, they found that human THE 1 elements are not repetitive in galago.[126] We have isolated the galago homologs to human THE 1 repeats as possible examples of identifiable proretroviral precursors.[97] Although our results are not yet conclusive, rearranged elements of the human THE 1 are identifiable in the galago sequence.[160] Hopefully, the human THE 1 repeat will be a recognizable mosaic of this ancestrally related sequence and provide an example of Temin's proposal that mobile sequences could give rise to retrovirus.[97,127]

E. TANDEMLY ORGANIZED REPEATS
1. Minisatellites Are Interspersed with Single-Copy DNA

Minisatellite DNAs are also interspersed with single-copy sequences, but, unlike retroposons, they do not appear to be inserted by way of RNA intermediates. The best-studied example is "poly(CA)", long tandem runs of alternating CA residues (Table 1).[128] The presence of this repetitive sequence in a wide variety of eukaryotic DNAs was immediately recognized and it was dubbed an evolutionarily conserved family.[128] The reason for its presence in distantly related species is not known. Similarly, we do not know the functions of other identified interspersed tandem repeats, i.e., minisatellites.[23,128-130,136]

However, DNA secondary and tertiary structures are not monotonous, but depend strongly on base sequence.[131-134] Alternating purine/pyrimidine sequences such as poly(CA) are the favored candidates to form left-handed Z DNA.[131] Although not proven, it is easy to imagine that poly(CA) serves a structural role and that other tandem repeats might also adopt unusual structures which are functionally important. In accordance with this view, poly(CA) has been implicated as a hot spot for DNA recombination, perhaps as a result of an unusual secondary structure.[135]

Minisatellites are accordians that can expand and contract in relatively brief evolutionary periods.[67,129,136,137] One possible mechanism for the generation of minisatellites is slippage, which might affect either DNA replication or recombination.[138] Runs of AT pairs have been found to have a novel secondary structure which provides an alternate rationale for understanding slippage.[133] Each A is paired with two T residues: its normal Watson/Crick partner and the neighboring T. In effect, the pairing in this duplex has built-in slippage. Simple sequences are likely to have other unrecognized novel structural features which will be exploited in their biochemical function. On the other hand, it has been observed that in the newt, RNA transcribed from an interspersed satellite undergoes self-cleavage *in vitro*.[139] Transcripts of this satellite are known to exist in the cytoplasm and resemble infectious plant RNAs in several details.[139,140] In this model, the origin of this sequence would be via an RNA intermediate. These two theories can be reconciled if retroposition were to account for the interspersion and replication/structure slippage for the length variability.

A practical consequence of frequent length mutations is that minisatellite-length polymorphisms are useful for mapping applications.[129] Human DNA in general and minisatellites in particular could become the ultimate in fingerprinting the identity and pedigree of individuals.[141]

2. New Techniques May Now Reveal Satellite DNA Structure

Satellites were the first mammalian repeats to be identified. Subsequently, their sequences have been determined in excruciating detail, prompting the insightful criticism that structure alone will not shed any additional light on satellite DNAs.[12] Quite probably, Miklos and John are correct in this criticism; however, the structure of satellite DNAs remains, in one sense, to be determined.

The practical definition of satellite DNAs has been that the length of their tandem organization so greatly exceeded any DNA fragment size which could be conveniently studied that their covalent linkage to other kinds of DNA was undetectable.[142] Satellite DNAs were considered to be essentially homogeneous in sequence and corresponding physical chemical properties. This has now changed.[142] The introduction of megabase electrophoretic methods will, for the first time, permit the structural characterization of complete tandem arrays of satellite DNA. It has long been postulated that satellite DNAs, because of their localization in centromeric heterochromatin and simplicity of sequence organization, were determinants of chromosome structure.[10,12,36,37,143] Interestingly, there are subtle differences between related satellite sequences on different chromosomes.[37] Since the complete structure of a satellite tract may only recently be determined, perhaps their integration into the overall sequence organization of a chromosome can now be elucidated.

Not only has the function of satellite DNA been elusive, but there are also few effects attributable to either satellite DNA or variations in its abundance or sequence.[10,12] Rather, closely related species may differ markedly in their satellite DNA content, e.g., 25% of African green monkey DNA consists of α-satellite DNA, whereas about 10% of the DNA in several macaques consists of this satellite.[10] The relatively rapid expansion/contraction of satellite DNA is not unlike that of minisatellite DNA and may be ascribed to unequal crossing-over.[155] It is noteworthy that in the only two human individuals examined, the length of Y chromosome satellite sequences differs by 100 kb![142]

Nakamura et al. have detected yet another class of satellite called a "midisatellite".[157] This midisatellite is hypervariable in the human population. It differs from other satellites in that it is found only at a single locus near the telomere of chromosome 1, suggesting a chromosome function. Conceivably, other chromosomes are also marked by their individual midisatellites which have not yet been identified.

The sequence simplicity and centromeric organization of satellite DNA have almost forced the notion that they perform a structural role. However, there is now a new appreciation that DNA structure occurs at several different levels. Radic et al. have demonstrated that mouse satellite DNA assumes a bent conformation.[143] Perhaps the structure-function relationship of satellite DNA will only be recognizable and ultimately appreciated at the level of the three-dimensional folding of the complex formed from satellite DNA and specific chromosomal recognition proteins.

III. SUMMARY AND PROSPECTUS

As mentioned, there is now serious consideration of physically mapping the human genome and some investigators are contemplating the complete sequence determination of human DNA. Assuming that some day the DNAs of several individual humans, as well as DNAs from several other species, are completely sequenced, it is worth speculating on what will be learned by analyzing this mass of data. In one sense, we already know the base sequence of 20% of human DNA. Additionally, the overall structural integration of these repetitive sequences into human DNA is also known, albeit at a superficial level. Thus, with respect to repetitive DNAs, we can make some educated guesses as to what will be revealed by comparing complete genomic sequences. Clearly, the mere accumulation of more sequenced members of repetitive DNA families will not in itself be informative. Moreover, it is probable, as reported here, that much of a genome is excess baggage including dead genes. Evolutionary comparisons of genomic sequences would certainly reveal mutations caused by retroposons. Although most retroposons have been fixed in the human genome for long evolutionary periods, there are exceptions; the sheer number of retroposons can turn even a relatively small number of exceptions into a large number of events. Similarly, mutations caused by homologous recombination between retroposons will also be apparent. Such structural comparisons alone would not, of course, identify all of the functional changes which might be attributable to the insertion of repetitive elements.

Another issue is whether the long-range ordering of repetitive sequences in the complete genomic sequence will provide new insights into chromosome structure. At present, there is no way to predict whether this will happen as the rules for chromosome folding have yet to be formulated. One might imagine finding phylogenetically conserved higher-order structures, but the presumptive role of some classes of repetitive sequences in chromosome folding is only indirectly inferred. Conceivably, it is here that complete physical maps would be most informative and provide insight into what is a purely structural problem.

The point that the search for repetitive DNA function may be a sterile exercise is well taken. This critical view should not, of course, discourage experimental tests of well-founded hypotheses. However, we must additionally distinguish between functions and effects. For example, proretrovirus transposons may be the parasitic DNA progeny of previous retroviral infections. Whether they have a cellular function is arguable, but they certainly affect gene structure and expression in a variety of interesting ways. Given that 20% of human DNA consists of an assortment of repetitive sequences, this result is not surprising. By analogy to what is known about the importance of repetitive sequences in *Drosophila* and yeast, we are barely beginning to appreciate these effects in the case of mammalian DNAs. The optimistic view is that a good beginning has now been made.

ACKNOWLEDGMENTS

Our research on repetitive human DNA sequences has been generously supported by

USPHS grant GM21346. Greg Matera is currently supported by a NIH Cellular and Molecular Biology predoctoral traineeship and a University of California biotechnology training grant. We acknowledge Ms. E. Bogren for her expertise in preparing this manuscript.

NOTE ADDED IN PROOF

The primate ID/*Alu* monomer transcript described in Section II.D.3 has been found to be a misidentification of 7SL RNA (Hintz, M. F. and Schmid, C. W., unpublished observations).

REFERENCES

1. **Smith, C. L. and Cantor, C. R.,** Preparation and manipulation of large DNA molecules: advances and applications, *Trends Biochem. Sci.,* 12, 288, 1987.
2. **Ivanov, I. G. and Markov, G. G.,** Heterogeneity of mouse unique deoxyribonucleic acid, *FEBS Lett.,* 47, 323, 1974.
3. **Fox, G. M. and Schmid, C. W.,** Related single copy sequences in the human genome, *Biochim. Biophys. Acta,* 609, 349, 1980.
4. **Ohno, S.,** *Evolution by gene duplication,* Springer Verlag, New York, 1970.
5. **Feeney, R. E. and Allison, R. G.,** *Evolutionary Biochemistry of Proteins,* Wiley Interscience, New York, 1969.
6. **Fornance, A. J., Cummings, D. E., Comeau, C. M., Kant, J. A., and Crabtree, G. R.,** Single-copy inverted repeats associated with regional genetic duplications in γ fibrinogen and immunoglobulin genes, *Science,* 224, 161, 1984.
7. **Britten, R. J. and Kohne, D. E.,** Repeated sequences in DNA, *Science,* 161, 529, 1968.
8. **Brutlag, D. L.,** Molecular arrangement and evolution of heterochromatic DNA, *Annu. Rev. Genet.,* 14, 121, 1980.
9. **Singer, M. F.,** Highly repeated sequences in mammalian genomes, *Int. Rev. Cytol.,* 67, 112, 1982.
10. **Miklos, G. L. G.,** Localized highly repetitive DNA sequences in vertebrate and invertebrate genomes, in *Molecular Evolutionary Genetics,* MacIntyre, R. J., Ed., Plenum Press, New York, 1985.
11. **Davidson, E. H. and Britten, R. J.,** Organization, transcription and regulation in the animal genome, *Q. Rev. Biol.,* 48, 565, 1973.
12. **John, B. and Miklos, G. L. G.,** Functional aspects of satellite DNA and heterochromatin, *Int. Rev. Cytol.,* 58, 1, 1979.
13. **Doolittle, W. F. and Sapienza, C.,** Selfish genes: the phenotypic paradigm and genome evolution, *Nature (London),* 284, 601, 1980.
14. **Orgel, L. E. and Crick, F. H. C.,** Selfish DNA: the ultimate parasite, *Nature (London),* 284, 604, 1980.
15. **Schmid, C. and Shen, C.-K.J.,** The evolution of interspersed repetitive DNA sequences in mammals and other vertebrates, in *Molecular Evolutionary Genetics,* MacIntyre, R. J., Ed., Plenum Press, New York, 1985.
16. **Weiner, A. M., Deininger, P. L., and Efstratiadis, A.,** Nonviral retroposons: genes, pseudogenes and transposable elements generated by the reverse flow of genetic information, *Annu. Rev. Biochem.,* 55, 631, 1986.
17. **Roger, J.,** The origin and evolution of retroposons, *Int. Rev. Cytol.,* 93, 187, 1985.
18. **Fanning, T. G. and Singer, M. F.,** Line 1: a mammalian transposable element, *Biochim. Biophys. Acta,* 910, 203, 1987.
19. **Marks, J., Shaw, J.-P., Perez-Stable, C., Hu, W.-S., Ayres, T. M., Shen, S., and Shen, C.-K.J.,** The primate α-globin gene family: a paradigm of the fluid genome, *Cold Spring Harbor Symp. Quant. Biol.,* 51, 499, 1986.
20. **Proudfoot, N. J. and Maniatis, T.,** The structure of a human α-globin pseudogene and its relationship to α-globin gene duplication, *Cell,* 21, 537, 1980.
21. **Lauer, J., Shen, C.-K.J., and Maniatis, T.,** The chromosomal arrangement of human α-like globin genes: sequence homology and α-globin gene deletions, *Cell,* 20, 119, 1980.
22. **Hardison, R. C., Sawada, I., Cheng, J.-F., Shen, C.-K. J., and Schmid, C. W.,** A previously undetected pseudogene in the human alpha globin gene cluster, *Nucleic Acids Res.,* 14, 1903, 1986.
23. **Proudfoot, N. J., Gil, A., and Maniatis, T.,** The structure of the human zeta-globin gene and a closely linked nearly identical pseudogene, *Cell,* 31, 553, 1982.

24. **Sawada, I. and Schmid, C. W.**, Primate evolution of the α-globin gene cluster and its Alu-like repeats, *J. Mol. Biol.*, 192, 693, 1986.
25. **Michelson, A. M. and Orkin, S. H.**, Boundaries of gene conversion within the duplicated human α globin genes, *J. Biol. Chem.*, 258, 15245, 1983.
26. **Zimmer, E., Martin, S., Beverly, S., Kan, Y., and Wilson, A.**, Rapid duplication and loss of genes coding for the α chains of hemoglobin, *Proc. Natl. Acad. Sci. U.S.A.*, 77, 2158, 1980.
27. **Dover, G. A. and Flavell, R. B.**, Molecular coevolution: DNA divergence and the maintenance of function, *Cell*, 38, 622, 1984.
28. **Gerbi, S. A.**, Evolution of ribosomal DNA, in *Molecular Evolutionary Genetics*, MacIntyre, R. J., Ed., Plenum Press, New York, 1985, 419.
29. **Sawada, I., Beal, M. P., Shen, C.-K. J., Chapman, B., Wilson, A. C., and Schmid, C. W.**, Intergenic DNA sequences flanking the pseudo alpha globin genes of human and chimpanzee, *Nucleic Acids Res.*, 11, 8087, 1983.
30. **Vanin, E. F.**, Processed pseudogenes: characteristics and evolution, *Biochim. Biophys. Acta*, 782, 231, 1984.
31. **Wagner, M.**, A consideration of the origin of processed pseudogenes, *Trends Genet.*, 2, 134, 1986.
32. **Linial, M.**, Creation of a processed pseudogene by retroviral infection, *Cell*, 49, 93, 1987.
33. **Boeke, J. D., Garfinkel, D. J., Styles, C. A., and Fink, G. R.**, Ty elements transpose through an RNA intermediate, *Cell*, 42, 507, 1985.
34. **Soares, M. B., Schon, E., Henderson, A., Karathanasis, S. K., Cate, R., Zeitlin, S., Chirgwin, J., and Efstratiadis, A.**, RNA mediated gene duplication: the rat preproinsulin I, *Mol. Cell. Biol.*, 5, 2090, 1985.
35. **Schmid, C. W. and Deininger, P. L.**, Sequence organization of the human genome, *Cell*, 6, 345, 1975.
36. **Jones, K. W.**, Satellite DNA, *J. Med. Genet.*, 10, 273, 1973.
37. **Willard, H. F. and Waye, J. S.**, Hierarchical order in chromosome specific human alpha satellite DNA, *Trends Genet.*, 3, 192, 1987.
38. **Cooke, H. J. and Hindley, J.**, Cloning of human satellite III DNA: different components are on different chromosomes, *Nucleic Acids Res.*, 10, 3177, 1979.
39. **Singer, M. F.**, Sines and Lines: highly repeated short and long interspersed sequences in mammalian genomes, *Cell*, 28, 433, 1982.
40. **Varmus, H.**, Form and function of retroviral provirus, *Science*, 216, 812, 1982.
41. **Varmus, H.**, Retrovirus, in *Mobile Genetic Elements*, Shapiro, J. A., Ed., Academic Press, New York, 1983, chap. 10.
42. **Roeder, G. S. and Fink, G. R.**, Transposable elements in yeast, in *Mobile Genetic Elements*, Shapiro, J. A., Ed., Academic Press, New York, 1983, chap. 6.
43. **Rubin, G. M.**, Dispersed repetitive DNAs in *Drosophila*, in *Mobile Genetic Elements*, Shapiro, J. A., Ed., Academic Press, New York, 1983, chap. 8.
44. **Fink, G. R.**, Pseudogenes in yeast?, *Cell*, 49, 5, 1987.
45. **Lee, M. G.-S., Loomis, C., and Cowan, N. J.**, Sequence of an expressed human tubulin gene containing ten Alu family members, *Nucleic Acids Res.*, 12, 5823, 1984.
46. **Friezner Degen, S. J., Rajput, B., and Reich, E.**, The human tissue plasminogen-activator gene, *J. Biol. Chem.*, 261, 6972, 1986.
47. **Wiginton, D. A., Kaplan, D. J., States, J. C., Akeson, A. L., Perme, C., Bilyk, I. J., Vaughin, A. J., La Hier, D. L., and Hutton, J. J.**, Complete sequence and structure of the gene for human adenosine deaminase, *Biochemistry*, 25, 8234, 1986.
48. **Deininger, P. L. and Daniels, G. R.**, The recent evolution of mammalian repetitive DNA elements, *Trends Genet.*, 2, 76, 1986.
49. **Jelinek, W. R. and Schmid, C. W.**, Repetitive sequences in eukaryotic DNA and their expression, *Annu. Rev. Biochem.*, 51, 813, 1982.
50. **Konarska, M. M., Padgett, R. A., and Sharp, P. A.**, Trans-splicing of mRNA precursors *in vitro, Cell*, 42, 165, 1983.
51. **Clemens, M. J.**, A potential role for RNA transcribed from B2 repeats in the regulation of mRNA stability, *Cell*, 49, 157, 1987.
52. **Siegel, V. and Walter, P.**, Removal of the Alu structural domain from signal recognition particle leaves its protein translocation activity intact, *Nature (London)*, 320, 81, 1986.
53. **Paulson, K. E. and Schmid, C. W.**, Transcriptional inactivity of Alu repeats in HeLa cells, *Nucleic Acids Res.*, 14, 6145, 1986.
54. **Watson, J. B. and Sutcliffe, J. G.**, Primate brain-specific cytoplasmic transcript of the Alu repeat family, *Mol. Cell. Biol.*, 7, 3324, 1987.
55. **Haynes, S. R. and Jelinek, W. R.**, Low molecular weight RNAs transcribed *in vitro* by RNA polymerase III from Alu-type dispersed repeats in Chinese hamster DNA are also found *in vivo, Proc. Natl. Acad. Sci. U.S.A.*, 78, 6130, 1981.

56. **Young, P. R., Scott, R. W., Hamer, D. W., and Tilghman, S. M.,** Construction and expression *in vivo* of an internally deleted mouse α-fetoprotein gene: presence of a transcribed Alu-like repeat within the first intervening sequence, *Nucleic Acids Res.,* 10, 3099, 1982.

57. **Kamerov, D. A., Lekakh, I. V., Samarina, O. P., and Ryskov, A. P.,** The sequences homologous to major interspersed repeats B1 and B2 of mouse genome are present in mRNA and small cytoplasmic poly(A)$^+$RNA, *Nucleic Acids Res.,* 10, 7477, 1982.

58. **Kamerov, D. A., Tillib, S. V., Ryskov, A. P., and Georgiev, G. P.,** Nucleotide sequence of small polyadenylated B2 RNA, *Nucleic Acids Res.,* 13, 6423, 1985.

59. **Trabuchet, G., Chebloune, Y., Savatier, P., Lachuer, J., Faure, C., Verdier, G., and Nigon, V. M.,** Recent insertion of an Alu sequence in the beta-globin gene cluster of the gorilla, *J. Mol. Evol.,* 25, 288, 1987.

60. **Singh, K., Carey, M., Saragosti, S., and Botchan, M.,** Expression of enhanced levels of small RNA polymerase III transcripts encoded by the B2 repeats in simian virus 40 — transformed mouse cells, *Nature (London),* 314, 553, 1985.

61. **Glaichenhaus, N. and Cuzin, F.,** A role for ID repetitive sequences in growth and transformation dependent regulation of gene expression in rat fibroblasts, *Cell,* 50, 1081, 1987.

62. **Sutcliffe, J. G., Milner, R. J., Gottesfeld, J. M., and Lerner, R. A.,** Identifier sequences are transcribed specifically in brain, *Nature (London),* 308, 237, 1984.

63. **Owens, G. P., Chaudhari, N., and Hahn, W. E.,** Brain identifier sequence is not restricted to brain: similar abundance in nuclear RNA of other organs, *Science,* 229, 1263, 1985.

64. **Lone, Y., Simon, M. P., Kahn, A., and Marie, J.,** Sequences complementary to the brain-specific "identifier" sequences exist in L-type pyruvate kinase mRNA (a liver-specific messenger) and in transcripts especially abundant in muscle, *J. Biol. Chem.,* 261, 1499, 1986.

65. **Sapienza, C. and St. Jacques, B.,** Brain-specific transcription and evolution of the identifier sequence, *Nature (London),* 319, 418, 1986.

66. **Willard, C., Nguyen, H. T., and Schmid, C. W.,** Existence of at least three distinct Alu subfamilies, *J. Mol. Evol.,* 26, 180, 1987.

67. **Sawada, I., Willard, C., Shen, C.-K. J., Chapman, B., Wilson, A. C., and Schmid, C. W.,** Evolution of Alu family repeats since the divergence of human and chimpanzee, *J. Mol. Evol.,* 22, 316, 1985.

68. **Maeda, N., Bliska, J. B., and Smithies, O.,** Recombination and balanced chromosome polymorphism suggested by DNA sequences 5′ to the human δ-globin gene, *Proc. Natl. Acad. Sci. U.S.A.,* 80, 5012, 1983.

69. **Koop, B. F., Miyamoto, M. M., Embury, J. E., Goodman, M., Czelusniak, J., and Slightom, J. L.,** Nucleotide sequence and evolution of the orangutan epsilon globin gene region and surrounding Alu repeats, *J. Mol. Evol.,* 24, 94, 1986.

70. **Daniels, G. R. and Deininger, P. L.,** A second major class of Alu family repeated DNA sequences in a primate genome, *Nucleic Acids Res.,* 11, 7595, 1983.

71. **Daniels, G. R., Fox, G. M., Lowensteiner, D. L., Schmid, C. W., and Deininger, P. L.,** Species-specific homogeneity of the primate Alu family of repeated DNA sequences, *Nucleic Acids Res.,* 11, 7579, 1983.

72. **Fukumaki, Y., Collins, F., Kole, R., Stoeckert, C. J., Jagadeeswaran, P., Duncan, C. H., and Weissman, S. M.,** Sequence of human repetitive DNA, *Cold Spring Harbor Symp. Quant. Biol.,* 47, 1079, 1983.

73. **Economou-Pachnis, A. and Tsichlis, P. N.,** Insertion of an Alu Sine in the human homologue of the Mlvi-2 locus, *Nucleic Acids Res.,* 13, 8379, 1985.

74. **Hwu, H. R., Roberts, J. W., Davidson, E. H., and Britten, R. J.,** Insertion and/or deletion of many repeated DNA sequences in human and higher ape evolution, *Proc. Natl. Acad. Sci. U.S.A.,* 83, 3875, 1986.

75. **Caras, I. W., Davitz, M. A., Rhee, L., Weddell, G., Martin, D. W., and Nussenzweig, V.,** Cloning of decay accelerating factor suggests novel use of splicing to generate two proteins, *Nature (London),* 325, 545, 1987.

76. **Sharma, S., Mehta, S., Morgan, J., and Maize, A.,** Molecular cloning and expression of a human B-cell growth factor gene in *E. coli, Science,* 235, 1489, 1987.

77. **Kress, M., Barra, Y., Seidman, J. G., Khoury, G., and Jay, G.,** Functional insertion of an Alu type 2 (B2 Sine) repetitive sequence in murine class 1 gene, *Science,* 226, 974, 1984.

78. **Raghow, R.,** Regulation of messenger RNA turnover in eukaryotes, *TIBS,* 12, 358, 1987.

79. **Lehrman, M. A., Goldstein, J. L., Russell, D., and Brown, M. S.,** Duplication of seven exons in LDL receptor gene caused by Alu-Alu recombination in a subject with familial hypercholesterolemia, *Cell,* 48, 827, 1987.

80. **Barsh, G. S., Seeburg, P. H., and Gelinas, R. E.,** The human growth hormone gene family: structure and evolution of the chromosomal locus, *Nucleic Acids Res.,* 11, 3939, 1983.

81. **Rouyer, F., Simmler, M.-C., Page, D. C., and Weissenbach, J.**, A sex chromosome rearrangement in a human XX male caused by Alu-Alu recombination, *Cell*, 51, 417, 1987.

82. **Hess, J. F., Fox, M., Schmid, C. W., and Shen, C.-K. J.**, Molecular evolution of the human adult α-globin-like region: insertion and deletion of Alu family repeats and non-Alu DNA sequences, *Proc. Natl. Acad. Sci. U.S.A.*, 80, 5970, 1983.

83. **Schimenti, J. C. and Duncan, C. H.**, Ruminant globin gene structures suggest an evolutionary role for Alu-type repeats, *Nucleic Acids Res.*, 12, 1641, 1984.

84. **Skowronski, J. and Singer, M. F.**, Expression of a cytoplasmic Line-1 transcript is regulated in a human teratocarcinoma cell line, *Proc. Natl. Acad. Sci. U.S.A.*, 82, 6050, 1985.

85. **Skowronski, J., Fanning, T. G., and Singer, M. F.**, Unit length Line-1 transcripts in human teratocarcinoma cells, *Mol. Cell. Biol.*, in press.

86. **Loeb, D. D., Padgett, R. W., Hardies, S. C., Shehee, W. R., Comer, M. W., Edgell, M. H., and Hutchison, C. A.**, The sequence of a large LIM δ element reveals a tandemly repeated 5′ end and several features found in retroposons, *Mol. Cell. Biol.*, 6, 168, 1986.

87. **Singer, M. F., Thayer, R. E., Grimaldi, G., Lehrman, M. I., and Fanning, T.**, Homology between the Kpn primate and Bam H1 (MIF) rodent families of long interspersed repeated sequences, *Nucleic Acids Res.*, 11, 5739, 1983.

88. **Skowronski, J. and Singer, M. F.**, The abundant Line-1 family of repeated DNA sequences in mammals: genes and pseudogenes, *Cold Spring Harbor Symp. Quant. Biol.*, 51, 457, 1986.

89. **Sakaki, Y., Hattori, M., Fujita, A., Yoshioka, K., Kuhara, S., and Takenaka, O.**, The Line-1 family of primates may encode a reverse transcriptase-like protein, *Cold Spring Harbor Symp. Quant. Biol.*, 51, 465, 1986.

90. **Kole, L. B. N., Haynes, S. R., and Jelinek, W. R.**, Discrete and heterogenous high molecular weight RNAs complementary to a long dispersed repeat family (a possible transposon) of human DNA, *J. Mol. Biol.*, 165, 257, 1983.

91. **Jubier-Maurin, V., Winker, P., Cuny, G., and Roizes, G.**, The relationship between the 5′ end repeats and largest members of the L1 interspersed repeated family in the mouse genome, *Nucleic Acids Res.*, 15, 7395, 1987.

92. **Jubier-Maurin, V., Dod, B. J., Bellis, M., Piechaczyk, M., and Roizes, G.**, Comparative study of the L1 family in the genus mus: possible role of retroposition and conversion events in its concerted evolution, *J. Mol. Biol.*, 184, 547, 1985.

93. **Rogers, J.**, Long interspersed sequences in mammalian DNA: properties of newly identified specimens, *Biochim. Biophys. Acta*, 824, 113, 1985.

94. **Brown, S. D. M. and Dover, G.**, Organization and evolutionary progress of a dispersed repetitive family of sequences in widely separated rodent genomes, *J. Mol. Biol.*, 150, 441, 1981.

95. **Martin, S. L., Voliva, C. F., Hardies, S. C., Edgell, M. H., and Hutchison, C. A.**, Tempo and mode of concerted evolution in the L1 repeat family of mice, *Mol. Biol. Evol.*, 2, 127, 1985.

96. **Lakshmikumaran, M. S., D'Ambrosio, E., Laimins, L. A., Lin, D. T., and Furano, A. V.**, Long interspersed repeated DNA (Line) causes polymorphism at the rat insulin locus, *Mol. Cell. Biol.*, 5, 2197, 1985.

97. **Temin, H. M.**, Reverse transcription in the eukaryotic genome: retrovirus, pararetrovirus, retrotransposons and retrotranscripts, *Mol. Biol. Evol.*, 2, 455, 1985.

98. **Clare, J. K. and Farabaugh, P.**, Nucleotide sequence of a yeast Ty element: evidence for an unusual mechanism of gene expression, *Proc. Natl. Acad. Sci. U.S.A.*, 82, 2829, 1985.

99. **Saigo, K., Kugimiya, W., Matsuo, Y., Inouye, S., Yoshioka, K., and Yuki, S.**, Identification of the coding sequence for a reverse transcriptase-like enzyme in a transposable genetic element in *Drosophila melanogaster*, *Nature (London)*, 312, 659, 1984.

100. **Mount, S. and Rubin, G.**, Complete nucleotide sequence of the *Drosophila* transposable element Copia: homology between Copia and retroviral proteins, *Mol. Cell. Biol.*, 5, 1630, 1985.

101. **Garfinkel, D. J., Boeke, J. D., and Fink, G. R.**, Ty element transposition: reverse transcriptase and virus-like particles, *Cell*, 42, 507, 1985.

102. **Paulson, K. E., Deka, N., Schmid, C. W., Misra, R., Schindler, C. W., Rush, M. G., Kadyk, L., and Leinwand, L.**, A transposon-like element in human DNA, *Nature (London)*, 316, 359, 1985.

103. **Lueders, K. K. and Kuff, E. L.**, Sequences associated with intracisternal A particles are reiterated in the mouse genome, *Cell*, 12, 963, 1977.

104. **Ono, M., Cole, M. D., White, A. T., and Huang, R. C. C.**, Sequence organization of cloned intracisternal A particle genes, *Cell*, 21, 465, 1980.

105. **Mager, D. L. and Henthorn, P. S.**, Identification of a retrovirus-like repetitive element in human DNA, *Proc. Natl. Acad. Sci. U.S.A.*, 81, 7510, 1984.

106. **Steele, P. E., Rabson, A. B., Bryant, T., and Martin, M. A.**, Distinctive termini characterize two families of human endogenous retroviral sequences, *Science*, 225, 943, 1984.

107. **O'Connell, C., O'Brien, S., Nash, W. G., and Cohen, M.**, ERV3, a full-length human endogenous provirus: chromosomal location and evolutionary relationships, *Virology*, 138, 225, 1984.

108. **Repaske, R., Steele, P. E., O'Neil, R. R., Rabson, A. B., and Marti, M. A.,** Nucleotide sequence of a full-length human endogenous retroviral segment, *J. Virol.,* 54, 764, 1985.

109. **Wirth, T., Gloggler, K., Baumrucker, T., Schmidt, M., and Horak, I.,** Family of middle repetitive DNA sequences in the mouse genome with structural features of solitary retroviral long terminal repeats, *Proc. Natl. Acad. Sci. U.S.A.,* 80, 3327, 1983.

110. **Schmidt, M., Gloggler, K., Wirth, T., and Horak, I.,** Evidence that a major class of mouse endogenous long terminal repeats (LTRs) resulted from recombination between exogenous retroviral LTRs and similar LTR-like elements (LTR-IS), *Proc. Natl. Acad. Sci. U.S.A.,* 81, 6696, 1984.

111. **Chen, S. J., Holbrook, N. J., Mitchell, K. F., Vallone, C. A., Greengard, J. S., Crabtree, G. R., and Lin, Y.,** A viral long terminal repeat in the interleukin 2 gene of a cell line that constitutively produces interleukin 2, *Proc. Natl. Acad. Sci. U.S.A.,* 82, 7284, 1985.

112. **Tamura, T., Noda, M., and Takano, T.,** Structure of the baboon endogeneous virus genome: nucleotide sequences of the long terminal repeats, *Nucleic Acids Res.,* 9, 6615, 1981.

113. **Bonner, T. I., O'Connell, C., and Cohen, M.,** Cloned endogenous retroviral sequences from human DNA, *Proc. Natl. Acad. Sci. U.S.A.,* 79, 4709, 1982.

114. **Nada, M., Kurihara, M., and Takano, T.,** Retrovirus-related sequence in human DNA detection and cloning of sequences which hybridize with the long terminal repeat of baboon endogenous virus, *Nucleic Acids Res.,* 10, 2865, 1982.

115. **Callahjan, R., Chiu, I.-M, Wong, J. F. H., Tronick, S. R., Roe, B. A., Aaronson, S. A., and Schlom, J.,** A new class of endogenous human retroviral genomes, *Science,* 228, 1208, 1985.

116. **Probst, F. and Van de Woude, G.,** A novel transposon-like repeat interrupted in the mouse c-mos locus, *Nucleic Acids Res.,* 12, 8381, 1984.

117. **Paulson, K. E., Matera, A. G., Deka, N., and Schmid, C. W.,** Transcription of a human transposon-like sequence is usually directed by other promoters, *Nucleic Acids Res.,* 15, 5199, 1987.

118. **Misra, R., Rush, M., Wong, E., and Schmid, C. W.,** Cloned extrachromosomal circular DNA copies of the human transposable element THE-1 are related predominantly to a single type of family member, *J. Mol. Biol.,* 196, 1, 1987.

119. **Krolewski, J. J., Bertelsen, A. H., Humayun, M. Z., and Rush, M. G.,** Members of the Alu family of interspersed repetitive DNA sequences are in the small circular DNA populations of monkey cells grown in culture, *J. Mol. Biol.,* 154, 399, 1982.

120. **Hisajima, H., Nishida, Y., Nakai, S., Takahashi, N., Ueda, S., and Honjo, T.,** Structure of the human immunoglobulin Cε2 gene, a truncated pseudogene: implications for its evolutionary origin, *Proc. Natl. Acad. Sci. U.S.A.,* 80, 2995, 1983.

121. **Wirth, T., Schmidt, M., Baumraker, T., and Horak, I.,** Evidence for the mobility of a new family of mouse middle repetitive DNA elements (LTR-IS), *Nucleic Acids Res.,* 12, 3603, 1984.

122. **Deka, N., Paulson, K. E., Willard, C., and Schmid, C. W.,** Properties of a transposon-like human element, *Cold Spring Harbor Symp. Quant. Biol.,* 51, 473, 1986.

123. **Jelinek, W. R., Toomey, T. P., Leinwand, L., Duncan, C., Biro, P. A., Choudary, P. V., Weissman, C. M., Rubin, C. M., Houck, C. M., Deininger, P. L., and Schmid, C. W.,** Ubiquitous interspersed repeated sequences in mammalian genomes, *Proc. Natl. Acad. Sci. U.S.A.,* 77, 1398, 1980.

124. **Johnson, E. M. and Jelinek, W. R.,** Replication of a plasmid bearing a human Alu-family repeat in monkey cos-7 cells, *Proc. Natl. Acad. Sci. U.S.A.,* 83, 4660, 1986.

125. **Wichman, H. A., Potter, S. S., and Pine, D. S.,** Mys:a family of mammalian transposable elements isolated by phylogenetic screening, *Nature (London),* 317, 77, 1985.

126. **Loyd, J. A., Lamb, A. N., and Potter, S. S.,** Phylogenetic screening of the human genome: identification of differently hybridizing repetitive sequence families, *Mol. Biol. Evol.,* 4, 85, 1987.

127. **Temin, H. M.,** Origin of retroviruses from cellular moveable genetic elements, *Cell,* 21, 599, 1980.

128. **Miesfeld, R., Krystal, M., and Arnheim, N.,** A member of a new repeated sequence family which is conserved throughout eukaryotic evolution is found between the human δ and β globin genes, *Nucleic Acids Res.,* 9, 5931, 1981.

129. **Jeffreys, A. J., Wilson, V., and Thein, S. L.,** Hypervariable minisatellite regions in human DNA, *Nature (London),* 314, 67, 1985.

130. **Gebhard, W. and Zachau, H. G.,** Simple DNA sequences and dispersed repetitive elements in the vicinity of mouse immunoglobulin κ light chain genes, *J. Mol. Biol.,* 170, 567, 1983.

131. **Zimmerman, S. B.,** The three dimensional structure of DNA, *Annu. Rev. Biochem.,* 51, 395, 1982.

132. **Trifonov, E. N.,** Curved DNA, *CRC Crit. Rev. Biochem.,* 19, 89, 1985.

133. **Nelson, H. C. M., Finch, J. T., Luisi, B. F., and Klug, A.,** The structure of an oligo(dA):oligo(dT) tract and its biological implications, *Nature (London),* 330, 221, 1987.

134. **Zahn, K. and Blattner, F. R.,** Direct evidence for DNA bending at the lambda replication origin, *Science,* 236, 416, 1987.

135. **Slightom, J. L., Blechl, A. F., and Smithies, O.,** Human fetal Gγ- and Aγ-globin genes: complete nucleotide sequences suggest that DNA can be exchanged between these duplicated genes, *Cell,* 21, 627, 1980.

136. **Willard, C. W., Wong, E., Hess, J. F., Shen, C.-K. J., Chapman, B., Wilson, A. C., and Schmid, C. W.,** Comparison of human and chimpanzee ζ globin genes, *J. Mol. Evol.,* 22, 309, 1985.

137. **Goodbourn, S. E. Y., Higgs, D. R., Clegg, J. B., and Weatherall, D. J.,** Molecular basis of length polymorphism in the human ζ-globin gene complex, *Proc. Natl. Acad. Sci. U.S.A.,* 80, 5022, 1983.

138. **Hentschel, C. C.,** Homocopolymer sequences in the spacer of a sea urchin histone gene repeat are sensitive to S1 nuclease, *Nature (London),* 295, 714, 1982.

139. **Epstein, L. M. and Gall, J. G.,** Self-cleaving transcripts of satellite DNA from the newt, *Cell,* 48, 535, 1987.

140. **Epstein, L. M., Mahon, K. H., and Gall, J. G.,** Transcription of a satellite DNA in the newt, *J. Cell Biol.,* 103, 1137, 1986.

141. **Jeffreys, A. J., Brookfield, J. F. Y., and Semeonoff, R.,** Positive identification of an immigration test-case using human DNA fingerprints, *Nature (London),* 317, 818, 1985.

142. **Tyler-Smith, C. and Brown, W. R. A.,** Structure of the major block of alphoid satellite DNA on the human Y chromosome, *J. Mol. Biol.,* 195, 457, 1987.

143. **Radic, M. Z., Lundgren, K., and Hamkalo, B. A.,** Curvature of mouse satellite DNA and condensation of heterochromatin, *Cell,* 50, 1101, 1987.

144. **Grossman, Z., Mietz, J. A., and Kuff, E. L.,** Nearly identical members of the heterogeneous IAP gene family are expressed in thymus of different mouse strains, *Nucleic Acids Res.,* 15, 3823, 1987.

145. **Kuff, E. L., Mietz, J. A., Trounstine, M. L., Moore, K. W., and Martens, C. L.,** cDNA clones encoding murine IgE-binding factors represent multiple structural variants of intracisternal A-particle genes, *Proc. Natl. Acad. Sci. U.S.A.,* 83, 6583, 1986.

146. **Ono, M., Kawakami, M., and Takezawa, T.,** A novel human nonviral retroposon derived from endogenous retrovirus, *Nucleic Acids Res.,* 15, 8725, 1987.

147. **Harada, F., Tsukada, N., and Kato, N.,** Isolation of three kinds of human endogenous retrovirus-like sequences using tRNA(pro) as a probe, *Nucleic Acids Res.,* 15, 9153, 1987.

148. **Deka, N., Willard, C., Wong, E., and Schmid, C. W.,** Human transposon-like elements insert at a preferred target site: evidence of a retrovirally mediated process, *Nucleic Acids Res.,* 16, 1143, 1988.

149. **Colicelli, J. and Goff, S. P.,** Mutants and pseudorevertants of Moloney murine leukemia virus with alterations at the integration site, *Cell,* 42, 573, 1985.

150. **Panganiban, A. T. and Temin, H. M.,** Circles with two tandem LTRs are precursors to integrated retrovirus DNA, *Cell,* 26, 673, 1984.

151. **Manuelidis, L.,** Repeating restriction fragments of human DNA, *Nucleic Acids Res.,* 3, 3063, 1976.

152. **Rinehart, F. P., Ritch, T. G., Deininger, P. L., and Schmid, C. W.,** Renaturation rate studies of a single family of interspersed repeated sequences in human deoxyribonucleic acid, *Biochemistry,* 20, 3003, 1981.

153. **Sun, L., Paulson, K. E., Schmid, C. W., Kadyk, L., and Leinwand, L.,** Non Alu family interspersed repeats in human DNA and their transcriptional activity, *Nucleic Acids Res.,* 12, 2669, 1983.

154. **Hamada, H., Petrino, M. G., and Kakunga, T.,** A novel repeated element with Z-DNA-forming potential is widely found in evolutionary diverse eukaryotic genomes, *Proc. Natl. Acad. Sci. U.S.A.,* 79, 6465, 1982.

155. **Waye, J. S. and Willard, H. F.,** Molecular analysis of a deletion polymorphism in alpha satellite of human chromosome 17: evidence for homologous unequal crossing-over and subsequent fixation, *Nucleic Acids Res.,* 14, 6915, 1986.

156. **Prossner, J., Fromm, M., Paul, C., and Vincent, P. C.,** Sequence relationships of three human satellite DNAs, *J. Mol. Biol.,* 187, 145, 1986.

157. **Nakamura, Y., Julier, C., Wolff, R., Holm, T., O'Connell, P., Leppert, M., and White, R.,** Characterization of a human midisatellite sequence, *Nucleic Acids Res.,* 15, 2537, 1987.

158. **Uhlenbeck, O.,** personal communication.

159. **Deka, N., Wong, E., Matera, A. G., Kraft, R., Leinwand, L., and Schmid, C. W.,** Repetitive nucleotide sequence insertions into a novel calmodulin-related gene and its processed pseudogene, *Gene,* 71, 123, 1988.

160. **Schmid, C. W.,** unpublished observations.

Chapter 2

SEQUENCE-SPECIFIC DNA-BINDING PROTEINS INVOLVED IN GENE TRANSCRIPTION

Graham H. Goodwin, Geoffrey A. Partington, and Neil D. Perkins

TABLE OF CONTENTS

I. INTRODUCTION

The transcriptional activation of a fully repressed eukaryotic gene involves a number of regulatory steps. The first probably requires that histones be dissociated from specific DNA regions near or within the gene, uncovering DNA sequences which are required for transcriptional activity of the gene. The sequence elements, termed promoters and enhancers, are composed of multiple short-sequence motifs which bind sequence-specific nonhistone proteins. These proteins, once bound to their sequences, promote the binding of RNA polymerase and its associated transcription factors to form a stable initiation complex. In some cases (for example, genes expressed in specific tissues), activation of the gene is accompanied by the demethylation of 5-methyl cytosine bases, though it is not known what role this plays in gene expression. Demethylation could prevent the binding of inhibitory proteins or it could allow the binding of ubiquitous DNA-binding factors required for gene transcription.[1] Most of the regulatory DNA sequences so far described are positively acting elements (that is, they stimulate transcription of the gene), but there are a number of examples of inhibitory or negative regulatory sequences. For example, within the immunoglobulin enhancer region, there are sequences which inhibit its activity in fibroblasts, presumably contributing to the total repression of this gene in nonlymphoid cells.[2] In the case of the interferon gene, its activity is modulated by double-stranded RNA through dissociation of a repressor protein from a regulatory sequence.[3] In the last few years, our understanding of how transcriptional regulatory sequences function has dramatically advanced as a result of the development of techniques to detect and purify proteins which bind to these sequences. The purpose of this chaper is to review the protein factors that have been found binding to enhancers and promoters, the main emphasis being on proteins found in mammalian and avian cells binding to genes transcribed by RNA polymerases II and III. For descriptions of RNA polymerase I transcription factors, yeast transcription factors, and the *Drosophila* regulatory genes involved in development, the reader is referred to papers and reviews by Learned et al.,[4] Struhl,[5] and Gehring.[6] A full compilation of eukaryotic transcription factors is given by Wingender.[7]

A typical polymerase II transcribed gene often has a TATA-box motif (consensus sequence $TATA_T^A A_T^A$) approximately 30 bp upstream of the start of transcription (the Cap site). This region, bracketed by the TATA box and the Cap site, probably binds RNA polymerase as well as TATA-box DNA-binding factor(s) and will be termed the promoter. The TATA box is required for correct initiation at the Cap site. Most genes have, just upstream of the TATA box, one or more upstream promoter elements (UPE) which are required for efficient transcription. For example, the β-globin genes have CAAT-box and CACACCC motifs upstream of the TATA box. The sequence GGGGCGG which binds the protein Sp1 is often found near the TATA box, but is also frequently found upstream of those polymerase-II-transcribed genes that do not have a TATA box. Genes that are inducible by extracellular stimuli such as growth factors and genes that are expressed in a tissue-specific manner have additional elements, termed enhancers, which may reside near the two 5′ upstream promoter elements, but often occur at considerable distances 5′ or 3′ to the gene. These elements are responsible for the enhanced transcription of induced genes and for the tissue-specific activation of the genes. Many enhancers have multiple redundant functional motifs which bind different proteins, and it is likely that these proteins interact with promoter- and UPE-bound proteins to fully activate the genes. At the moment it is not clear whether enhancers and upstream promoter elements are truly functionally distinct elements since motifs found in the latter have been found in the former and vice versa. Nevertheless, multimerization of a single motif found in an enhancer produces an effective enhancer and yet multimers of the UPE CCAAT-box motif do not,[8] suggesting a functional difference (though this may simply reflect weaker interactions between the CCAAT-box factors and the transcriptional machinery). Also, enhancers may be required to initially activate the fully repressed gene, but once the gene is being transcribed, they may no longer be required to maintain the transcriptionally active state[9] (for a further discussion of the function of enhancers and promoters, see the review by Maniatis et al.[10]). A possible model of how these elements function is to consider the UPE as a unit which binds factors which, acting over short distances through protein-protein interactions with promoter factors (TATA-box factors and polymerase), enhance the binding of the promoter factors to their sequences. If the gene also has an enhancer, the enhancer as a unit binds multiple proteins which can interact over long distances with one or more of the UPE factors, and the combined enhancer-UPE nucleoprotein complex then enhances the binding of promoter factors more effectively than a UPE alone.

Eukaryotic viruses appear to transcribe their genomes by very much the same mechanisms used by cellular genes. Thus, DNA viruses such as SV40 and polyoma have promoter, UPE, and enhancer elements. Similarly, the proviral DNA of RNA retroviruses contain such elements in their long terminal repeats (LTRs). Both types of viruses use host DNA-binding factors to activate their transcription and, indeed, many of the factors first described as being utilized by eukaryotic viruses have subsequently been found to be used by cloned eukaryotic genes. Thus, this review will describe factors binding to viral regulatory sequences before going on to detail those found binding to cellular genes.

<u>SV40 UPSTREAM PROMOTER</u>

Sp1 BINDING SITES:

```
        VI              V             IV              III            II              I
      ------         ------                         ------         ------         ------
5'-TCCCGCCCCTAACTCCGCCCATCCCGCCCCCTAACTCCGCCCAGTTCCGCCCATTCTCCGCCCCA
```

```
Sp1 CONSENSUS:     GGGGCGGGGC
                   ‾‾         ‾‾‾‾
                   TA        TAAT
```

FIGURE 1A.   The six GC boxes in the SV40 upstream promoter, five of which are bound by Sp1 (dashed lines above sequence). The Sp1 consensus binding site is shown below.

<u>SV40 ENHANCER</u>

```
   ---GT-IIA-----                              ------NFκB-----
        -GT-IIB-                         ---------CBP--------
        ---GT-IIC--                      ----------AP2------------
5'-CGGAAAGAACCAGCTGTGGAATGTGTGTCAGTTAGGGTGTGGAAAGTCCCCAGGCTCCCCAG
            GT-II                           GT-I
```

```
             --------                  ---AP1--
CAGGCAGAAGTATGCAAAGCATGCATCTCAATTAGTCAGCAACCAG-3'
           O                          P
```

FIGURE 1B.   The SV40 enhancer, showing the GT-II, GT-I, octamer (O), and P motifs bound by protein (dashed lines above sequence show protein binding sites).

II. PROTEINS BINDING TO TRANSCRIPTION ELEMENTS IN VIRUSES

A. SV40 VIRUS

Probably the best understood set of eukaryotic transcriptional activating sequences are those of SV40. Upstream of the TATA-box motif there is an upstream promoter element (UPE) consisting of the so-called 21-bp repeats which have five binding sites for the cellular protein Sp1 (Figure 1A). Further upstream there is an enhancer residing within a duplicated region (the 72-base-pair repeats) plus some additional upstream sequences (Figure 1B). The UPE and the enhancer have been shown to bind multiple proteins, some of which differ in different cell types.

1. The Upstream Promoter

The transcription factor Sp1 was first detected in HeLa cell extracts due to its ability to bind to the multiple GC boxes of the 21-bp repeats in the SV40 UPE and stimulate transcription in an *in vitro* system.[11,12] *In vitro* DNAse I footprinting and dimethyl sulfate methylation protection experiments have demonstrated that Sp1 binds to only five of the six GC boxes contained in the three 21-bp repeats with varying degrees of affinity (Figure 1A). Binding to box V seemed to exclude binding from GC box IV and there was no evidence of cooperativity at any of the sites.[13] Sp1 makes protein-DNA contacts in the major groove of the DNA, and in SV40 each binding site occurs once per turn of the DNA helix such that the five bound promoters are all aligned on the same face of the DNA[14] In a monkey promoter containing multiple Sp1 binding sites, however, the Sp1 promoters bind to opposite faces of the DNA. The SV40 21-bp repeats have been shown to be able to stimulate transcription in the late gene direction as well as the early gene direction. Mutational analysis has shown that binding of Sp1 to GC boxes I, II, and III is primarily concerned with early gene transcription, whereas binding to GC boxes III, V, and VI affects transcription on the other strand in the late gene direction.[13] Since it has been found that inversion of the 21-bp repeats, despite their asymmetric nature, does not affect early gene transcription,[15] it would appear that the binding sites nearer the start of transcription have the greatest influence over RNA synthesis. This is supported by the fact that, in other genes, the Sp1 site that is closest to the start of RNA synthesis is usually located 40 to 70 bp upstream and that its effect on transcription will decrease if it is moved only a few tens of base pairs upstream.[16]

Sp1 has been found in a variety of different tissues[17] and has been found to bind to and activate transcription from the promoters of other genes containing similar GC boxes. These include the promoters of the herpes-simplex virus IE-3, IE 4/5,[18] and TK genes,[19] the HIV virus[20] LTR, the mouse dihydrofolate reductase gene,[21] the human metallothionein gene,[22] and two monkey genes.[23] From these promoters it has become apparent that only one GC box is necessary for Sp1 to bind and stimulate transcription. The affinity of Sp1 for these binding sites, which have wide sequence variation, can vary up to an order of magnitude. Nevertheless, comparison has enabled a decanucleotide consensus sequence to be determined[24] (Figure 1A). From the position of the Sp1 recognition sites it would appear likely that Sp1 is able to interact with other cellular transcription factors such as CTF, AP1, AP2, and TATA-box-bound factors. These interactions are critically dependent on the correct stereospecific alignment of the Sp1 binding sites on the DNA double helix relative to the other transcription factor binding sites.[16]

Purification of Sp1 by sequence-specific DNA affinity chromatography[25] yielded two polypeptides of apparent molecular weight 105 and 95 kDa, both of which could be renatured from an SDS-polyacrylamide gel to give Sp1-like activity. Subsequently, however, treatment with dilute ammonium hydroxide converted the two polypeptides to a single 105-kDa form, suggesting that the 95-kDa form was due to some form of post-translational modification.[26] When the gel is electrophoresed in the absence of reducing agents, a 190-kDa polypeptide is observed, indicating that Sp1 might be naturally present as a dimer. Recently, Kadonaga et al.[26] have cloned a cDNA by screening a HeLa cDNA library with synthetic oligonucleotides obtained from the amino acid sequence of Sp1 peptides. This cDNA encodes the 696 C-terminal amino acids of Sp1, amounting to 72 kDa of the total 105-kDa molecule. The DNA-binding activity of Sp1 has been located to the 168 C-terminal residues which contain the 71-amino-acid sequence of three putative DNA-binding Zn(II) finger motifs similar to those first identified in the RNA polymerase III transcription factor TFIIIA (see Section VI). Binding of Sp1 has been shown to be dependent on the presence of Zn(II).

Recently, another transcription factor has been reported, distinct from Sp1, that also binds to the GC boxes of the 21-bp repeats of the SV40 early promoter region.[27] This factor, named LSF, is present in HeLa cells in approximately five times less abundance than Sp1 and appears to stimulate transcription from the late SV40 initiation sites (and slightly from the early sites). It would seem likely that it is involved in the switch from early-gene to late-gene transcription during SV40 infection.

2. The Enhancer

Initially using DNAse I footprinting and methylation protection techniques, proteins in nuclear extracts from HeLa and lymphoid B cells were found to bind to four regions within the enhancer.[28,29] The footprint patterns obtained with the two cell extracts differed, suggesting that there could be different proteins in the two cell types binding to the same sequences. Since then, these DNA-binding activities have been better defined and shown to bind to sequences important for enhancer activity (the P, GT-I, GT-II, octamer, and Sph motifs).

a. The P Sequence

A protein of molecular weight 47 kDa, which binds to the P sequence shown in Figure 1B, has been purified from HeLa cells.[30] This protein, termed AP1, acts as a transcription factor *in vitro* and is probably the same as the protein PEA1 which binds to a similar sequence in the polyoma enhancer (see below). AP1 binds to sequences containing the motif TGACTCA and such binding sites have been found flanking a number of genes that are inducible by the phorbol ester, TPA. Thus, the metallothionein II_A, stromolysin, and collagenase enhancers contain TPA-responsive elements which bind the AP1 protein.[22,30,31] The SV40 and polyoma enhancers are both TPA responsive. It is likely that phorbol ester induction occurs through activation of protein kinase C which, in turn, phosphorylates AP1, resulting in an increase in DNA-binding activity, though as yet there is no direct evidence for phosphorylation of AP1.

The DNA sequences bound by Ap1 are very similar to that bound by the yeast transcription factor GCN4. Also, it has recently been discovered that there are considerable protein sequence similarities between the C-terminal DNA-binding domains of GCN4 and that of the JUN oncoprotein encoded by the v-*jun* oncogene found in the avian sarcoma virus ASV17.[32] Replacing the GCN4 DNA-binding domain with the JUN DNA-binding domain resulted in a hybrid protein which was able to act as a transcription factor in yeast.[33] This pointed to the likelihood that the cellular homolog of the viral *jun* oncogene encodes the gene of one of the genes of AP1. This has now been confirmed by the finding that the human cellular homolog c-*jun* encodes a protein with the same properties as AP1.[34,35]

b. The Octamer Motif

Gel retardation and footprinting experiments have revealed the presence of at least four proteins that bind to the same octamer sequence, ATGCAAAG[36-39] (Figure 1B). Using the nomenclature of Rosales et al.,[36] two of the factors, OctB1-B and OctB2 (or NFA2 of Staudt et al.[40]), are lymphoid-specific factors, while OctB1-A is found in some nonlymphoid cells such as HeLa and F9 cells. The Oct-B3 protein is found in all cell types. These factors also bind to the octamer motif found in the IgH immunoglobulin gene enhancer, the U2 snRNA enhancer, and the histone H2B and immunoglobulin V_H UPEs. An inhibitory factor which binds to the OctB2 protein is present in some lymphoid cells.[36] The OctB1-A factor binds with high affinity to the octamer sequence in the Ad2 adenovirus origin of replication and may be the same as the NFIII replication factor described by Pruijn et al.[37,38] and Rosenfeld et al.[41] As described below, the lymphoid-specific octamer binding factors may be at least partly responsible for the tissue-specific activation of the immunoglobulin genes. Three octamer-binding proteins have been purified: a 74-kDa ubiquitous protein,[42] the 58- to 62-kDa lymphoid-specific factor(s),[43] and the 90-kDa S-phase specific factor which is involved in regulating histone gene expression[44] (see Section IV.C).

c. The GT-II Motif

The "enhancer core" consensus sequence GTG $_{TTT}^{AAA}$G occurs twice in the SV40 enhancer as the GT-I and GT-II motifs. Although the two motifs are very similar, they appear to bind a quite different set of factors. The GT-II sequence and sequences immediately upstream (Figure 1B) in fact make up three partially overlapping sequence motifs which can bind four proteins *in vitro* — GT-IIA, GT-IIBα, GT-IIIBβ, and GT-IIC, in the nomenclature of Xiao et al.[45] None of these factors bind to the related GT-I motif. Factors GT-IIA, GT-IIBα, and

GT-IIBβ are present in all cells except differentiated F9 cells which lack the two GT-IIB proteins. Protein GT-IIC is present in all cells except lymphoid cells.

d. GT-I Motif

Three DNA-binding proteins have been detected binding to this region of the enhancer (Figure 1B). One such protein (AP2) has been purified from HeLa cells.[46] This 50 kDa protein also binds to similar sequence motifs in a number of other enhancer/UPEs of genes, including the human metallothionein IIA gene which has three such binding sites upstream of the TATA-box motif (Figure 1C). The AP2 binding sites are responsible for transcriptional activation by TPA and forskolin, an activator of adenylate cyclase that mimics the action of hormones that activate cAMP-dependent kinases. Thus, the AP2 activity is modulated by two independent pathways, the kinase C and cAMP kinase pathways, presumably by postsynthetic modifications. Unlike AP1, its DNA-binding activity does not appear to be activated by TPA or forskolin, and therefore these agents probably act by increasing its ability to interact with other proteins in the transcriptional machinery.

Protein AP2 has also been found to bind to the UPE and enhancers of the human growth hormone, human c-*myc* and histocompatibility H-2K^b genes, and the BPV enhancer (Figure 1C).

The GT-1 sequence also binds a second, small, basic heat-stable protein termed EBP20 or CBP.[47] This protein binds to a sequence mostly overlapping the AP2 binding site in the SV40 enhancer containing the enhancer core sequence GTGGAAAG (Figure 1B). It also binds the related enhancer core sequences in the polyoma enhancer (Figure 1D) and the murine sarcoma virus enhancer. Rather surprisingly, as discussed below (Section II.D), this protein also binds to the CAAT-box motifs of the murine sarcoma virus (MSV) LTR and the HSV thymidine kinase upstream promoter elements.

The third protein identified as binding to the GT-1 region is the factor NFκB, which is involved in activating the immunoglobulin κ-chain enhancer in B cells[48] and the human immunodeficiency virus (HIV) enhancer in T-cells[49-51] (see Sections II.H and III.A).

e. The Sph Motifs

Overlapping the octamer sequence is a duplication of the sequence AAGNATGCA. Using footprinting techniques, a protein (or proteins) has been described in HeLa cells[28,29] (but not B-cells) which binds to this duplicated sequence. However, this putative factor was not detected using the gel retardation method and has not been further characterized.

B. POLYOMA VIRUS

The 224-bp element residing upstream of the origin of replication and the TATA box of the early promoter contains an enhancer element that resembles that of the SV40 virus in that it has two independently functioning domains (A and B) which reside in a nucleosome-free region, and several DNA-binding proteins interact with motifs within the enhancer (Figure 1D).

Two factors in mouse 3T3 nuclear extracts, termed PEA1 and PEA2, have been found to bind two similar sequences partly overlapping one another[52] (Figure 1D). PEA1 also binds to the SV40 P-element (which has a similar sequence), and it is therefore possible that this protein is the same or related to the SV40-binding protein AP1 described above. PEA1 also binds to a similar sequence in the c-*fos* enhancer (see Section IV.A).

Both PEA1 and PEA2 DNA-binding activities are absent in undifferentiated F9 embryonal carcinoma cells (which do not support wild-type polyoma transcription and replication), but PEA1 is induced when the cells differentiate on exposure to retinoic acid.[53] Since polyoma transcription and replication can occur in the differentiated cells, the absence of PEA1 activity is probably one reason for the lack of transcription in undifferentiated cells.

AP2 BINDING SITES

```
5'-GGTGTGGAAAGTCCCCAGGCTCCCCAGCA-3'                          SV40


              -180        -170
5'-AGGGAACTGACCGCCCGCGGCCCGTGTG-3'
              -120        -110
5'-CCGGGTGTTTCGCCTGGAGCCGCAAGTG-3'        HUMAN METALLOTHIONEIN IIA
            -220        -210
5'-GCCGAGGCGTCCCCGAGGCGCAAGTGG-3'



            -270          -280
5'-TGGGTGCCCTCTGGCCGCAGGCCATGGT-3'
            -150        -160                HUMAN GROWTH HORMONE GENE
5'-CTGGGATCATGCCCCCTGGCTTGTCAT-3'



             610          620
5'-CTAATCTCCCGCCCACCGGCCCTTTA-3'
           650          660                HUMAN c-MYC
5'-CGGCTGAGGACCCCCGAGCTGTGCTGCT-3'



             7650        7660
5'-GATATAGGTTTGGGGCTCCCCAAGGGACTGCTGG-3'          BPV


            -170                -190
5'-AGATGGGGAATCCCCAGCCCTGGGCTTCCCCACCCCTGACCTCA-3'   H-2Kᵇ
```

Consensus: TCCCCANGCG
           C G   C CGC

FIGURE 1C.   Sequences bound by the protein AP2.

The polyoma enhancer is also strongly stimulated in F9 cells and in a myeloma cell line if the cells are co-transfected with c-Ha-*ras* oncogene or treated with TPA.[54] These results are consistent with a model in which the oncogene or TPA activates protein kinase C phosphorylation of PEA1 (AP1), thus inducing its DNA-binding activity.

A third protein, EFC, has been detected in nuclear extracts of F9, 3T3, HeLa, and L-cells which binds to a 6-bp inverted repeat downstream of the PEA1-PEA2 binding sites.[55,56] A similar sequence in the hepatitis B virus enhancer has been found to bind a liver protein[57] (see Section II.F below).

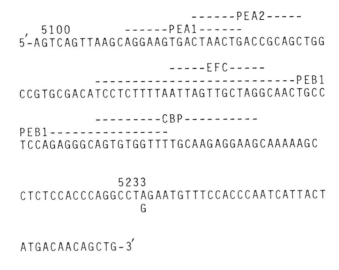

POLYOMA ENHANCER

```
                              ------PEA2-----
   , 5100            ------PEA1------
  5'-AGTCAGTTAAGCAGGAAGTGACTAACTGACCGCAGCTGG

                       -----EFC-----
              ---------------------------PEB1
  CCGTGCGACATCCTCTTTTAATTAGTTGCTAGGCAACTGCC

                    ---------CBP----------
  PEB1----------------
  TCCAGAGGGCAGTGTGGTTTTGCAAGAGGAAGCAAAAAGC


              5233
  CTCTCCACCCAGGCCTAGAATGTTTCCACCCAATCATTACT
                G


  ATGACAACAGCTG-3'
```

FIGURE 1D.   Five protein binding sites in the polyoma enhancer. The PyEC mutation at position 5233 is also shown.

A fourth protein, PEB1, binds over the same polyoma region.[58,59] This protein binds to an extensive 50-bp region which includes the 6-bp inverted repeat and a second GC-rich inverted repeat on the early side. Most of the contacts made by the protein are on a short region of one DNA strand. These DNA-binding features resemble that of the 5S gene transcription factor TFIIIA, and it is therefore likely that this protein has zinc fingers.

As described above, the CBP/EBP20 protein binds to the enhancer core sequence in the polyoma enhancer.[47]

Mutants of polyoma virus have been isolated which will grow in undifferentiated EC cells. In the case of the F441 mutation at base number 5233, which changes the sequence TAGAATGT to TGGAATGT, two factors have been found in undifferentiated and differentiated EC cells (as well as in other cell types) which bind to the second (mutant) sequence.[60] The binding of these ubiquitous factors could then possibly compensate for the lack of AP1 activity to activate polyoma transcription in undifferentiated cells. The mutant sequence resembles that of the SV40 GT-11 motif, and it is therefore possible that one or more of the factors that bind to that motif bind to the mutant polyoma sequence.

An additional feature which may be responsible for the inactivity of the wild-type enhancer in undifferentiated cells is that there may be an E1a-like repressor in these cells, which may not be able to act on the mutated enhancer.[61]

C. ADENOVIRUS
1. The Adenovirus-2 Major Late Promoter and Upstream Element

Upstream of the TATA-box element of the adenovirus major late promoter (MPL) resides an upstream transcriptional element with a dyad symmetry. This sequence binds a heat-stable protein termed MLTF, USF, or UEF, a 46-kDa monomeric protein present in uninfected HeLa cells (Figure 1E).[62-66] The partially purified factor stimulates *in vitro* transcription of this promoter by RNA polymerase II and partially purified transcription factors TFIIB, TFIID, and TFIIE.[62] TFIID is a TATA-box binding factor which gives a 10-bp footprint over the MPL TATA-box sequence when MPE is used as the DNA cleaving agent, but using DNAse I, two protected regions are seen: one stretches from bases $+1$ to about -40; the second footprint, from $+1$ to $+35$, is interrupted by cutting sites every 10 base pairs,

<u>ADENOVIRUS</u>

<u>MAJOR LATE PROMOTER</u>

```
     -70 -------MLTF-------              -40  -------------TFIID----------
  5'-TTATAGGTAGGCCACGTGACCGGGTGTTCCTGAAGGGGGGCTATAAAAGGGGGCGCGTTCGTC
```

```
              TFIID
     -------   --------  --------  ------- +40
  CTCACTCTCTTCCCCTCCATACCCTTCCTCCATCTATACCACC-3'
```

<u>EIa ENHANCER</u>

```
          ----------------
  5'-TGTAAGCGCCGGATGTGGTAAAAGTGACGTTTTTTGTGTGCGCCGGTGTACACAGGAAGTG
```

```
    ------E2F-----      -------------
  ACAATTTTCGCGCGGTTTTAGGCGGATGTTGTAGTAAATTTGGGCGTAACCGAGTAAGATTTG
```

```
    ------E2F------        -200
  GCCATTTTCGCGGGAAAACTGAATAAGAGG-3'
```

<u>EII EARLY UPSTREAM PROMOTER</u>

```
          ----EIIaE-C -- -----EIIaE-A----
  -150                                -110
    5'-GTAGGGCGTGGGAATTTCCTTGCTCATAATGGCGCTGACGACAGGTGCTGGCGCCGGG
```

```
                         ---E2F---            EIIaE-A/E2F
            ----EIIaE-B--------                ----------
  -90                              -60
      TGTGGCCGCTGGAGATGACGTAGTTTTCGCGCTTAAATTTGAGAAAGGGCGCGAAA-3'
```

FIGURE 1E. Protein binding sites in the adenovirus major late promoter (MLTF and TATA-box factor TFIID),
the EIa enhancer (E2F and two "enhancer core" binding proteins), and the EII early promoter (five protein binding
sites).

staggered by 4 base pairs on the two strands (Figure 1E). This suggests that the DNA double
helix is in contact with a protein surface, leaving one side of the double helix exposed. The
MLTF factor binds to a sequence whose center is 30 base pairs upstream of the TATA box,
suggesting that this factor binds to the same side of the DNA double helix and could therefore
interact with the TATA-box factor.

2. The EIa Enhancer

The EIa gene has an enhancer element 150 to 350 base pairs upstream of the promoter
which binds several factors (Figure 1E). Within this enhancer are two repeats of the sequence
TTTCGCGC which binds a host transcription factor, E2F,[67] originally detected by its ability
to bind to two of the same motifs in the EII upstream promoter element (see below). As
discussed below, the level of this factor is elevated by the EIa gene product. The EIa
enhancer also contains several "enhancer core" sequences which have been found to bind
factors which presumably correspond to one or more of the SV40 GT-I and GT-II binding
factors.[68]

3. The EII Early Upstream Promoter

The EII gene has three upstream promoter elements — A, B, and C — which bind several different factors (Figure 1E).[69-73] The promoter-proximal element A binds a factor, EIIaE-A,[72] which also binds weakly to a sequence within the most distal element C. A second factor, EIIaE-B[72] (or EII A-F),[73] binds to a sequence within the element B, and it probably also binds to the UPEs of the E1a and EIV genes. The third factor[72] (EIIaE-C) binds to a site within the C element adjacent to the EIIaE-A binding site. EII gene transcription is enhanced by the adenovirus E1a gene product, but this does not appear to be mediated by E1a increasing the DNA binding activity of any of these three factors. However, the activity of a fourth factor, E2F (or EIIF),[69-71] which binds to the two repeats of the sequence TTTCGCGC in the A and B elements, is elevated by E1a. It would appear, then, that E1a induces EII gene transcription by increasing the levels of the E2F factor or increasing its affinity for its DNA binding site. The E2F activity is present in undifferentiated F9 cells, but disappears when the cells are induced to differentiate,[70] which is consistent with the postulated presence of an E1a-like activity in undifferentiated cells which disappears on differentiation.[61]

4. The EII Late Upstream Promoter

Six binding sites have been described in the UPE of this gene (Figure 1F), three of which (sites II, IV, and VI) contain the CCAAT motif and one binds Sp1 (site I).[75] The proteins binding to the CCAAT boxes probably correspond to the NFY/CP1/CP2 factors which bind to a large number of CCAAT motifs (see Section II.D). A third protein binds to two stretches of six Gs in sites V and VI, but the function of these sequences is not known. The fourth protein binds to a sequence (site III) resembling that present in the UPE of the c-*fos* gene. This factor (NUSFII) binds to the c-*fos* UPE, binding to the same sequence as the factor AP1 (or PEA1) (see Figure 3A), but the two factors appear to be different.

5. The EIII Upstream Promoter

Four proteins have been identified binding to the promoter and UPE of this gene by footprinting techniques[76] (Figure 1F). One factor binds to the TATA box and the binding of this factor is stabilized by two other factors binding further upstream — AP1 and the factor E4F1, which binds to the EIV upstream promoter (see below). The fourth factor binds to an inverted repeat sequence homologous to that bound by the replication and transcription factor NF1 (see Section II.D for a fuller discussion on NF1 and its binding sites). The adenovirus E1a gene product does not increase the DNA binding activity of any of these factors.

6. The EIV Upstream Promoter

Three binding sites for the protein E4F1 have been described upstream of the TATA box of this gene.[77] One binding site was in an upstream promoter element and two others in an E1a-inducible enhancer between bases -138 and -200 (Figure 1F). The putative DNA-binding consensus sequence for this protein is ACGT(A/C)AC. The protein binds to similar sequences upstream of three other adenovirus early genes (E1a, EII, and EIII), but there is no evidence that the DNA-binding activity of the factor is elevated by the E1a gene product. This protein is probably the same as the 43-kDa cAMP regulatory element binding protein (CREB) which binds to the somatostatin gene[78] (see Section IV.H).

D. HERPES-SIMPLEX VIRUS (HSV) THYMIDINE KINASE GENE

The upstream promoter element of the HSV thymidine kinase gene has been extensively analyzed by deletion and linker scanning mutations. Five motifs have been delineated (Figure 1G). Two Sp1 binding sites in opposite orientations are found on either side of a CCAAT-

<u>ADENOVIRUS (CONTD)</u>

<u>EII LATE UPSTREAM PROMOTER</u>

```
                  ----------VI---------
                 -250
5'-CGGCTGCTGATAGGGCTGCGGCGGCGGGGGGGATTGGGTTGAGCTCC


      ----------V-----------------
             -200
TCGCCGGACCTGGGGGTCCAGGTAAACCCCCCGTCCCTTTCGTAGCAGAAA


         ------------IV------------    -------
                -150
CTCTGGCGGGCTTTGTTGATGGCTTGCAATTGGCCAAGGATGTGGCCCT


--------III----------          -------II-------
          -100
GGGTAATGACGCAGGCGGTAAGCTCCGCATTTGGCGGGCGGGATTGGTCT


          ----------I-------
               -50
TCGTAGAACCTAATCTGCTGGGCGTGGTAGTCCTCAGGT-3'
```

<u>EIII UPSTREAM PROMOTER</u>

```
          ----------NF1----------- -150
5'-TCCCCGTAGTTGGCCCGCTTCCCTGGTGTACCAGGAAAGTCCCGCTCCCACCACTGTGGTAC


         -110      ---------AP1----------
TTCCCAGAGACGCCCAGGCCGAAGTTCAGATGACTAACTCAGGGGCGCAGCTT


    --------E4F1---------    -40   -----------------------
GCGGGCGGCTTTCGTCACAGGGTGCGGTCGCCCGGGCAGGGTATAACTCACCTGAAA-3'
```

<u>EIV UPSTREAM PROMOTER</u>

E4F1 FACTOR BINDING SITES: -167 / -161   5'-ACGTAAC-3'

                               -146 / -140   5'-TCGTCAC-3'

                               -53 / -47   5'-ACGTAAC-3'

FIGURE 1F.   Protein binding sites in the adenovirus EII late promoter (six binding sites, three of which [II, IV, and VI] have CCAAT motifs and I has the Sp1 sequence), EIII promoter (binding NF1, AP1, E4F1, and a TATA-box factor). Binding sites for the E4F1 factor in the EIV upstream promoter regions is also shown.

box motif.[19,79] The distal Sp1 binding site plus the CCAAT box are functional in stimulating transcription in conjunction with the TATA-box element. The more proximal Sp1 binding site is a weaker site and is dependent on the presence of the distal Sp1 and CCAAT-box elements for activity. The fifth, most distal element is the octamer motif found in SV40, immunoglobulin, and other gene regulatory elements (Section II.A), and this element augments expression of the thymidine kinase promoter.[80]

HERPES VIRUS THYMIDINE KINASE
UPSTREAM PROMOTER

```
                  -120                    --Sp1--                    -80
5'-ATATTTGCATGTCTTTAGTTCTATGATGACACAAACCCCGCCCAGCGTCTTGTCATTGGCGAAT
          0
```

```
                      --Sp1--        -40
    TCGAACACGCAGATGCAGTCGGGGCGGCGCGGTCCCAGGTCCACTTCGCATATTAA
```

FIGURE 1G. HSV thymidine kinase upstream promoter showing the octamer (O), the CCAAT-box and the TATA-box motifs (underlined), and the two Sp1 binding sites.

A basic, heat-stable protein CBP was described that binds to the CCAAT motif of the HSV thymidine kinase gene and to the CCAAT motif of the Moloney murine sarcoma virus (MSV) upstream promoter.[79] In both cases the binding of this factor was increased when the CCAAT was mutated to GCAAT. Rather surprisingly, this mutation decreases the activity of this element *in vivo*. In retrospect, this is probably due to the fact that CBP is probably the same as the "enhancer core" binding protein EBP20.[47] Thus, while this protein can bind to the CCAAT box *in vitro*, this sequence is probably normally bound *in vivo* by one of the "true" CCAAT-box factors NFY, CP1, CP2, or possibly CTF (see below). Mutation to GCAAT then destroys the binding sites of the true CCAAT-box protein, accounting for the down mutation. Presumably, the binding of EBP20 (CBP) to the mutated GCAAT box fails to compensate for the loss of the correct CCAAT-box-bound protein.

At this stage it is necessary to review in more detail the various factors which have been described as binding to CCAAT motifs. Although there is a superficial similarity between the CCAAT regions from different genes,[81-84] it has recently become apparent that the surrounding nucleotides play a role in determining which of the multiple, ubiquitous factors that recognize this sequence actually bind.

The CCAAT transcription factor (CTF) is a protein (or group of proteins) shown to be responsible for the stimulation of transcription from the HSV thymidine kinase (TK),[19] the human *hsp70*,[85,86] and human α-globin promoters in an *in vitro* system.[87] CTF has also been reported binding to CCAAT elements in the MSV LTR and human β-globin promoter.[19] The similarity between the CCAAT sequence and half of the nuclear factor one (NF1) inverted-repeat recognition sequence (see below) led to speculation that these proteins were in some way related. Sequence-specific DNA affinity chromatography gave a family of polypeptides ranging from 52 to 66 kDa, the same as for NF1.[87] Jones et al. reported that CTF and NF1 were indeed indistinguishable and concluded that they were either the same protein or were both members of a family of closely related proteins.[87]

Recently, however, Chodosh et al.[88] reported the identification and characterization of three CCAAT-binding proteins: CP-1, CP-2, and NF-1. CP-1 bound to CCAAT sequences previously reported bound by CTF, but not, significantly, to the NF-1 inverted repeat. Their preparation of NF-1, on the other hand, bound to its previously characterized sites, but not to CCAAT domains bound by CP-1 and CTF. Thus, it is possible that the CTF/NF-1 reported by Jones et al.[87] is in fact composed of two distinct DNA-binding activities, i.e., CP-1 and NF-1. Several instances now exist where promoters contain separate binding sites for NF-1 and a CCAAT-binding factor,[89,90] e.g., the avian β^H-globin gene.[307]

CP-1 and CP-2 appear to be different proteins which recognize different CCAAT sequences (Figure 1H). They do, however, overlap in some instances, binding the same site with equal affinity.[88] Surprisingly, all three proteins (CP-1, CP-2, and NF-1) have been shown to be composed of heterologous subunits which, in the case of CP-1, are interchangeable with the HAP2 and HAP3 gene products from yeast.[91]

CP1, CP2 AND NF-1 CONSENSUS BINDING SEQUENCES

```
CP1:           CNNNNNNAACCAATCANCG
               T        GG        TT


CP2:           CAGCNNNARCCAATCNNNR
               T   T   G


NF-1:          NTTGGCNNNNNGCCAAN
```

FIGURE 1H.   Consensus DNA sequences bound by the CCAAT-box binding factors CP1 and CP2. Also shown is the NF1 consensus sequence.

A factor binding to a CCAAT region in the Y box of the MHC class IIE promoter, designated NF-Y, has also been shown binding to the human α-globin promoter, the MSV LTR, and to the rat albumin promoter.[92,93] It has been shown to be a metalloprotein with a protease-resistant DNA-binding core. Its DNA binding complex is 250 to 300 kDa and may be composed of multiple proteins.[94] A possibly similar, but distinct, protein (NF-Y*) has also been identified binding to the CCAAT region of the HSV *tk* promoter.[92]

CCAAT binding factors have also been identified binding to the mouse α-globin promoter,[95] the sea urchin sperm histone H2B promoter,[96] the human TK promoter,[97] the mouse α-2(I) collagen promoter,[98] and the adenovirus-2 EII-late promoter[75] (which also binds to the rat albumin and HSV *tk* promoters).

It is probable that many of the CCAAT-binding factors described above are either similar or identical to each other. It would seem, however, that there exist a number of distinct proteins capable, in some instances, of recognizing the same CCAAT regions. This, combined with the likely multi-subunit and heterologous nature of these proteins, would suggest that CCAAT-binding factors have a complex regulatory role in the control of gene expression.[99]

Nuclear factor 1 (NF-1) was first identified by Nagata et al.[100,101] as a protein in HeLa cell nuclear extracts which bound to an inverted repeat sequence within the adenovirus type 2 origin of replication. This was found to enhance the initiation reaction of viral replication both *in vitro* and *in vivo*.[41,102,103] Independently, Nowock and Sippel[104,105] characterized the TGGCA protein, the chicken homolog of NF-1, from chicken oviduct tissue. Both of these proteins recognize the same DNA sequences and both will enhance adenovirus replication in an *in vitro* system.[106,107] It has, however, not been proven that these proteins have any role in cellular DNA replication; indeed, their cellular function may be unrelated to DNA replication, having possibly been recruited by the adenovirus for its own ends. Jones et al.[87] also found that HeLa cell NF-1 is identical to CTF, a protein that binds to the CCAAT motifs of a number of genes (see above). Miksicek et al.[108] argue, however, that it is extremely unlikely that the TGGCA protein is a true CCAAT-binding factor. It is possible that these proteins may be members of a family of functionally related, but distinct, proteins.

NF-1/TGGCA protein is a ubiquitous protein found in many different cell extracts and has been shown to interact with the enhancers and promoters of many cellular and viral genes.[109-111] These include 5′ of the human c-*myc* gene,[112] a human IgM gene,[113] the mouse mammary tumor virus (MMTV) long-terminal repeat (LTR),[106,108,114] the hepatitus B virus S gene promoter,[115] the human papovirus BK enhancer,[106] the chicken lysozyme gene enhancer,[104,105] the rat albumin gene upstream promoter,[90] and the major immediate early promoters of simian and human cytomegalovirus (which contains at least 20 adjacent binding sites).[116-118] Study of the homology between these recognition sequences and single-base mutagenesis have produced a NF-1 consensus sequence of 5′-TGG(N)$_{6-7}$GCCAA-3′, while for the TGGCA-protein, it is 5′-PyTGGCA(N)$_3$TGGCAPu-3′.[107] Sequences bordering these

MMTV ENHANCER-PROMOTER

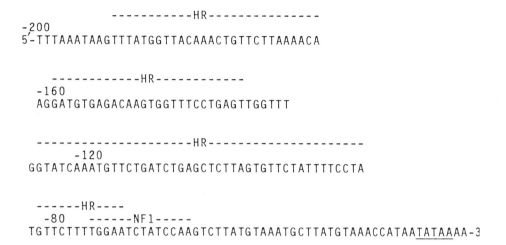

```
          - - - - - - - - - - - H R - - - - - - - - - - - - - -
-200
5'-TTTAAATAAGTTTATGGTTACAAACTGTTCTTAAAACA


         - - - - - - - - - - - - H R - - - - - - - - - - -
  -160
  AGGATGTGAGACAAGTGGTTTCCTGAGTTGGTTT


    - - - - - - - - - - - - - - - - - - H R - - - - - - - - - - - - - - - - - - -
       -120
  GGTATCAAATGTTCTGATCTGAGCTCTTAGTGTTCTATTTTCCTA


   - - - - - -HR- - - -
     -80     - - - - - -NF1- - - - -
  TGTTCTTTTGGAATCTATCCAAGTCTTATGTAAATGCTTATGTAAACCATAATATAAAA-3
```

FIGURE 1I.   The MMTV enhancer-promoter showing the four regions bound by steroid hormone receptors (HR) and by NF1. The TATA box is underlined.

motifs have, however, been shown to influence binding affinity.[119] NF-1/TGGCA protein has been shown to bind in the major groove of DNA, binding to one side of the double helix, possibly as a dimer (or tetramer).[120]

While the relative abundance, ubiquitous nature, and association of the NF-1/TGGCA protein with areas of DNAse I hypersensitivity-containing enhancer and promoter elements would seem to indicate that it plays some major role in the regulation of the basal functions of the eukaryotic cell, no clear mechanism as to how this might be brought about has so far been elucidated. The NF-1 binding site in the MMTV LTR (Figure 1I) has been shown to be essential for maximal glucocorticoid-induced transcription from this promoter *in vivo*.[108] Similarly, the NF-1 binding site is required for efficient activity of the hepatitis B virus S gene promoter.[115] NF-1/CTF has also been demonstrated to stimulate transcription from an α-globin promoter in an *in vitro* system.[87] However, in other cases, no role has been found for NF-1 in promoter function. Promoter constructs containing just NF-1/TGGCA protein sequences have been found not to stimulate transcription.[106] It is possible that the NF-1/TGGCA protein may have a general role in the regulation of gene expression, possibly via the activation of chromatin during the early stages of cell differentiation. Alternatively, NF-1 could act by interacting with other transcription factors, increasing their affinity for their own DNA recognition sequences. It will be interesting to see whether NF-1 stimulates the initiation of adenovirus replication via the same processes as those used in its role as a transcriptional regulator.

DNA affinity chromatography purification of NF-1 from HeLa cells by Rosenfeld and Kelly[121] produced a protein with a number of different polypeptides with apparent molecular weights ranging from 66,000 to 55,000 Da. Purification of CTF by Jones et al.[87] produced an identical pattern of proteins. Diffley and Stillman,[122] however, reported purified NF-1 to be 160 kDa, a result which Jones et al.[87] were not able to repeat. Nagata et al.[101] previously reported the presence of a major 47-kDa polypeptide in purified fractions containing NF-1, which Rosenfeld and Kelly suggest is either a contaminant or a proteolytic product of NF-1. Purification of chicken liver TGGCA protein by Rupp and Sippel[123] also revealed a family of polypeptides; however, these ranged from 36.8 to 29.8 kDa. Both Rosenfeld and Kelly[121] and Rupp and Sippel[123] present evidence that it is unlikely that these families of polypeptides result from proteolytic degradation from some larger polypeptide. Proteolysis of both these families of polypeptides reveals that they are structurally related to one another. It is possible that these proteins are the product of a gene family which encodes DNA-binding proteins with similar recognition sequences, but with a range of functions. Alternatively, they could be the product of differential mRNA splicing or different post-translational modifications, such as phosphorylation.

Support for the idea that the NF-1/TGGCA protein recognition sequence may bind a number of different protein species is reported by Lichsteiner et al.[124] Examining the different DNA-binding proteins binding to the mouse albumin gene upstream promoter, they found that purified HeLa NF-1 gave a footprint indistinguishable from proteins in their nuclear extract (see Section III.C). They found, however, that the major activity in their liver extract consisted of proteins with characteristics different from NF-1. This raises the possibility of tissue-specific NF-1/TGGCA protein-like activities.[124]

E. MOUSE MAMMARY TUMOR VIRUS (MMTV)

The LTR of MMTV is a steroid hormone-responsive enhancer and has been found to have four hormone-responsive elements (HRE) which correspond to binding sites for the glucocorticoid, progestin, and androgen receptor proteins[125,126] (Figure 1I). In addition, there is an NF-1 binding site just upstream of the TATA-box motif.[114,125]

Studies using mouse cells containing an episomal BPV vector carrying the LTR sequences have shown that, in the absence of glucocorticoid hormone, the LTR is covered with specifically positioned nucleosomes.[114,127] On induction with hormone, histones forming the nucleosome over the four HREs are dissociated (forming a DNAse I "hypersensitive site"), with the concomitant binding of NF-1 and TATA-box proteins to their sequences just downstream from the HREs. This occurs in the absence of episomal DNA synthesis, showing that access of transcription factors to their sequences does not require chromatin replication. Rather surprisingly, there was no evidence from exonuclease III protection experiments for the binding of the steroid hormone receptor to the nucleosome-free region, presumably due to the labile nature of the receptor DNA complex. A description of steroid hormone receptors and their mode of gene activation is given in Section V.

F. HEPATITIS B VIRUS (HBV)

An enhancer element in this virus has been described which is functionally most active in liver cells. A factor present in liver cells of different states of differentiation and also in HeLa cells has been detected binding to a sequence (the E site) within this enhancer (Figure 1J).[57] This sequence is very similar to the polyoma inverted repeat which binds to the ubiquitous factor EF-C (see Section II.B above). Two other factors present only in nuclear extracts of differentiated hepatoma cells bind to three sequences (UE1, UE2, and UE3) upstream of the enhancer.[57] The factor binding to UE1 is probably the same as that binding to UE3. Note that the footprint sequences UE2 and UE3 contain the CAAT-box sequence CCAAT.

G. HUMAN CYTOMEGALOVIRUS

Transcription from the immediate early gene IE1 is controlled by an enhancer element located between sequences -524 and -117. This enhancer is functional in a number of cell types and binds multiple nuclear proteins (Figure 1K).[118] Two of the binding sites (I and VII) are probably NF-1 binding sites. Footprint II has the cAMP response element motif TGACGTCA (see Section IV.H).

H. HUMAN IMMUNODEFICIENCY VIRUS (HIV)

Five protein-binding regions within the LTR of this virus have been shown to be important for transcription[49-51] (Figure 1L): the TATA-box region (site IV), three Sp1 binding sites

HEPATITIS B VIRUS ENHANCER

```
                           881                    900
UE3 SITE:                  5'-GGGTTATGTCATTGGATGTT-3'


                           972                    991
UE2 SITE:                  5'-CCTATTGATTGGAAGTATG-3'


                           1048                   1029
UE1 SITE:                  5'-TTTACACAATGTGGTTATCC-3'

                           -----------EF-C-----------
                           1187                       1212
E SITE:                    5'-AAGTGTTTGCTGACGCAACCCCCACTG-3'
```

FIGURE 1J.   Four sequences bound by protein within and bordering the hepatitis B viral enhancer. Sites UE2 and UE3 have the CCAAT motif. Site UE1 has the inverted repeat bound by factor EFC.

CYTOMEGALOVIRUS ENHANCER

```
  -500       -----------I-----------    -------II------
5'-AATGGCCCGCCTGGCTGACCGCCCAACGACCCCCGCCCATTGACGTCAA


-------------III-----------------   --------IV--------
TAATGACGTATGTTCCCATAGTAACGCCAATAGGGACTTTCCATTGACGTCAATGGGT


             -380       -----------V-----------
GGAGTATTTACGGTAAACTGCCCACTTGGCAGTACATCAAGTGTATCATATGCCAAGTAC


-------VI---------         ---------VII----------
GCCCCCTATTGACGTCAATGACGGTAAATGGCCCGCCTGGCATTATGCCCAGTACATGAC


  -270       -----------VIII------------
CTTATGGGACTTTCCTACTTGGCAGTACATCTACGTATTAGTCATCGCTATTACCATGG


---------IX-----------     -180
TGATGCGGTTTTGGCAGTACATCAATGGGCGTGGATAG-3'
```

FIGURE 1K.   Protein binding sites within the cytomegalovirus enhancer. Binding sites I and VII have the NF1 binding sequences; site II has the c-AMP response element.

just upstream of the TATA box, a duplicated GGGACTTTCC sequence (found also in the SV40 and immunoglobulin κ-chain enhancers), a negative regulatory sequence (site I), and a sequence downstream of the TATA box (site V). HeLa nuclear protein extracts contain all five DNA binding proteins,[49] although only one of the GGGACTTTCC repeats is fully bound (site III). This is in contrast to the lymphoid factor NFκB (which binds to this motif in the immunoglobulin κ-chain enhancer),[48] which is capable of binding to both repeats of

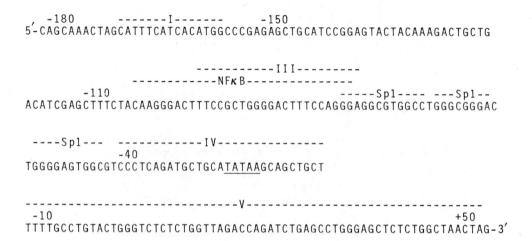

FIGURE 1L.   Protein binding sites in the LTR of the human immunodeficiency virus. Three sites are bound by Sp1, one by a TATA-box factor (IV), and one by a factor binding downstream of the TATA-box (V). The lymphoid factor NFκB binds to both repeats of the sequence GGGACTTTCC, but the HeLa factor III to only one repeat.

the HIV enhancer, suggesting that the HeLa and lymphoid factors are different. The HeLa factor could be the protein AP2 since AP2 binds to a similar motif in the H-2K^b gene enhancer (Figure 1C).

As discussed later in connection with the immunoglobulin light-chain enhancers (Section III.A), NFκB activity is normally found only in B cells, but its activity can be induced in other cells by treating the cells with phorbol esters and phytohemagglutinin. Thus, when Jurkat T-cells are stimulated with these agents, NFκB DNA-binding activity is induced, as is the activity of the enhancer.[50] This stimulation may be responsible for the activation of latently infected HIV provirus in T cells.

A more detailed analysis of the lymphoid proteins binding to the GGGACTTTCC sequence has been carried out using two-dimensional gel electrophoresis techniques,[51] which have revealed the existence of up to seven proteins of similar molecular weight (78 to 86 kDa), the levels of one of which is elevated in Jurkat cells by treatment with TPA and PHA treatment.

I. ROUS SARCOMA VIRUS (RSV)

The LTR of this virus has a typical TATA-box promoter, upstream promoter element, and enhancer. All three elements reside in the U3 region[128,129] (Figure 1M). The upstream promoter has the CCAAT motif which binds a protein (FII) present in avian erthryroid cells and fibroblasts.[130,131] Methylation of the two Gs of the CCAAT box inhibit binding of this factor. In addition, the methylation of a G approximately 10 bases upstream in the enhancer region also prevent binding, which is consistent with data suggesting the binding of an additional factor which binds cooperatively with the CCAAT-box factor. This CCAAT-binding factor is not the same as NF1/CTF or the CBP factor and is probably the same as CP1, CP2, or the NFY factor described above (Section II.D).

Two proteins have been found to bind sequences within the enhancer region upstream of the CCAAT box (Figure 1M). One protein binds to a sequence containing the element GCAAT[130,131] and may therefore be related to the CBP factor, which binds to the thymidine kinase CCAAT box more strongly when it is mutated to GCAAT (see Section II.D above). As discussed above, CBP may be the same factor EBP20, which binds to the "enhancer

```
ROUS SARCOMA VIRUS ENHANCER - UPSTREAM PROMOTER

  -229 ------CBP-----                      -----------FI-----------
   5'-AATGTAGTCTTATGCAATACTCTTGTAGTCTTGCAACATGGTAACGATGAGTTAGCAACATGCC


        -150        --------FII----------        -110
   TTACAAGGAGAGAAAAAGCACCGTGCATGCCGATTGGTGGAAGTAAGGTGGTACGATCGTGCCTT-3'
```

FIGURE 1M.   Three protein binding sites in the avian sarcoma virus LTR. Protein CBP binds to one site, FII to a CCAAT-box motif (underlined), and the third (F1) to an inverted repeat sequence.

core'' sequence of various enhancers such as those of MSV and SV40. The second protein (FI in Figure 1M) binds to an inverted repeat TGGTAACGATAGTTAGCAAC.[131] The polyoma virus enhancer also appears to bind this factor, although it does not have the inverted repeat. However, the polyoma enhancer does have the inverted repeat AGTTGCT-AGGCAACT, which binds the ubiquitous protein EFC (see Figure 1D). Thus, the proteins which bind to the two enhancers may be the same, binding to a common inverted-repeat sequence, GTNNC----GCAAC, with a flexible distance between the two repeats.

The avian sarcoma virus has two other enhancer elements: one just upstream of the 3' LTR of the provirus, the other within the *gag* gene near the 5' LTR. Nothing is known about proteins binding to the 3' enhancer, but two footprints have been detected in the *gag* enhancer using nuclear protein extracts.[132] Both binding sites have GCAAT motifs and therefore probably bind the same protein (CBP) which binds to the GCAAT of the LTR enhancer.

The avian sarcoma virus naturally causes sarcomas arising from tissues of mesenchymal origin. However, none of the factors described so far are specific for these tissues, having been found in erythroid cells as well as in fibroblasts.

J. MOLONEY MURINE LEUKEMIA VIRUS (Mo-MuLV)

Like many animal retroviruses, the U3 region of the LTR contains promoter and enhancer elements typical of eukaryotic genes. Upstream of the TATA and CAAT box there is a 75-bp repeated unit which acts as a typical enhancer in a wide variety of cell types (though the virus only causes thymic leukemia and the disease specificity is conferred by the U3 region). Six factors have been identified that bind to each enhancer repeat from a variety of cell types, though none of the factors are T-cell specific[133] (Figure 1N). One of the factors (GR) binds to a sequence homologous to the glucocorticoid responsive element (GRE) found in a number of hormone-responsive enhancers (see Section V). A second factor binds to an NF1 binding sequence adjacent to the GRE. A third factor binds to an "enhancer core" motif and may therefore be related to one of the factors described above that binds to the SV40 enhancer. The three remaining factors (LVa, b, and c) bind to sequences which have not been previously identified in other enhancers as being protein-binding sites.

A CCAAT motif resides between the enhancer and the TATA motif of the Mo-MuLV and the similar amphoteric virus A-MuLV. A factor present in retrovirus nonpermissive EC cells, but not in a permissive cell line, binds to this CCAAT motif.[134] This factor would appear to be different from the other CCAAT-box-binding factors described above (since methylation of the two Gs does not inhibit binding of this factor), and it has been suggested that this protein may be responsible for the repression of the retrovirus expression in EC cells. This factor could therefore be related to the CAAT-box displacement factor, an inhibitory factor found in somatic cells which inhibits testis-specific histone gene expression (see Section IV.C).

MOLONEY MURINE LEUKEMIA VIRUS ENHANCER

```
      LVa              NF1      LVb              CORE
     ------          -------  -------         --------
,        7940
5'-GAACAGCTGAATATGGGCCAAACAGGATATCTGTGGTATGCA


      LVc            NFI          GR
     -----        ------------ ----------
                      8000
GTTCCTGCCCCGGCTCAGGGCCAAGAACAGATG-3'
```

FIGURE 1N.   Protein-binding sites in the Moloney murine leukemia virus enhancer. Two sequences are bound by NF1, one by the glucocorticoid-receptor complex (GR), and a third by an "enhancer core" protein (CORE). The other three sequences are bound by proteins LVa, b, and c.

GIBBON APE LEUKAEMIA VIRUS ENHANCER

5'-AGAAATAGATGAGTCAACAGC-3'

FIGURE 1O.   Sequence in the Gibbon ape leukemia viral enhancer bound by a nuclear protein.

K. GIBBON APE LEUKEMIA VIRUS

The U3 region of the LTR of the retrovirus contains a 48-base-pair repeated element which acts as an enhancer. This element contains the binding site for a protein factor present in a number of different cell lines (Figure 1O).[135]

III. PROTEINS BINDING TO TRANSCRIPTION ELEMENTS OF TISSUE-SPECIFIC GENES

A. THE IMMUNOGLOBULIN GENES

The lymphoid B-cell-specific gene transcription of the rearranged heavy-chain gene is regulated by three sequence elements: a 5' upstream promoter element, an enhancer in the intron separating the variable and constant-region exons, and a third intragenic element. The UPE upstream of the TATA box has the so-called octamer motif which, as discussed above in Section II.A, is also found in the SV40 enhancer, where it has been found to bind at least four different proteins, two of which are lymphoid specific.[28,36,37,39,136-140] The motif is also present in the μ heavy-chain enhancer, and it is probable that the binding of the lymphoid-specific octamer-binding factors to these motifs of the UPE and enhancer is necessary for B-cell-specific gene transcription. Bases surrounding the octamer motif may be important for selectivite binding of the lymphoid-specific factors, rather than the ubiquitous octamer-binding factor.[36] The lymphoid-specific factor(s) have been purified and shown to consist of three peptides of 58 to 62 kDa.[43]

Methylation protection experiments have demonstrated that in the intact B cell or nucleus there are five or more proteins bound to the heavy-chain enhancer.[141] One protein is bound to the octamer motif and four proteins to the E motifs, as shown in Figure 2A. Gel retardation

IMMUNOGLOBULIN HEAVY CHAIN ENHANCER

```
      NFμE1                                              IgPE1
      --------                                    -----------------
5'-GTTGAGTCAAGATGGCCGATCAGAACCAGAACACCTGCAGCAGCTGGCAGGAA
         E1                                              E2
IgPE1
------  NFμE3
        ----------
GCAGTCATGTGGCAAGGCTTTTGGGGAAGGGAAAATAAAACCACTAGGTAAAC
     E3

                                              -------
TTGTAGCTGTGGTTTGAAGAAGTGGTTTTGAAACACTCTGTCCAGCCCCACCAA

   IgPE2
   --------------                        -------
ACCGAAAGTCCAGGCTGACAAAACACCACCTGGGAATTGCATTTCAAAAT-3'
                     E4                       O
```

IMMUNOGLOBULIN κ-CHAIN ENHANCER

```
            ------NFκB---------
        5'-CAGAGGGGACTTTCCGAGAGG-3'
```

FIGURE 2A.   Part of the immunoglobulin heavy chain enhancer showing the E1-E4 and octamer (O) motifs complexed with protein *in vivo* (underlined). Dashed lines show sequences bound by factors *in vitro*. Also shown below is the sequence in the immunoglobulin κ-light chain enhancer bound by the factor NFκB.

techniques have shown that nuclear extracts from lymphoid and nonlymphoid cells contain two proteins, NFμE1 and NFμE3, which bind to the E1 and E3 motifs, respectively.[48,142] Footprinting studies using proteins from lymphoid and nonlymphoid cells have revealed a complex set of binding proteins.[143-146] Factors binding over the E2, E3, and octamer motifs (but not the E4 motif) have been detected in both types of cells. Also, a factor (IgPE-2) has been detected[143] which binds to a sequence between the E3 and E4 motifs which corresponds to an *in vivo* methylation-protected sequence.[141] Although extracts from lymphoid and non-lymphoid cells have proteins which bind to all the above motifs, the individual footprints obtained with the two types of extracts are somewhat different (i.e., the exact sequences protected from digestion differ), suggesting that proteins binding to the same regions of the enhancer may differ in the two types of cells.

Two negative regulatory elements surrounding the heavy-chain enhancer and a third negative element within the enhancer may contribute to maintaining the repressed state of the immunoglobulin gene in nonlymphoid cells.[2] Although two factors have been detected binding to the flanking negative elements,[146] these have been found in lymphoid and non-lymphoid cells, and it is therefore not clear whether they are responsible for repression of the enhancer.

Like the heavy-chain gene, the κ light-chain UPE has a conserved octamer motif, ATTTGCAT, although it should be noted that the immunoglobulin conserved sequence is in fact more likely to be the decanucleotide sequence TNATTTGCAT,[140] and nuclear factors have been detected binding to it which presumably are the same as the heavy-chain and SV40-binding factors[136,139] discussed above.

The κ-chain enhancer is located within the major intron and two proteins have been detected binding to its sequences *in vitro*.[48] One protein is probably NFμE3 since it binds to a sequence in the κ-chain enhancer which is similar to its binding sequence in the heavy-chain enhancer. A second protein, NFκB, binds to a sequence present in the κ-chain enhancer (Figure 2A) and to a similar sequence in the SV40 and HIV enhancers (see Sections II.A and II.H). The DNA-binding activity of NFκB is detectable only in B cells which are expressing the κ-chain gene. Agents which induce transcription of this gene in B cells, such as bacterial lipopolysaccharide and phorbol esters, also induce the DNA-binding activity of NFκB, presumably via a mechanism involving protein kinase C phosphorylation.[147,148] Phorbol esters can also induce its activity in non-B cells.[148] The data suggest that the lymphoid B-cell-specific activity of the κ-chain enhancer is dependent on the presence of NFκB- DNA binding activity. However, this enhancer activity may only be transiently required for activation of the κ-chain gene during B-cell differentiation.

B. THE GLOBIN GENES

The various members of the α- and β-globin gene family are expressed in different erythroid cell lineages appearing in the blood during embryonic development. In the case of the β-globin genes, this developmental regulation is mediated via sequences at the 5' and 3' ends of the genes and, in the case of the human adult β-globin gene, sequences within the gene itself.[149] These sequences are marked by the appearance of nuclease-hypersensitive sites within the nuclear chromatin during red blood cell development.

The chicken β-globin genes have been extensively investigated for sequence-specific DNA binding proteins. The chicken β-globin family is composed of two embryonic genes, ρ and ε, and two adult genes, β^H and β^A. During embryonic development, the earliest red blood cells in the blood islands express the ρ and ε globin genes. After about day 7 of embryonic development, red blood cells emerging from the bone marrow express the two adult hemoglobin genes β^H and β^A. In the adult chicken, only β^A globin is transcribed. These changes in gene expression are reflected by the appearance of the nuclease-hypersensitive sites around the genes. For example, the nuclease-hypersensitive site at the 5' end of the β^H globin gene is present in nuclei from 15-day-old embryonic red blood cells, but is absent in adult chicken erythrocytes when the gene is not expressed.[307] In contrast, the 5' hypersensitive site of the β^A-globin gene is present at both stages. Multiple protein factors have been detected binding to these two hypersensitive regions.

Three proteins have been found to bind to sequences upstream of the CCAAT and TATA boxes of the β^A-globin gene (Figure 2B).[150,151] One protein binds to a stretch of 16 to 18 guanine bases, the second binds to an inverted repeat similar to that bound by the protein NF-1, and the third protein binds to the CACCC motif found in the upstream promoters of a number of β-globin genes.

Three proteins have been found to bind to the upstream promoter element of the β^H-globin gene.[151,307] One binds to the CCAAT motif; a second, NF-1, binds to an inverted repeat in the middle of the hypersensitive region. A third protein binds to a sequence contiguous to the NF-1 binding site (Figure 2B). This third factor, termed EF1, is an erythroid-specific factor, which also binds to two sequences within the enhancer downstream of the β^A-globin gene.[307] This enhancer binds five factors present in erythroid and nonerythroid nuclear extracts.[152] One of the factors (factor I in Figure 2B) found in both cell types binds to a sequence similar to that bound by NF-1. Two of the sequences (II and IV) bind erythroid-specific factors. Of particular interest is the large footprint IV. This footprint is produced by the binding of EF1 to the enhancer; it is, in fact, probably two contiguous footprints produced by the binding of two molecules of the protein.[307]

Two proteins have been found to bind to three sequences within the nuclease-hypersensitive region in the second intron of the mouse β^{major}-globin gene[153] (Figure 2B). One of

CHICKEN β-GLOBIN GENES

β^A-GLOBIN UPSTREAM PROMOTER ELEMENT

```
  -200    --------------------
5-GGGAATCGGGGGGGGGGGGGGGGGCGGGTGGTGGTGTGGCCCACC

           --------------------------------
-----------NF1------------                    -110
GATCTGGGCACCTTGCCCTGAGCCCCACCCTGATGCCGCGTTCCCTCCCCC-3
```

β^H-GLOBIN UPSTREAM PROMOTER ELEMENT

```
                       ------ EF1------
           ----------NF1------------
5-CTTTTCCCCTTGGACAGGATCCAAATTTGCTATCTTTCCTTCTGC-3
```

β^A-GLOBIN ENHANCER

```
   1820       ---------I----------   ------------II----------
5-GATCATTTCTGGCATTCAGCCTCCCCGAAAGGAGCTGACTCATGCTAGCCCAGCAG


       ---III---     ---------------IV---------------
CCAGCTGGGTGGGGGCAGGTTGCAGATAAACATTTTGCTATCAAGACTTGCACAGACCT


                                                    -----
TGTTTATGCACTTCTTCACCCTACGCTGCCCATTCTGCTGCTCTGCGTGAGGGAAGAGAG


--V----------  2010
GGGGTTAATCCTGTCAATA-3
```

MOUSE α-MAJOR GLOBIN INTRON

```
   533   -----B2------  -------B1----------
5-CAACCTTCCTATCAGAAAAAAAGGGGAAGCGATTCTAG-3
```

CHICKEN α^D-GLOBIN UPSTREAM PROMOTER

```
                    ----B-----
-79        ----C-----               -------A--------   -30
5-GCCCCTCCGTGCGGATAAGATAAGGCCGGGGCGGGTGTACAGGGAGCTATAAGA-3
```

FIGURE 2B.   Protein binding sites associated with chicken and mouse globin genes. The β^A-globin gene UPE binds three factors: NF1, a polyG binding factor, and a CACCC binding factor. The β^H-globin UPE binds NF1 and the erythroid-specific factor EF1, which also binds to the β^A-globin enhancer (site IV) and possibly to the mouse β^{major}-globin gene (site B2) and the chicken α^D-globin gene (site B).

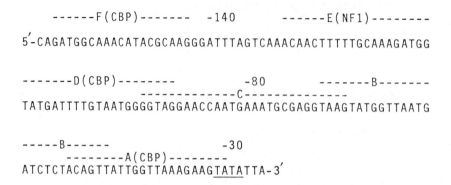

FIGURE 2C. Protein binding sites within the mouse albumen upstream promoter. These sites are bound by CBP, NF1 a CCAAT-box factor (C), and a liver factor (B).

these proteins (B2) is an erythroid-specific factor and binds to a sequence resembling that bound by the chicken EF1. Next to the binding site of this factor is a binding site for a second factor (B1) which is found in erythroid and nonerythroid hemopoietic cells. This second factor also binds to a second site 100 base pairs away.

An erythroid-specific factor binds to a sequence, AAGATAAGG, between the CCAAT and TATA boxes of the chicken α^D-globin gene (Footprint B, Figure 2B).[154] Again, the sequence resembles those bound by EF1, and hence the two factors may be the same. Two other factors, one of which may be Sp1, are seen binding on either side of this factor when high protein concentrations are used.

The sequence AAGATAAGG is also found upstream of the ρ globin gene[154] and the erythroid nuclear factor binds to this sequence. Thus, it can be seen that the same erythroid-specific factor (EF1) may be binding to the β^A enhancer and the upstream promoter elements of the β^H-, ρ- and α^D- globin genes and is therefore likely to be involved in erythroid-specific activation of these globin genes.

The UPE of the human γ-globin gene contains two CCAAT motifs, a CACCC motif, and, at −180, the octamer motif ATTTGCAT, which binds several proteins.[155] One of these proteins is the ubiquitous octamer-binding factor; the other two are present only in uninduced erythroid K562 cells and bind to the γ-globin octamer motif with high affinity. These octamer binding factors disappear when K562 cells are induced to differentiate, suggesting that they could be negative regulatory proteins. A fourth A-rich protein-binding region is present further upstream of the octamer motif.[155]

C. THE ALBUMEN GENE

The upstream promoter element of the rat and mouse albumen genes contains a CCAAT-box motif and a region from bases −93 to −157 containing two conserved distal elements, which are at least partially responsible for hepatocyte-specific gene expression. Six protein-binding sites have been described in these regions[90,124,156] (Figure 2C). (1) Rat liver nuclear extracts contain several NF-1-like proteins which bind to the site E sequence TGGCA (which is half of the inverted repeat-consensus sequence bound by the protein); some of these NF-1 proteins are more abundant in liver than other tissues. (2) A protein resembling NF-Y binds to the CCAAT box (site C). (3) A third, heat-stable protein binds to three sites (A, D, and F) with different affinities; this protein is probably the enhancer core-binding protein CBP and would appear to be more abundant in liver. (4) A protein(s) binds to site B in Figure 2C, which resembles the sequence of site A.

INSULIN ENHANCER

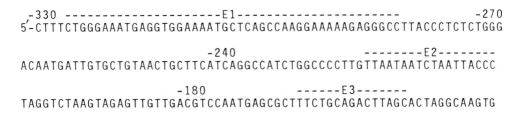

```
 -330 --------------------E1--------------------          -270
5'-CTTTCTGGGAAATGAGGTGGAAAATGCTCAGCCAAGGAAAAAGAGGGCCTTACCCTCTCTGGG

                         -240                 --------E2--------
ACAATGATTGTGCTGTAACTGCTTCATCAGGCCATCTGGCCCCTTGTTAATAATCTAATTACCC

                     -180              ------E3-------
TAGGTCTAAGTAGAGTTGTTGACGTCCAATGAGCGCTTTCTGCAGACTTAGCACTAGGCAAGTG
```

FIGURE 2D.   Three protein binding sites within the insulin enhancer.

HUMAN GROWTH HORMONE GENE

```
               SITE 3(AP2)
    ----------------------------------- -250
5'-CACCATGGCCTGCGGCCAGAGGGCACCCACGTGACCCTTAAAGAGAGGACAAGTTGGGTGGT


                       -200
ATCTCTGGCTGACACTCTGTGCACAACCCTCACAACACTGGTGACGGTGGGAAGGGAAAGATGA

-------AP2-------------            -----------SITE 2a---------
             -150          ------SITE 2-------------
CAAGCCAGGGGGCATGATCCCAGAGCATGTGTGGGAGGAGCTTCTAAATTATCCATTAGCACAA

               SITE 1
    --------------------------- -60
GCCCGTCAGTGGCCCCATGCATAAATGTACACAGAAACAGGTGGGGG-3'
```

FIGURE 2E.   Protein binding sites in the human growth hormone gene UPE. The factor GHF1 binds to sites 1 and 2a. Underlined is a glucocorticoid response element.

A comparison of the footprint patterns obtained with liver and nonhepatic nuclear extracts reveals that proteins are present in nonhepatic tissues[90,156] which bind over the same regions as the liver proteins, but there are qualitative and quantitative differences in the protection patterns,[90] supporting the idea that albumen gene expression in liver cells is controlled by the binding of positive regulators and that repression in nonhepatic cells is achieved by competitive binding of repressor molecules.

D. THE INSULIN GENE

The 5' flanking sequence of the insulin gene contains two regions important for cell-specific expression, the most distal of which binds three proteins present in nuclear extracts of insulin-secreting cells[157] (Figure 2D). Two of the proteins (E2 and E3) are also present in noninsulin producing cells.

E. THE GROWTH HORMONE GENE FAMILY

The 5' flanking region of the human growth hormone gene contains a tissue-specific enhancer which binds three proteins *in vitro*, one of which is specific to pituitary cells[158] (Figure 2E). The pituitary-specific factor GHF1 binds to a proximal element (site 1) just upstream of the TATA box and also more weakly to a second site further upstream (site 2a). Another uncharacterized factor binds to site 2 and there are also two AP2 binding sites.[46] Similar factors have been described binding to the rat growth hormone gene.[159-161]

THYROID HORMONE RECEPTOR BINDING SITES

```
                         - - - - - - - - - - - - - - - - -
           - - - - - - - - - - - - - - - - - -
        -190                                    -160
      5'-TGGAAAGGTAAGATCAGGGACGTGACCGCAGGA-3'
```

FIGURE 2F.   Two binding sites upstream of the rat growth hormone gene
that are bound by the thyroid hormone receptor (c-*erb*-A).

PROLACTIN UPSTREAM PROMOTER FOOTPRINT SEQUENCES

```
                    III
        -163                -145
      5'-CCTGAAAATGAATAAGAAA-3'


            II
        -131          -120
      5'-GATGTTTAAAAT-3'


                    I
        -67                  -48
      5'-CCTGATTATATATATATTCA-3'
```

FIGURE 2G.   Three sequences bound by protein in the UPE of the prolactin gene.

The growth hormone gene is regulated by glucocorticoids (there is a GRE in the 5′
flanking sequences)[158] and also by thyroid hormone. Purified thyroid hormone receptor has
been found to bind to a sequence centered at −180 base pairs upstream of the rat growth
hormone gene[161] (Figure 2F). The sequence is comprised of a direct repeat of the sequence
AGGNANG. Another study describes a thyroid hormone receptor footprint over the sequences −163 to −178, i.e., just 3′ and overlapping one of the direct repeats[160] (Figure
2F). Thus, there may be two binding sites for this receptor. The human c-*erb*A gene product
synthesized *in vitro* also bound to this sequence, confirming the identity between the c-*erb*A
gene product and the thyroid hormone receptor[160] (see Section V).

The prolactin gene is similarly expressed in pituitary cells and three footprint regions
have been defined in the 5′ flanking regulatory region of this gene[162,163] (Figure 2G). One
footprint (footprint II) is produced with protein extracts from pituitary and nonpituitary cells,
while the other two (footprints I and III) are pituitary cell specific. Footprints I and III have
similar sequences and may therefore be bound by the same factor. This factor may also be
responsible for epidermal factor and phorbol ester-mediated transcriptional enhancement.

Although there are resemblences between the footprint sequences of the prolactin and
growth hormone upstream elements (for example, several have AT-rich motifs), it is not
known how the various DNA binding activities are related.

F. THE TYROSINE-AMINOTRANSFERASE (TAT) GENE

The TAT gene is transcribed in parenchymal cells of the liver. Basal level transcription
can be increased by glucocorticoids or by cAMP. Three nuclease-hypersensitive sites in the
nuclear chromatin mark important regulatory regions,[164,165] the most distal of which (site
III) is 2.5 kb upstream of the promoter and contains two glucocorticoid-responsive ele-

TYROSINE AMINOTRANSFERASE UPSTREAM ELEMENTS

HYPERSENSITIVE SITE III

```
               ---------GR---------        -2480
5'-ACCAATCTCTGCTGTACAGGATGTTGTAGCTACTTTATTTGCAATAGAAAATCTGAAA

      --------------GR-------------------        -2420
GTTTCCCCATGTCCAACAAGACTAGAACAAACAAGTCCTGCGTAGTCGCCTGTCGGTTTT

      ---------  -2390
CTGGGTGTGGTGGTATAG-3'
```

HYPERSENSITIVE SITE II

```
     -1100                -----------------------        -------------
5'-GCAGCGTGGTGGTGGGAGCACTGCACATGCGCAGAGACGCTACTATGCAAATAATAGTCTA

------------------------------------------        -1000
GCGCCTCTTGTGGACGGTGTTGTAGCTGCGGTCTGTGCCTGTTGGA-3'
```

HYPERSENSITIVE SITE I

```
                 ------        -200
5'-ACCAAATGACTAGAAAGAGTTAACAGGATTCCAGATACTTGATGTAAG

      ----------------------- -150          ---------------
GACAAATCCCAGATTGGAAGGTGGCCCAGGGTTGGGGTGAGAAACAGCAGAGTGGGGGGTGGGG

---------   ----------------------        -80    --------------------
TATGGGGGTAGGTCCGGGGGAGGGACTTAGTTCTCACTCTCAACCAATAGCACGAAGGCTTCGG

-----    ---------------------------------- -10
GCCCAACGCCATTGGCTGAAACTATTTCAAGGGTCAGGACTGCAC-3'
```

FIGURE 2H.   *In vivo* and *in vitro* protein binding sites within the three hypersensitive sites (I, II, and III) of the TAT gene. Note that in site III there are two sites bound by glucocorticoid receptor (GR) and one binding to the CACCC motif. In site I, there are three protein-bound sequences which have CCAAT motifs (underlined).

ments.[164] Genomic footprinting with dimethyl sulphate has been used to define *in vivo* protein-bound regions in the hypersensitive sites.[164,165] Hormone-dependent binding of the glucocorticoid receptor (GR) to the two GREs in site III is observed in hepatoma cells (Figure 2H). Additional protein-binding sites are also observed on neighboring sequences, e.g., a CACCC motif is protected from DMS *in vivo*, suggesting that a factor similar to that bound to the chicken β^A-globin gene (Figure 2B) binds near the glucocorticoid receptor.

Protein binding within the hypersensitive sites I and II is not dependent on glucocorticoids.[165] Two proteins are seen bound to the hypersensitive site II *in vivo* and at least four proteins to the hypersensitive I region. Similar proteins are detected *in vitro* by DNAse I footprinting of nuclear extracts of hepatoma and nonhepatoma cells. Thus, in the promoter region, hypersensitive site I *in vivo* and *in vitro* protein binding is observed over the TATA box, three sequences containing CCAAT motifs and three other sequences, one of which has protein(s) binding over a long GC-rich sequence, reminiscent of the protein-binding site of the chicken β^A-globin 5' flanking region.[150,151] (Figure 2B).

PROTEIN BINDING SITES UPSTREAM OF ACTIN GENES

```
                                      -150
       CARDIAC MUSCLE α-ACTIN:        5'-TCCATGAATGG-3'

                                      -110
                                      5'-ACCAAATAAGG-3'


                                      -90
       SKELETAL MUSCLE α-ACTIN:       5'-CCAAATATGGCG-3'


       XENOPUS ACTIN:          5'-AGATGCCCATATTTGGCGATCT-3'
```

FIGURE 2I.   Sequences upstream of actin genes which bind nuclear protein(s).

G. THE ACTIN GENES

Two protein factors of different molecular weight bind to the same sequence just upstream of the TATA box of the skeletal muscle actin gene[166] (Figure 2I). The higher molecular weight form is present in nonmuscle cells, while the lower molecular weight protein predominates in myoblasts and myotubes. The cardiac β-actin gene has two similar sequence motifs upstream of the TATA box which bind the same transcription factor.[167] The sequences of the two actin genes bound by protein resemble one another strongly, suggesting that a common muscle-specific transcription factor activates these genes.

Note that the *Xenopus* actin 5'-flanking sequence has a similar sequence and this has been shown to be homologous to the serum response element of the c-*fos* gene (see Section IV.A). However, the relationship between the serum response factor and the two factors binding to the cardiac and skeletal muscle actin genes is not clear.

H. THE ADIPOCYTE P2 GENE (aP2)

This gene has a 13-base-pair negative regulatory element situated 108 to 124 bp upstream of the start of transcription. Nuclear extracts from preadipocytes give a smeared gel-retarded band when mixed with an oligonucleotide of the regulatory element, but differentiated adipocyte extracts give a discrete band with a somewhat higher mobility.[168] Both extracts give a similar footprint over the regulatory element (Figure 2J). Both gel-retarded complexes are disrupted by antibodies to the c-*fos* protein. This evidence, together with cross-linking data, indicate that c-*fos* is bound to the regulatory DNA element.

I. THE THYROGLOBULIN GENE

The thyroglobulin gene is only expressed in thyroid cells and approximately 170 base pairs 5' flanking this gene are involved in this tissue-specific gene expression. A factor found in rat thyroid cells binds to a sequence some 60 to 70 base pairs upstream of the start of transcription[169] (Figure 2J).

J. THE FIBRINOGEN GENES

The UPE of the γ-fibrinogen gene binds the transcription factor MLTF,[170] the 46-kDa protein originally discovered as a protein binding to the adenovirus major late promoter (see Section II.C.1 above). Since this protein is not liver specific, it is probably involved with other liver-specific transcription factors to activate transcription of this gene in the liver.

The β-fibrinogen gene is also a liver-expressed gene and a sequence in the 5' flanking

ADIPOCYTE P2 GENE UPSTREAM PROMOTER

```
 -124---------------
5'-AAACATGACTCAGAGGAAAACATAC-3'
```

THYROGLOBULIN UPSTREAM PROMOTER

```
       -80 --------------- -60
5'-GAGCCACTGCCCAGTCAAGTGTTCTTGA-3'
```

c-MOS ENHANCER

```
 -------       ------  -------
5'-TGGTTTGACTCCAACCACAGTGTTTTAA-3'
```

OVALBUMIN UPSTREAM PROMOTER

```
     ----COUP------      -65
5'-ATGGTGTCAAAGGTCAAACTTCTG-3'
```

FIGURE 2J.   Sequences upstream of various tissue-specific genes which bind nuclear proteins.

β-FIBRINOGEN UPSTREAM PROMOTER

```
 -150            -------------------
5'-TGCTGGGAAGATGTTGCTCAAATGATAAAAACGGTTCAGCCAA
```

```
       -----------HNF1------------- -70
CAAGTGAACCAAACTGTCAAATATTAACTAAAGGGAGGTAA-3'
```

FIGURE 2K.   Two protein binding sites in the β-fibrinogen UPE. HNF1 is a liver-specific factor.

UPE has been identified that is important for liver-specific gene transcription. A liver-specific factor HNFI binds to this region (Figure 2K) and contains the sequence motif ATTAAC which is also found in the UPEs of the α-fibrinogen and α₁-antitrypsin genes.[171] The albumen UPE also binds a similar liver-specific protein (footprint B of Figure 2C). Further upstream of the β-fibrinogen UPE binding site there is a second protein binding site (Figure 2K). The factor that binds to this sequence is not specific to liver.

K. THE C-*MOS* GENE

The c-*mos* proto-oncogene is only expressed in testis and ovaries. This tissue-specific restriction is probably brought about by a repressor DNA element situated 1.6 to 1.8 kb

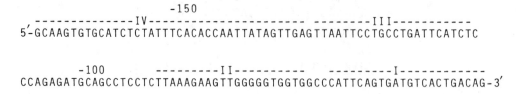

```
                            -150
   ---------------IV---------------------- --------III-----------
5'-GCAAGTGTGCATCTCTATTTCACACCAATTATAGTTGAGTTAATTCCTGCCTGATTCATCTC

          -100        ---------II---------- ----------I------------
CCAGAGATGCAGCCTCCTCTTAAAGAAGTTGGGGGTGGTGGCCCATTCAGTGATGTCACTGACAG-3'
```

FIGURE 2L.   Protein-binding sites in the UPE of the T-cell receptor β-chain variable-region gene.

upstream of the coding region. Just upstream of the repressor element is an enhancer element which is functional in nongerm cells and has three protein-binding sites (Figure 2J), two of which resemble the SV40-enhancer core motifs GT-I and II.[172] The third protein binds to an AT-rich sequence.

L. THE T-CELL RECEPTOR β-CHAIN VARIABLE-REGION GENE

The Ti-β chain gene expression is switched on in lymphoid cells in the thymus. Footprinting analysis of the 5' flanking region of the gene using nuclear proteins from various cell lines revealed the presence of five footprinting activities, one of which (binding to footprint II) is lymphoid specific[173] (Figure 2L).

M. THE OVALBUMIN GENE

The ovalbumin gene is expressed in the chicken oviduct and is regulated in a complex fashion by negative and positive regulatory elements, including those binding steroid hormones.[174] Two transcription factors, present in HeLa cells, have been extensively purified and shown to be required for *in vitro* transcription of the ovalbumin gene.[175] One of the factors (COUP) is a DNA-binding protein which binds to a specific sequence in the upstream promoter of the ovalbumin gene (Figure 2J). Although it binds specifically to this sequence, it requires a second factor (S300-11) of molecular weight 39 kDa to activate transcription *in vitro*. This second factor does not bind to ovalbumin sequences. This factor stabilizes the binding of COUP to its DNA-binding sequence by lowering the rate of dissociation of the COUP-DNA complex. S300-11 also stimulates *in vitro* transcription from the MMTV and lysozyme upstream promoter/enhancer, presumably by interacting with transcription factors present in the crude transcription extracts which bind to these elements.

IV. PROTEINS BINDING TO TRANSCRIPTIONAL ELEMENTS OF GROWTH-ASSOCIATED AND INDUCIBLE GENES

A. THE C-*FOS* GENE

Transcription of the c-*fos* proto-oncogene is rapidly induced by exposure of cells to growth factors and mitogens. Two sequence elements have been defined in the 5' flanking regions of the gene which are responsible for the rapid induction of the gene by serum and v-sis-conditioned culture medium.[176,177] The serum-responsive element (SRE), some 300 bp upstream of the start of transcription, binds a 67-kDa protein which has recently been purified by affinity chromatography[177-181] (Figure 3A). The activity of this factor in A431 cells is induced by epidermal growth factor,[179] but in HeLa cells it is detectable even before serum stimulation,[177] suggesting that transcriptional activation of the HeLa gene by serum is mediated by postsynthetic modification of the preexisting factor (e.g., via protein kinase C phosphorylation). The serum-responsive element is also present in the β-actin gene and the *Xenopus* skeletal actin gene, both of which are serum inducible. The second c-*fos* regulatory

```
                         c-FOS UPSTREAM PROMOTER

                                                              ---AP1--
     -350 ----IV-----      -330         ----------SRF-----------
5'-AGCAGTTCCCGTCAATCCTCCCCCCTTACACAGGATGTCCATATTAGGACATCTGCGTCAGCA-3'
                                          SRE

     -150    ---II---       -130
5'-GGGTTGAAAGCCTGGGGCGTAGAGTTGACGAC-3'


    110  ----II-----    -90
5'-AGAGGGCCTTGGGGCGCGCTTC-3'


     -80             --I---            -50
5'-CCTTCCAGTTCCGCCCAGTGACGTAGGAAGTCCATC-3'
```

FIGURE 3A.   Protein-binding sites in the c-*fos* 5'-flanking sequences. The serum response element (SRE) is underlined.

element, just upstream of the serum-responsive element, binds another factor which is induced by conditioned medium, but not by phorbol esters, insulin, or EGF.[176]

Just downstream of the SRE is an AP1 binding site[52] and three more protein-binding elements are located near the promoter, two of which bind the same factor.[181] The promoter proximal sequences are probably involved in basal level transcription of the gene, although footprint I has an element homologous to the cAMP response element CRE (see Section IV.H) and may therefore be involved in cAMP stimulation of the gene.

B. THE C-*MYC* GENE

Transcription of the c-*myc* gene is elevated when cells are stimulated to divide and repressed in quiescent cells and when cells terminally differentiate. Terminal differentiation is accompanied by the loss of multiple nuclease-hypersensitive sites in the 5' region of the c-*myc* gene. Both negative and positive regulatory DNA elements are present upstream of the coding region,[182,183] but these have not been correlated with the multiple protein-binding sites that have been described. Thus, an NF1-binding site[112] and two AP2-binding sites (Figure 1C) are present in the 5' flanking region, but the functional significance of these sites is not known. A factor, *myc*-PCF, present in plasmacytomas (which do not express c-*myc*) and absent in B-cells (which express c-*myc*) binds to a sequence approximately 300 bp upstream of the start of transcription[184] (Figure 3B). This factor could therefore be a repressor, though there is no evidence that it binds to a negative regulatory element.

The 5' flanking region of the chicken c-*myc* gene is GC-rich and would be classified as a "CpG-rich island", which may characterize important regulatory regions.[1] This region binds multiple proteins present in expressing and nonexpressing erythroid cells.[185] All the footprint sequences are very GC-rich, some of which may correspond to Sp1 and AP2 binding sites (Figure 3B).

C. HISTONE GENES

Transcription of the histone genes is regulated during the cell cycle. Also, specific members of the histone gene family are expressed during embryonic development and in different tissues (for example, chicken histone H5 is expressed only in differentiated erythrocytes).

The 5' flanking regions of the histone genes contain elements necessary for S-phase-specific transcription of the histone genes. Two proteins, H4TF1 and H4TF2, bind to the 5' flanking regulatory regions of the human histone H4 gene.[186] H4TF1 binds to sequences

```
                    CHICKEN c-MYC UPSTREAM PROMOTER

 , 20  ---------I----------      50
5'-GCAGCCGCTCCCCCCGCAGCCTCCTCCTCCCGTTTAATCCTCCGGGATAACGAAGCAGCG

     -------II-------          110
ACACGGGCGGGGGTGCGCGAGCTACGGACGCTCCTTTGTGCCGGTAGGGTAGCCGGCAAC

140 --------III---------
CGCCCCGCCCGCAGCCGCGTTACGGGT-3'


                             ----------------V'-------------
 ,     -----IV----- 460    ---------------V--------------
5'-AGTACTCGGGGGGGGGCACGGAGCCCCTCGGCCGCCCCCTCGCGGCGCGCCCTCCCCGCT

 ----------VI-----------                            550
CACGGAGCCCGCGCGGAGCCGGGGGCGAGCGGGAGGGAGATGAAGCGGCGACGCGCACCG

  ----------VII---------    590       ---------VIII---------
CGAGAGCGCGCACTCGCGGGGCCCCGCCGTGCCGCTCGTGCTCCCGCCCCCGCTGCATCTCCC-3'

                    MURINE c-MYC UPSTREAM PROMOTER


         -300    -------------------------        -260
          5'-ACAACCGTACAGAAAGGGAAAGGACTAGCGCGCGAGAAGAG-3'
```

FIGURE 3B. Protein-binding sites in the 5′-flanking regions of the chicken and mouse c-*myc* genes. Footprints I—III, V′, and VI—VIII are observed with chicken erythrocyte nuclear extracts. Footprint V is only obtained with extracts from proliferating erythroid precursor cells. The numbering of the bases is as given in Lobanenkov et al.[185]

between −80 and −110, which contain the sequence GATTTC which is related to the sequence GACTTTC present in several histone genes. The factor H4TF2 binds just upstream of the TATA box to a region containing the histone-conserved sequence GGTCC.

Genomic footprinting of the human histone H4 5′-flanking region has delineated two regions (I and II) which are complexed with protein within the DNAse I nuclease-hypersensitive site[187] (Figure 3C). Site II encompasses the TATA box and sequences 30 base pairs upstream which include the GGTCC motif, presumably bound by the factor H4TF2. Site I lies −117 to −150 bp upstream and possibly binds two factors, Sp1 and a factor which binds to an AT-rich sequence also present in the H3 UPE.[188] These proteins do not vary with the cell cycle.

Octamer-binding proteins bind to a number of histone genes and are probably involved in stimulating S-phase gene transcription; for example, the recently purified 90-kDa OTF-1 factor, which binds to the histone H2B 5′-flanking octamer sequence, is present only in S-phase cells.[44] Also, there is a testis-specific octamer-binding protein which binds to two sequences just upstream of the sea urchin sperm histone H2B-1 gene.[96] Two other factors bind to this region.[96] One is a CCAAT-box factor and the second is a factor present in embryonic extracts which binds over a sequence overlapping the CCAAT box (Figure 3C). This factor (CDP) is thought to act as a repressor in somatic cells, preventing the expression of this sperm histone gene by excluding the binding of the CCAAT-box protein. CDP is

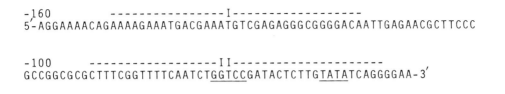

A   HUMAN HISTONE H4 UPSTREAM PROMOTER

-160   ----------------I-------------------
5'-AGGAAAACAGAAAAGAAATGACGAAATGTCGAGAGGGCGGGGACAATTGAGAACGCTTCCC

-100   -------------------II---------------------
GCCGGCGCGCTTTCGGTTTTCAATCTGGTCCGATACTCTTGTATATCAGGGGAA-3'

B   SPERM H2B1 UPSTREAM PROMOTER

                 --CBF---
   -110   ----------------------CDP----------------------
5'-CAGCTGTGCGTTGATTGGTAGATTATCATGTCGCGCTGATAAGCAGAGGATGCACTGAACCC-3'

FIGURE 3C.   A — Genomic footprint sequences in the 5'-flanking sequences of the histone H4 gene. The histone conserved sequences GGTCC and the TATA-box are underlined. B — 5'-flanking sequences of the sperm histone H2B1 gene bound by a CCAAT-box factor (CBF) and the CAAT displacement factor (CDP).

                HUMAN HSP70 UPSTREAM PROMOTER

           -110 --------------------------
           5'-AACCCCTGGAATATTCCCGACCTGGCAGCCTC-3'

FIGURE 3D.   Heat shock element of the *hsp70* gene bound by a nuclear protein.

also present in vertebrate cells and binds to a similar sequence overlapping the distal CCAAT box of the human γ^A-globin gene. This binding may be responsible for the repression of this gene in adult red blood cells. This is supported by the observation that the Greek type of hereditary persistence of fetal anemia is associated with a mutation of the TGA sequence just upstream of the distal CCAAT box to TAA, which would prevent binding of the repressor.

D. THE HEAT SHOCK PROTEIN 70 GENE

A symmetrical heat-shock element centered at -100 bp upstream of the human *hsp70* gene is responsible for heat shock-induced transcription of this gene. A protein that binds over this element has been detected at high levels in nuclear extracts from cells incubated at 43°C[189] (Figure 3D). This induction of DNA binding activity does not require protein synthesis, suggesting that the protein is modified by the heat-shock treatment.

This protein is probably the human homolog of the *Drosophila* heat-shock transcription factor (HSTF) which has recently been purified.[190]

E. THE HISTOCOMPATIBILITY AND β2-MICROGLOBULIN GENES

A 60-kDa factor, KBF1, which binds to the 5' enhancer element of the H-2K^b and β2-microglobulin genes has been purified[191-193] (Figure 3E). This factor also appears to bind to the same sequence in the SV40 enhancer that binds the protein AP2 (Figures 1B and 1C). Since purified AP2 binds to the same sequence on the *H-2K^b* gene, it would appear that the two proteins are the same.

The class II major histocompatibility genes are expressed predominantly in lymphoid

KBF1 BINDING SITES

```
                                          -170
H-2Kᵇ ENHANCER:                           5′-CTGGGGATTCCCCA-3′

                                          -120        -130
β2-MICROGLOBULIN ENHANCER:                5′-AGGGACTTTCCCATTT-3′
```

FIGURE 3E.   Sequences bound by the transcription factor KBF1.

HLA-DQ2   UPSTREAM PROMOTER

```
-150                                                                           -90
5′-AGGCACTGGATTCAGAACCTTCACAAAAAAAAAAATCTGCCCAGAGACAGATGAGGTCCTTCAG
           W                                       X


                                   -60
CTCCAGTGCTGATTGGTTCCTTTCCAAGG-3′
          Y
```

FIGURE 3F.   Upstream promoter sequences of the HLA-DQ2 gene showing the conserved boxes X, Y, and W which bind nuclear proteins. Note box Y has the CCAAT motif.

MOUSE METALLOTHIONEIN I METAL REGULATORY ELEMENTS

```
                            -150
        MREd:               5′-CTCTGCACTCCGCCCGA-3′

                            -132
        MREc:               5′-AAGTGCGCTCGGCTCTG-3′

                            -56            -72
        MREb:               5′-GTTTGCACCCAGCAGGC-3′

                            -54
        MREa:               5′-CTTTGCGCCCGGACTCG-3′
```

FIGURE 3G.   Metal regulatory elements upstream of the mouse metallothionein I gene.

cells and are induced by γ-interferon and interleukin-4, and repressed by prostaglandins. Several conserved DNA elements in the 5′ flanking region have been found to be important for gene expression (the X, Y, and W boxes in Figure 3F). The third box, W, may be involved in γ-interferon induction. Factors binding over these elements have been detected in nuclear extracts.[194]

F. THE METALLOTHIONEIN GENES

The mouse metallothionein I gene has four functional metal regulatory elements (MRE) upstream of the TATA-box element (Figure 3G). One of these (MREd) is the strongest such element and has been shown to bind a factor present in nuclear extracts.[195] Extracts exposed to EDTA do not bind this element, but can be reactivated with cadmium. It is of interest to note that the MREd also contains an Sp1 consensus sequence, suggesting that the cadmium-binding factor could be bound next to an Sp1 protein *in vivo* to activate transcription.

Genomic footprinting of the rat metallothionein I upstream promoter element demonstrates the cadmium-induced binding of factors to five MRE elements, together with the constitutive binding of Sp1 next to the most distal MRE.[196]

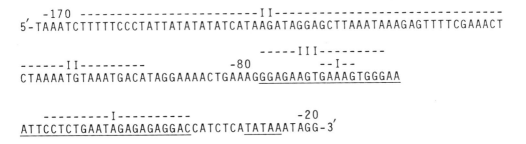

β-INTERFERON UPSTREAM PROMOTER

```
     -170 ------------------------II--------------------------------
5'-TAAATCTTTTTCCCTATTATATATATCATAAGATAGGAGCTTAAATAAAGAGTTTTCGAAACT

                                    -----III---------
------II---------            -80          --I--
CTAAAATGTAAATGACATAGGAAAACTGAAAGGGAGAAGTGAAAGTGGGAA

        ---------I----------              -20
ATTCCTCTGAATAGAGAGAGGACCATCTCATATAAATAGG-3'
```

FIGURE 3H. Genomic footprints in the β-interferon upstream regulatory elements. In the uninduced state, protein II and protein I (possibly two molecules) are bound to the upstream regulatory element of the β-interferon gene. On induction, these are replaced by factor III. The TATA-box and constitutive-enhancer elements are underlined.

SOMATOSTATIN UPSTREAM PROMOTER

```
                    ----------CREB----------
        -60                                  -32
5'-CTGGGGGCGCCTCCTTGGCTGACGTCAGAGAGAGAG-3'
```

FIGURE 3I. Sequence bound by the CREB protein on the cAMP response element (underlined) of the somatostatin gene.

The 5' flanking sequences of the human metallothionein IIA_A gene has four MREs, an Sp1 binding site, an AP1 binding site, and three AP2 binding sites.[22] Like the metallothionein I gene, it also has a far upstream glucocorticoid-responsive element.

G. THE β-INTERFERON GENE

Human β-interferon gene expression is induced by viral infection or by poly(I)-poly(C). Genomic footprinting experiments have been used to investigate the changes in proteins associated with the *cis*-acting DNA elements in the 5' region of the gene.[3] The data suggest that before induction there are two proteins (possibly repressors) bound to the 5' flanking region, one bound to a distal negative regulatory element (factor II, Figure 3H) and the second over the constitutive enhancer element (IRE) (factor I). On induction, the two putative repressor molecules are replaced by an activator protein III binding onto the IRE.

H. THE SOMATOSTATIN GENE AND cAMP RESPONSE ELEMENTS

Somatostatin biosynthesis is regulated by cAMP in the hypothalamic cells. Transcription of the gene is also induced by cAMP and this is thought to occur via phosphorylation of a 43-kDa nuclear protein by cAMP-dependent kinase II.[78,197] This 43-kDa protein (CREB) has been purified[78] and shown to be a phospho-protein which binds to a cAMP-responsive element (CRE) in the UPE of the somatostatin gene, 30 to 60 base pairs upstream from the start of the transcription (Figure 3I). The CRE contains the palindrome TGACGTCA which appears upstream of a number of other cAMP-regulated genes, e.g., the phosphoenolpyruvate carboxykinase,[198] proenkephalin,[199] and c-*fos* (Section IV.A) genes. The cytomegalovirus enhancer also has this motif (Section II.G). The human glycoprotein α-subunit protein gene has two repeats of this sequence plus one which functions in concert with an upstream enhancer element which binds a placental protein factor.[200]

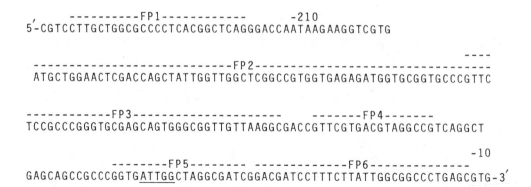

FIGURE 3J.   Six protein-binding sites in the HMG-CoA reductase upstream promoter. FP5 has a CCAAT-box element; FP4 has the sequence CGTGAC bound by factor MLTF.

I. THE HMG-CoA REDUCTASE GENE

The expression of this gene is controlled by end-product repression; transcription is substantially repressed by cholesterol and other mevalonate-derived products. *Cis*-acting elements required for transcription and repression are located at the 5' end of the gene and its 5' flanking sequences. Nuclear protein extracts give six footprints over the 5' flanking region of the gene,[201] two of which, FP4 and FP6, are associated with sequences important for transcription (Figure 3J). FP4 contains the sequence CGTGAC, which is present in the adenovirus major late promoters and is recognized by the transcription factor MLTF. FP6 includes the sequence TATT, which probably serves as a TATA box for the HMG Co-A reductase gene, and FP5 has a CAAT-box element.

V. NUCLEAR HORMONE RECEPTORS

A large number of genes are known to be activated by steroid-hormone receptor complexes by a mechanism in which the steroid hormone is translocated to the nucleus and the hormone receptor binds to the hormone-responsive enhancers or upstream promoter elements. The genes for the glucocorticoid,[202,203] mineralocorticoid,[204] estrogen,[205,207] and progesterone[208-210] receptors have been cloned and, also recently, a member of the c-*erb*-A gene family has been shown to code for the thyroid hormone receptor.[211,212] All four receptor proteins have similar structures (reviewed by Green and Chambon[213]), having two highly conserved domains: one central DNA-binding domain of 66 to 68 amino acids has the potential to form two zinc fingers and the other C-terminal domain of 220 to 250 amino acids is required for hormone binding. These two domains are joined by a basic hydrophilic-hinge region. The N-terminal domains of the different receptors are variable in length and sequence. The DNA-binding domain acts autonomously in that it can bind alone to its DNA recognition sequence in the absence of the hormone-binding domain and the N-terminal domain. The N-terminal domain and the central DNA-binding domain are both important for transcriptional activation.[214,215] Deletion analysis of the glucocorticoid and estrogen receptors suggests that the C-terminal domain may contain a region which masks the DNA-binding domain and that, when the hormone binds, it causes a conformational change in the protein, unmasking the DNA-binding domain.[215,216] These results are consistent with *in vivo* footprinting data[164] which show that glucocorticoid hormone binding to the receptor is required for binding of the receptor to its recognition sequence in the nucleus. However, *in vitro,* the receptor can bind to DNA sequences in the absence of hormone;[217,218] this may

LYSOZYME UPSTREAM PROMOTER

```
          --------------------PR-----------------
          -------------GR---------------                    -150
5'-AGATATTGCAACAGACTATAAAATTCCTCTGTGGCTTAGCCAATGTGGTACTCCCACATT


                                                -100
GTATAAGAAATTTGGCAAGTTTAGAGCAATGTTTGAAGTGTTGGGAAATTTCTGTATACTCAA


          ------------GR--------------         -40
GAGGGCGTTTTTGACAACTGTAGAACAGAGGAATCAAAAGGGGGTGG-3'
```

A

VITELLOGENIN ESTRADIOL RESPONSE ELEMENT

```
      ----NHP-1----        -----NHP-2----
         -620                    -600
5'-TCCTGGTCAGCGTGACCGGAGCTGAAAGAACACATTGAT-3'
```

B

FIGURE 4.   (A) Progesterone-receptor (PR) and glucocorticoid-receptor (GR) binding sites in the lysozyme upstream promoter. (B) Binding sites for the two proteins NHP-1 and NHP-2 to the estradiol response element of the avian vitellogenin gene. The palindrome sequence thought to be bound by the estradiol receptor is underlined.

be due to alteration in the protein during its purification or to loss of an inhibitory cofactor (e.g., the heat-shock protein).[218]

The glucocorticoid and progesterone receptors bind to very similar sequences based on the palindromic motif GGTACA NNN TGTTCT.[219,220] For example, the UPE of the chicken lysozyme gene has two binding sites for the glucocorticoid receptor, the more distal site having the lower affinity[221] (Figure 4A). The distal site also binds the progesterone receptor, although the contact points differ between the two proteins.

The interaction of the estrogen receptor with a number of hormone-responsive genes has been studied (e.g., the vitellogenin[222] and prolactin genes[223]). The sequences bound by the unpurified receptor resemble that of the glucocorticoid- and progesterone-responsive elements and are based on the palindromic consensus GGTCA NNN TGACC. Although preparations of the estrogen receptor have been found to bind to estrogen-responsive elements by nitrocellulose filter assays, the purified protein has not yet been demonstrated to bind directly to the estrogen-response element using footprinting techniques.[224] This may be because it requires additional factors for tight binding. Two such proteins, NHP-1 and NHP-2, have recently been found which bind to the avian vitellogenin estrogen-response element some 600 bp upstream of the gene and these enhance binding of the estrogen-receptor complex.[224] One of the factors (NHP-1) binds to the symmetrical estrogen palindromic element and the other (NHP-2), just downstream (Figure 4B).

Studies carried out by transfecting the cloned gene have suggested that full activation of transcription by the estrogen receptor requires the C-terminal hormone-binding domain of the protein as well as the N-terminal and DNA-binding domains.[215] The requirement of the C-terminal domain may be due to this domain being involved in dimerization of the receptor.[225] The basic hydrophilic-hinge region has recently been found to be involved in negative regulation of prolactin gene transcription by estrogen receptors.[226] This is probably caused by this region interacting with other transcription factors, preventing their correct binding.

The thyroid hormone triiodothyronine binds to a receptor present in the nucleus of virtually all mammalian cells. Binding of hormone to the receptor activates transcription of a number of genes. For example, the thyroid hormone receptor binds to thyroid hormone-responsive regions − 160 to − 190 base pairs upstream of the rat growth-hormone gene (see Section III.E above for a description of this binding site). The thyroid hormone receptor is encoded by one of the c-*erb*-A genes and has a structure very similar to the steroid hormone receptors.[211,212] It has a putative two-zinc-finger DNA-binding domain and a C-terminal hormone-binding domain.

The protein product of the chicken v-*erb*-A gene present in the avian erythroblastosis virus differs functionally from the cellular chicken c-*erb*-A protein in that the former does not bind hormone, which may account for its effect on erythroid cell differentiation and in transformation. The v-*erb*-A protein may be constitutively active in the absence of hormone and may compete for DNA binding sites normally bound by c-*erb*-A protein and thereby block the activation of genes normally activated by c-*erb*-A in erythroid cells. (These may include δ-aminolevulinic acid synthase, band 3, and the globin genes.[227]

Retinoic acid may be the natural morphogen involved in the development of the chick limb bud. It also induces F9 cells to differentiate and this is accompanied by changes in the transcription of a number of genes. These effects are likely to be brought about by retinoic acid binding to a protein receptor. The gene coding for this receptor has recently been cloned and shown to have a similar structure to the thyroid and steroid hormone receptors.[228] As yet, the DNA sequences to which it binds have not been identified. Similarly, the vitamin D3 receptor gene has been cloned and found to resemble the steroid hormone receptors.[229]

VI. PROTEINS BINDING TO GENES TRANSCRIBED BY RNA POLYMERASE III

Genes transcribed by polymerase III in eukaryotic cells can be grouped into three classes with respect to the promoter structure utilized and the requirement for particular ancillary transcription factors. The adenovirus VA1 genes, Epstein-Barr virus, tRNA genes, and the Alu family repeats form one group. The promoters of these genes have a split internal promoter in which the protein-binding regions, designated box A and box B, are separated from one another by 30 to 60 bases (for reviews see References 230, 231). At least two factors, designated TFIIIB and TFIIIC, are required to bind these regions to establish transcriptional activity.[232-234] A second group of genes comprised of the *Alu*, 7SL, 7SK, and U6 snRNA genes have an internally coded box A sequence together with upstream elements.[235-237] A third group, the 5S rRNA genes, also contains an internal promoter or internal control region (ICR), consisting of two boxes designated A and C, which bind the factor TFIIIA.[238,239] The 5S A box is homologous to the tRNA A box. As discussed below, TFIIIC and TFIIIB subsequently bind to promote transcriptional activation.

A. 5S RNA GENES

Although 5S RNA genes are ubiquitous in eukaryotic cells, the transcription of 5S genes has been most intensively studied in the frog *Xenopus*. There are two major classes of 5S genes in amphibians, the oocyte (about 20,000 copies per haploid genome) and somatic types (about 400 copies per haploid genome). Both types are transcribed at very high levels during oogenesis, whereas after blastulation only the somatic type is expressed. Oocyte 5S genes differ from somatic genes by six base changes within the 5' part of the internal control region (for a review, see Reference 240). The effect of this change is to weaken the binding of the transcription factor TFIIIA to the oocyte 5S ICR by about fourfold in comparison to the somatic gene.[241] The binding site for TFIIIA has been located to a region of about 50 bp between nucleotides 45 to 96 of the 5S gene.[240] The site has been defined by deletion mapping[242,243] and refined using linker substitution mutants.[244]

The purification of TFIIIA,[245] the first eukaryotic transcription factor to be purified to homogeneity, was aided by the finding that the protein is stored in large amounts, complexed with the gene product 5S RNA in oocytes of *Xenopus laevis*.[246,247] The availability of large amounts of TFIIIA enabled a detailed study to be undertaken to elucidate the interaction of the protein with the 5S gene. The stoichiometry of binding of the protein to the ICR was derived by both DNA footprinting and gel exclusion chromatography of the DNA-protein complex. In these experiments, it was demonstrated that each 5S gene ICR binds one molecule of TFIIIA.[248,249] Using limited proteolytic cleavage of TFIIIA, it was shown[248] that the 38-kDa protein consists of three structural domains — a large 20-kDa domain at one end and two smaller 10-kDa domains at the other. The large domain contains the sequence-specific DNA-binding regions which recognize key nucleotides on the 3′ side of the internal control region (i.e., the C-box region, Figure 6 below). In conjunction with the adjacent 10-kDa domain, binding is extended into the 5′ region of the ICR (i.e., the A-box region). This 30-kDa TFIIIA fragment can support transcription weakly. The third domain at the carboxyl end does not bind to the ICR. It is, however, responsible for most of the transcription-enhancing activity of the intact protein, probably because it interacts with the other transcription factors required to form the active complex.

The development of the cell-free transcription systems has enabled a detailed understanding to be made of the molecular mechanisms involved in 5S gene transcription and regulation (e.g., References 250, 251). The proteins which mediate transcription of 5S genes were chromatographically resolved into three fractions designated TFIIIA, TFIIIB, and TFIIIC.[232-234,252] As discussed previously, TFIIIA is capable of binding to the DNA template itself with a 1:1 stoichiometry to form a metastable complex. Subsequent binding of TFIIIC stabilizes this complex. The rate-limiting step is the addition of TFIIIB, a step requiring both ATP and Mg^{2+}. The addition of TFIIIB is necessary to induce maximal transcription rates.[253,254] This complex of factors is stable to challenge by competing templates and is maintained through many cycles of transcription by RNA polymerase III.[255,256]

Since polymerase III can initiate transcription on a naked template — albeit with little specificity, but with accurate and efficient termination of transcription — the role of these factors must be limited to initiation and possibly elongation.[257]

A cDNA clone of TFIIIA has been isolated from a cDNA library prepared from polyA$^+$ RNA from mature *Xenopus* ovaries.[258] From this sequence it was possible to establish the complete amino acid sequence of the protein. Using these data, Miller et al.[259] suggested that the 30-kDa DNA-binding domain of the protein comprised nine, imperfect, tandemly repeated regions, each comprised of about 30 amino acid residues. It was proposed that each of these linearly arranged, independently folding domains, or "fingers", was stabilized by a tetrahedral arrangement of zinc ligands coordinated to the two invariant pairs of cysteine and histidine residues in each unit (Figure 5). Furthermore, these workers confirmed an earlier report[260] that TFIIIA contained Zn^{2+} and estimated that there are 7 to 11 Zn^{2+} atoms in each molecule. Subsequently, Rhodes and Klug[261] revealed that within the TFIIIA binding site there is a 5 $^1/_2$-bp periodicity of guanine residues which are in contact with the protein. Thus, a total of nine such repeats in the internal control region of the 5S gene corresponded to the nine binding fingers of TFIIIA. Furthermore, it was suggested that the periodicity of half or double turns of the DNA, together with the same periodicity of the fingers, would result in each finger having the same type of contact with the DNA and, hence, interact with the same groove of the double helix, most probably the major groove.[262]

Recently Brown and co-workers have extended their original characterization of the functional domains of TFIIIA.[263] In this study, protein was prepared from deletion mutants.

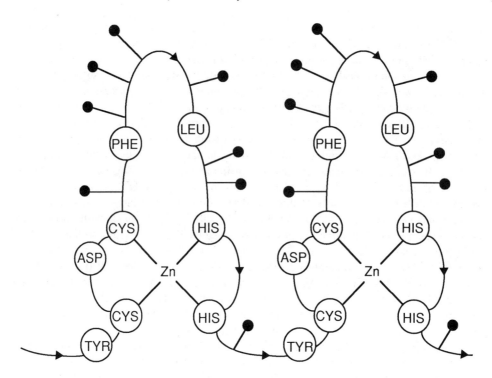

FIGURE 5.   TFIIIA zinc-finger. Ringed circles are amino acids that are conserved in the TFIIIA fingers; the black circles are DNA-binding amino acids (Lys, Arg, His, Asn, Gln, Thr). (Adapted from Miller, J., McLachlan, A. D., and Klug, A., *EMBO J.*, 4, 1609, 1985.)

Using SP6 polymerase, mRNA was transcribed from each construct *in vitro* and this was used to synthesize the protein in a wheat germ *in vitro* translation system. The effect of the deletions was then assayed for binding of the protein to the ICR region by DNAse I and hydroxyl radical footprinting. These results were compared with the ability of each mutant to support 5S RNA transcription. In summary, the data suggested the following model of the interaction of TFIIIA with the ICR (Figure 6). The greatest contribution to the binding energy is provided by the interaction of the N-terminal fingers 1 and 2 to the 3' end of the ICR (box C). Probably, the protein is in close contact in the major groove with most of the nucleotides through at least one turn of the DNA helix. The protein emerges from the major groove at about nucleotide residue 80 in the central region of the ICR to lie on one side of the DNA for about 20 base pairs. In this region (box M), fingers 5 and 6 are in contact with the minor groove of the DNA, but also simultaneously make contacts with nucleotides in the major groove. Binding to the 5' end of the ICR (box A) is mediated by fingers 8 and 9 as well as a short polypeptide region on the carboxyl side of finger 9.

Especially noteworthy is the information from the footprinting data that clusters of fingers bind to three regions of the ICR, rather than adjacent fingers binding to precisely adjacent nucleotides along the ICR, as predicted by Fairall et al.[262] This observation is in good agreement with the results of Pieler et al.[264,265] who by analysis of point mutations and deletions defined almost identical regions as being important for TFIIIA function.

The availability of semipurified and purified transcription factors of the *Xenopus* 5S genes has prompted an examination of their role in the control of the developmental regulation of the 5S genes. A major aim has been to determine how the selective inhibition of oocyte 5S transcription during development is brought about. The fact that TFIIIA has a fourfold higher affinity for the somatic 5S gene relative to the oocyte gene has provided the basis for models to explain the differential expression of oocyte and somatic genes during de-

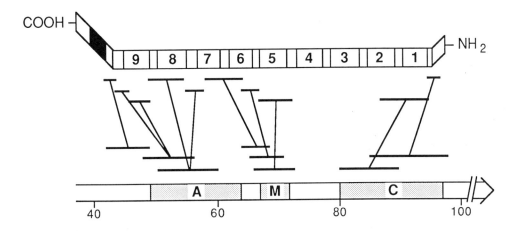

FIGURE 6. Model of the interaction of the nine TFIIIA zinc fingers (1—9) with the A, M, and C elements of the 5S gene internal promoter. Horizontal bars denote interacting domains of the protein and DNA. The C-terminal transcriptional activating domain of the protein is shaded. (From Vrana, K. E., Churchill, M. E. A., Tullius, T. D., and Brown, D. D., *Mol. Cell. Biol.*, 8, 1684, 1988. With permission.)

velopment since the levels of TFIIIA fall during development (for reviews, see References 266, 267). However, oocyte 5S genes can be inactive, even in the presence of excess TFIIIA.[268] Furthermore, it has recently been reported[269] that in an extract prepared from whole ovaries of *Xenopus*, exogenous somatic genes were transcribed 100-fold more efficiently than the oocyte type. However, not only were the active somatic 5S genes shown to be complexed with TFIIIA in the extract, but so were the inactive oocyte genes. The study of which component of the transcription complex is rate limiting in this *in vitro* system has been extended by Wolfe,[270] who has shown that it is factor TFIIIC which is present in limiting amounts. This limiting amount selectively stabilizes the TFIIIA complex with the somatic genes, hence leading to a stimulation of transcription from these genes. McConkey and Bogenhagen[271] have reexamined the relative affinity of the somatic- and oocyte-type 5S genes and reported that while a low abundance oocyte-type 5S gene has indeed a reduced affinity for TFIIIA in comparison to the somatic gene, the major oocyte-type 5S gene binds TFIIIA with an affinity equal to that of the somatic gene.

Evidence is also accumulating from other types of experiments to show that TFIIIC plays a crucial role in modulating the activity of polymerase III genes. A number of reports have shown that the adenovirus EIa gene product can enhance the transcription of polymerase III genes both *in vivo* and in extracts from EIa-expressing cells.[272-274] This enhancement of polymerase III transcription has been attributed to the ability of EIa proteins to increase the effective concentration of factor TFIIIC.[272,274] Conversely, infection of cells with poliovirus resulted in a severely reduced activity of factor TFIIIC and, to some extent, TFIIIB as well.[275] This inhibition of TFIIIC activity did not involve destruction of its DNA-binding domain.[275]

TFIIIC binds to the 5S ICR only in the presence of TFIIIA,[254,276] suggesting that protein-protein interactions are important for binding. However, analysis of point mutations of the ICR and mutants with an altered spacing of box A and C regions suggests that TFIIIC, once bound, may also be interacting with DNA elements in both the A and C boxes.[277] This model has to be qualified by the recent finding[278,279] that the human TFIIIC can be chromatographically separated into two forms (see below), and these two forms could explain the observed bipartite binding to box A and box C in the *Xenopus* 5S gene.

B. tRNA GENES

The highly conserved intragenic promoter of tRNA genes has been shown to be split into two regions, each of about 10 nucleotides known as the A and B blocks. These sequences have been identified as being important for specific initiation of RNA synthesis by deletion and point-mutation analysis. The distance between the A and B blocks is variable in natural tRNAs and can be altered on a particular tRNA species without abolishing promoter activity (for reviews, see Reference 230, 231). Sequences 5' of the coding region have been shown to have a modulating role in transcriptional efficiency in a species-specific manner (see Reference 280 and references therein).

Chromatographic fractionation of crude cellular extracts for reconstituted transcription systems from yeast, *Drosophila*, *Xenopus*, and human cells has shown that the primary event required for template commitment is the binding of factor TFIIIC. Factor TFIIIB then binds to stabilize the complex, but has not been observed to bind to the ICR region in the absence of TFIIIC.[232,234,245,281,282] When analyzed by the DNAse I footprinting technique, TFIIIC was shown to bind more strongly to the B block region at the 3' end of the ICR, with weaker binding over the A block region.[283] It should be noted that, although TFIIIC binds more strongly to the B block region, mutations in the A block lower overall binding efficiency and lower promoter strength.[284,285]

Attempts to purify factors TFIIIB and TFIIIC have been more successful with yeast than with mammalian systems. TFIIIB prepared from yeast appears to be a single monomeric protein of about 60 kDa,[285] while TFIIIC has been reported to be a multimeric protein of about 30 kDa.[287]

C. ADENOVIRUS VA GENES

The two tandemly located adenovirus genes, designated VA1 and VA2, which are differentially regulated during the course of viral infection, each code for small cytoplasmic RNA species.[288,289] These genes are transcribed by polymerase III[290] and have internal promoters.[291] Mutational analysis has demonstrated that the promoters are divided into two domains, the A and B blocks, which have sequence homologies with the corresponding domains of tRNA genes.[291-293] In addition to RNA polymerase III, it has been shown that factors TFIIIB and TFIIIC are required to obtain *in vitro* transcription of these genes, with binding of TFIIIC at the B block element being the first step for commitment of transcription.[234,254,276] A more detailed study has recently been published on the protein-binding sites on both genes.[294] Using proteins partially purified from HeLa cell nuclear extracts, the binding sites were determined using DNAse I and MPE footprinting techniques. A total of five protected regions were found on a DNA segment containing both the VA1 and VA2 genes. Further analysis characterized two binding sites of TFIIIC, one centered over the B block region of the VA1 gene and the other to an analogous, but lower affinity, site on the VA2 gene. A second protein which bound to two sites immediately downstream of the two VA1 RNA termination sites was identified as NF-1 (see Section I.D). A previously undescribed protein, designated VPB, was shown to bind to only one site, 5' to the B block element of the VA1 gene. The function of these proteins was analyzed in a reconstituted *in vitro* transcription system. As would be expected, TFIIIC and TFIIIB were necessary to obtain transcription and the addition of VPB altered the relative ratios of transcription of VA1 and VA2. Addition of NF-1 to the transcription assays appeared to have no effect.

Evidence has recently been presented that factor TFIIIC prepared from human cells can be resolved into two components, TFIIIC1 and TFIIIC2.[278,279] Both are required for *in vitro* transcription of the adenovirus VA1 gene.[278,279] DNAse I footprinting experiments have shown that TFIIIC2 protects the internal B block and the addition of TFIIIC1 extends the footprint to the A block region. Furthermore, it was demonstrated that TFIIIC2 is the limiting component necessary for stable transcription complex formation and hence may be the target of EIa-mediated transactivation.[278,279]

D. THE U6 GENES

The U1-U6 snRNA gene family codes for capped, abundant, small nuclear RNAs involved in the processing of mRNA. These genes are thought to be transcribed by RNA polymerase II (see Reference 295 and references contained therein). However, recently there have been several reports that the U6 RNA genes are, as assessed by several criteria, transcribed by polymerase III.[296-299] Most intriguingly, the promoter elements of these genes are located 5' to the coding region and most closely resemble those of polymerase-II-type genes.[296,298-302] The U6 genes appear to be ubiquitous (for references see Reference 297) but, to date, the most detailed studies have been confined to U6 RNA genes from human, mouse, and *Xenopus* cells.[298,300-302] These genes also contain an internal characteristic tRNA-type A box element. However, it has been reported that mutants of mouse and *Xenopus* U6 genes lacking this element can be transcribed efficiently both *in vitro* and in injected *Xenopus* oocytes.[300,301]

Upstream of the initiation site of RNA synthesis, human, mouse, and *Xenopus* U6 genes have sequences homologous to the distal sequence element (DSE) and the proximal sequence element (PSE) of the U1 and U2 snRNA genes.[298-302] The DSE has structural homology to "enhancers" found in pol II genes and contains an "octamer" motif, first characterized as being part of the transcriptional control elements of immunoglobulin genes (see Section III.A). Close inspection of the *Xenopus* U6 gene sequence revealed that it contains two adjacent octamer motifs arranged as an inverted repeat and, in addition, two Sp1-binding homologies upstream of the DSE and another internal "octamer" sequence just 3' to the internal box A sequence.[301] The DSE has been shown to enhance U6 gene transcription up to tenfold, to bind factors as assayed by *in vitro* DNAse I footprinting, and to functionally substitute for the DSE of the polymerase II-transcribed U2 gene.[298,300-302] The PSE of the U6 genes has both sequence and functional homology to those of the U1 and U2 RNA genes, but it has not yet been possible to detect any protein binding to this element.[300] In addition to the DSE and PSE, the U6 genes include a TATA-like box at about -30, a feature characteristic of many polymerase-II-transcribed genes. At present, it is not clear whether this element has a major or minor functional role.[300,302]

VII. CONCLUSION

It is apparent from this review that a large number of eukaryotic DNA-binding proteins have been detected and a few have been purified. In a number of cases, these proteins have been shown to be transcription factors *in vitro*, and it is anticipated that the rest of the factors will be so characterized. It is likely that the majority of the common transcription factors utilized by many genes have now been identified and the stage is set for the study of how these factors function to activate transcription. The important aspect here will be the study of how the various factors interact with one another; that is, the focus of attention will be directed toward protein-protein interactions. This will involve investigating both how the DNA-binding proteins interact with one another and, equally important, how these factors bind other non-DNA-binding transcription factors (for example, factors TFIIB and TFIIE which, together with the TATA-box factor TFIID, are involved in directing polymerase II to initiate transcription from the cap site[303-305]). In the case of 5S gene transcription, the importance of interactions between factors TFIIIA, B, and C in forming a transcription complex has already been demonstrated. The study of the interaction of transcription factors will require the molecular cloning of the genes for mutational analysis and for the production of large amounts of protein for structural studies.

Finally, another important line of research will be the characterization of postsynthetic modifications of the proteins (such as phosphorylation) since many of the regulatory events will involve changes in either protein-DNA binding or protein-protein interactions brought about by changes in the state of protein modification.

REFERENCES

1. **Bird, A. P.**, CpG-rich islands and the function of DNA methylation, *Nature (London)*, 321, 209, 1986.
2. **Imler, J.-L., Lemaire, C., Wasylyk, C., and Wasylyk, B.**, Negative regulation contributes to tissue specificity of the immunoglobulin heavy-chain enhancer, *Mol. Cell. Biol.*, 7, 2558, 1987.
3. **Zinn, K. and Maniatis, T.**, Detection of factors that interact with human β-interferon regulatory region in vivo by DNAase 1 footprinting, *Cell*, 45, 611, 1986.
4. **Learned, R. M., Learned, T. K., Haitiner, M. M., and Tjian, R. T.**, Human rRNA transcription is modulated by the coordinate binding of two factors to an upstream control element, *Cell*, 45, 847, 1986.
5. **Struhl, K.**, Promoters, activator proteins, and the mechanism of transcriptional initiation in yeast, *Cell*, 49, 295, 1987.
6. **Gehring, W. J.**, Homeo boxes in the study of development, *Science*, 236, 1245, 1987.
7. **Wingender, E.**, Compilation of transcription regulating proteins, *Nucleic Acids Res.*, 16, 1879, 1988.
8. **Blenz, M. and Pelham, H. R. B.**, Heat shock regulatory elements function as an inducible enhancer in the *Xenopus hsp70* gene and when linked to a heterologous promoter, *Cell*, 45, 753, 1986.
9. **Wang, F. and Calame, K.**, SV40 enhancer-binding factors are required at the establishment but not the maintenance step of enhancer-dependent transcriptional activation, *Cell*, 47, 241, 1986.
10. **Maniatis, T., Goodbourn, S., and Fischer, J. A.**, Regulation of inducible and tissue-specific gene expression, *Science*, 236, 1237, 1987.
11. **Dynan, W. S. and Tjian, R.**, Isolation of transcription factors that discriminate between different promoters recognized by RNA polymerase II, *Cell*, 32, 669, 1983.
12. **Dynan, W. S. and Tjian, R.**, The promoter-specific transcription factor Sp1 binds to upstream sequences in the SV40 early promoter, *Cell*, 35, 79, 1983.
13. **Gidoni, D., Kadonaga, J. T., Barrera-Saldana, H., Takahashi, K., Chambon, P., and Tjian, R.**, Bidirectional SV40 transcription mediated by tandem Sp1 binding interactions, *Science*, 23, 511, 1985.
14. **Gidoni, D., Dynan, W. S., and Tjian, R.**, Multiple specific contacts between a mammalian transcription factor and its cognate promoters, *Nature (London)*, 312, 409, 1984.
15. **Everett, R. D., Baty, D., and Chambon, P.**, The repeated GC-rich motifs upstream from the TATA box are important elements of the SV40 early promoter, *Nucleic Acids Res.*, 11, 2447, 1983.
16. **Takahashi, K., Vigneron, M., Matthes, H., Wildeman, A., Zenke, M., and Chambon, P.**, Requirement of stereospecific alignments for initiation from the simian virus 40 early promoter, *Nature (London)*, 319, 121, 1986.
17. **Briggs, M. R., Kadonaga, J. T., Bell, S. P., and Tjian, R.**, Purification and biochemical characterization of the promoter-specific transcription factor, Sp1, *Science*, 234, 47, 1986.
18. **Jones, K. A. and Tjian, R.**, Sp1 binds to promoter sequences and activates herpes simplex virus "immediate-early" gene transcription *in vitro*, *Nature (London)*, 317, 179, 1985.
19. **Jones, K. A., Yamamoto, K. R., and Tjian, R.**, Two distinct transcription factors bind to the HSV thymidine kinase promoter in vitro, *Cell*, 42, 559, 1985.
20. **Jones, K. A., Kadonaga, J. T., Luciw, P. A., and Tjian, R.**, Activation of the AIDS retrovirus promoter by the cellular transcription factor, Sp1, *Science*, 232, 755, 1986.
21. **Dynan, W. S., Sazer, S., Tjian, R., and Schimke, R. T.**, Transcription factor Sp1 recognizes a DNA sequence in the mouse dihydrofolate reductase promoter, *Nature (London)*, 319, 246, 1986.
22. **Lee, W., Haslinger, A., Karin, M., and Tjian, R.**, Activation of transcription by two factors that bind promoter and enhancer sequences of the human metallothionein gene and SV40, *Nature (London)*, 325, 368, 1987.
23. **Dynan, W. S., Saffer, J. D., Lee, W. S., and Tjian, R.**, Transcription factor Sp1 recognizes promoter sequences from the monkey genome that are similar to the simian virus 40 promoter, *Proc. Natl. Acad. Sci. U.S.A.*, 82, 4915, 1985.
24. **Kadonaga, J. T., Jones, K. A., and Tjian, R.**, Promoter-specific activation of RNA polymerase II transcription by Sp1, *Trends Biochem.*, 11, 20, 1986.
25. **Kadonaga, J. T. and Tjian, R.**, Affinity purification of sequence-specific DNA binding proteins, *Proc. Natl. Acad. Sci. U.S.A.*, 83, 5889, 1986.
26. **Kadonaga, J. T., Carner, K. R., Maziarz, F. R., and Tjian, R.**, Isolation of cDNA encoding transcription factor Sp1 and functional analysis of the DNA binding domain, *Cell*, 51, 1079, 1987.
27. **Kim, C. H., Heath, C., Bertuch, A., and Hansen, U.**, Specific stimulation of simian virus 40 late transcription *in vitro* by a cellular factor binding the simian virus 40 21-base-pair repeat promoter element, *Proc. Natl. Acad. Sci. U.S.A.*, 84, 6025, 1987.
28. **Davidson, I., Fromental, C., Augereau, P., Wildeman, A., Zenke, M., and Chambon, P.**, Cell-type specific protein binding to the enhancer of simian virus 40 in nuclear extracts, *Nature (London)*, 323, 544, 1986.

29. **Wildeman, A. G., Zenke, M., Schatz, C., Wintzerith, M., Grundstrom, T., Matthes, H., Takahashi, K., and Chambon, P.,** Specific protein binding to the simian virus 40 enhancer in vitro, *Mol. Cell. Biol.,* 6, 2098, 1986.

30. **Lee, W., Mitchell, P., and Tjian, R.,** Purified transcription factor AP-1 interacts with TPA-inducible enhancer elements, *Cell,* 49, 741, 1987.

31. **Angel, P., Imagawa, M., Chiu, R., Stein, B., Imbra, R. J., Rahmsdorf, H. J., Jonat, C., Herrlich, P., and Karin, M.,** Phorbol ester-inducible genes contain a common *cis* element recognized by a TPA-modulated *trans*-acting factor, *Cell,* 49, 729, 1987.

32. **Vogt, P. K., Bos, T. J., and Doolittle, R. F.,** Homology between the DNA-binding domain of the GCN4 regulatory protein of yeast and the carboxyl-terminal region of a protein coded for by the oncogene, *jun, Proc. Natl. Acad. Sci. U.S.A.,* 84, 3316, 1987.

33. **Struhl, K.,** The DNA-binding domains of the jun oncoprotein and the yeast GCN4 transcriptional activator protein are functionally homologous, *Cell,* 50, 841, 1987.

34. **Bohmann, D., Bos, T. J., Admon, A., Nishimura, T., Vogt, P. K., and Tjian, R.,** Human proto-oncogene c-*jun* encodes a DNA binding protein with structural and functional properties of transcription factor AP-1, *Science,* 238, 1386, 1987.

35. **Bos, T. J., Bohmann, D., Tsuchie, H., Tjian, R., and Vogt, P. K.,** v-*jun* encodes a nuclear protein with enhancer binding properties of AP-1, *Cell,* 52, 705, 1988.

36. **Rosales, R., Vigneron, M., Macchi, M., Davidson, I., Xiao, J. H., and Chambon, P.,** *In vitro* binding of cell-specific and ubiquitous nuclear proteins to the octamer motif of the SV40 enhancer and related motifs present in other promoters and enhancers, *EMBO J.,* 6, 3015, 1987.

37. **Pruijn, G. J. M., van Driel, W., and van der Vliet, P. C.,** Nuclear factor III, a novel sequence-specific DNA-binding protein from HeLa cells stimulating adenovirus DNA replication, *Nature (London),* 322, 656, 1986.

38. **Pruijn, G. J. M., van Driel, W., van Miltenburg, R. T., and van der Vliet, P. C.,** Promoter and enhancer elements containing a conserved sequence motif and recognized by nuclear factor III, a protein stimulating adenovirus DNA replication, *EMBO J.,* 6, 3771, 1987.

39. **Bohmann, D., Keller, W., Dale, T., Scholer, H. R., Tebb, G., and Mattaj, I. W.,** A transcription factor which binds to the enhancers of SV40 immunoglobulin heavy chain and U2 snRNA genes, *Nature (London),* 325, 268, 1987.

40. **Staudt, L. M., Singh, H., Sen, R., Wirth, T., Sharp, P. A., and Baltimore, D.,** A lymphoid-specific protein binding to the octamer motif of immunoglobulin genes, *Nature (London),* 323, 640, 1986.

41. **Rosenfeld, P. J., O'Neill, E. A., Wide, R. J., and Kelly, T.,** Sequence-specific interactions between cellular DNA-binding proteins and the adenovirus origin of DNA replication, *Mol. Cell. Biol.,* 7, 875, 1987.

42. **Wang, J., Nishiyama, K., Araki, K., Kitamura, D., and Watanabe, T.,** Purification of an octamer sequence (ATGCAAAT)-binding protein from human B cells, *Nucleic Acids Res.,* 15, 10105, 1987.

43. **Schedereit, C., Heguy, A., and Roeder, R. G.,** Identification and purification of a human lymphoid-specific octamer-binding protein (OTF-2) that activates transcription of an immunoglobulin promoter in vitro, *Cell,* 51, 783, 1987.

44. **Fletcher, C., Heintz, N., and Roeder, R. G.,** Purification and characterization of OTF-1, a transcription factor regulating cell cycle expression of a human histone H2b gene, *Cell,* 51, 774, 1987.

45. **Xiao, J. H., Davidson, I., Ferrandon, D., Rosales, R., Vigneron, M., Macchi, M., Ruffenach, F., and Chambon, P.,** One cell-specific and three ubiquitous nuclear proteins bind *in vitro* to overlapping motifs in the domain B1 of the SV40 enhancer, *EMBO J.,* 6, 3005, 1987.

46. **Imagawa, M., Chiu, R., and Karin, M.,** Transcription factor AP-2 mediates induction by two different signal-transduction pathways: protein kinase C and cAMP, *Cell,* 51, 251, 1987.

47. **Johnson, P. F., Landschulz, W. H., Graves, B. J., and McKnight, S. L.,** Identification of a rat liver nuclear protein that binds to the enhancer core element of three animal viruses, *Genes Dev.,* 1, 133, 1987.

48. **Sen, R. and Baltimore, D.,** Multiple nuclear factors interact with the immunoglobulin enhancer sequences, *Cell,* 46, 705, 1986.

49. **Garcia, J. A., Wu, F. K., Mitsuyasu, R., and Gaynor, R. B.,** Interactions of cellular proteins involved in the transcriptional regulation of the human immunodeficiency virus, *EMBO J.,* 6, 3761, 1987.

50. **Nabel, G. and Baltimore, D.,** An inducible transcription factor activates expression of human immuno-deficiency virus in T cells, *Nature (London),* 326, 711, 1987.

51. **Franza, B. R., Jr., Josephs, S. F., Gilman, M. Z., Ryan, W., and Clarkson, B.,** Characterization of cellular proteins recognizing the HIV enhancer using a microscale DNA-affinity precipitation assay, *Nature (London),* 330, 391, 1987.

52. **Piette, J. and Yaniv, M.,** Two different factors bind to the α-domain of the polyoma virus enhancer, one of which also interacts with the SV40 and c-*fos* enhancers, *EMBO J.,* 6, 1331, 1987.

53. **Kryszke, M.-H., Piette, J., and Yaniv, M.,** Induction of a factor that binds to the polyoma virus A enhancer on differentiation of embryonal carcinoma cells, *Nature (London),* 328, 254, 1987.

54. **Wasylyk, C., Imler, J. L., Perez-Mutul, J., and Wasylyk, B.,** The c-Ha-*ras* oncogene and a tumor promoter activate the polyoma virus enhancer, *Cell,* 48, 525, 1987.

55. **Fujimura, F. K.,** Nuclear activity from F9 embryonal carcinoma cells binding specifically to the enhancers of wild-type polyoma virus and PyEC mutant DNAs, *Nucleic Acids Res.,* 14, 2845, 1986.

56. **Ostapchuk, P., Diffley, J. F. X., Bruder, J. T., Stillman, B., Levine, A. J., and Hearing, P.,** Interaction of a nuclear factor with the polyomavirus enhancer region, *Proc. Natl. Acad. Sci. U.S.A.,* 83, 8550, 1986.

57. **Shaul, Y. and Ben-Levy, R.,** Multiple nuclear proteins in liver cells are bound to hepatitis B virus enhancer element and its upstream sequences, *EMBO J.,* 6, 1913, 1987.

58. **Piette, J. and Yaniv, M.,** Molecular analysis of the interaction between an enhancer binding factor and its DNA target, *Nucleic Acids Res.,* 14, 9595, 1986.

59. **Bohnlein, E. and Gruss, P.,** Interaction of distinct nuclear proteins with sequences controlling the expression of polyomavirus early genes, *Mol. Cell. Biol.,* 6, 1401, 1986.

60. **Kovesdi, I., Satake, M., Furukawa, K., Reichel, R., Ito, Y., and Nevins, J. R.,** A factor discriminating between the wild-type and a mutant polyomavirus enhancer, *Nature (London),* 328, 87, 1987.

61. **Hen, R., Borrelli, E., Fromental, C., Sassone-Corsi, P., and Chambon, P.,** A mutated polyoma virus enhancer which is active in undifferentiated embryonal carcinoma cells is not repressed by the adenovirus-2 E1A products, *Nature (London),* 321, 249, 1986.

62. **Sawadogo, M. and Roeder, R. G.,** Interaction of a gene-specific transcription factor with the adenovirus major late promoter upstream of the TATA box region, *Cell,* 43, 165, 1985.

63. **Moncollin, V., Miyamoto, N. G., Zheng, X. M., and Egly, J. M.,** Purification of a factor specific for the upstream element of the adenovirus-2 major late promoter, *EMBO J.,* 5, 2577, 1986.

64. **Carthew, R. W., Chodosh, L. A., and Sharp, P. A.,** An RNA polymerase II transcription factor binds to an upstream element in the adenovirus major late promoter, *Cell,* 43, 439, 1985.

65. **Chodosh, L. A., Carthew, R. W., and Sharp, P. A.,** A single polypeptide possesses the binding and transcription activities of the adenovirus major late transcription factor, *Mol. Cell. Biol.,* 6, 4723, 1986.

66. **Lennard, A. C. and Egly, J. M.,** The bidirectional upstream element of the adenovirus-2 major late promoter binds a single monomeric molecule of the upstream factor, *EMBO J.,* 6, 3027, 1987.

67. **Kovesdi, I., Reichel, R., and Nevins, J. R.,** Role of an adenovirus E2 promoter binding factor in E1A-mediated coordinate gene control, *Proc. Natl. Acad. Sci. U.S.A.,* 84, 2180, 1987.

68. **Barrett, P., Clark, L., and Hay, R. T.,** A cellular protein binds to a conserved sequence in the adenovirus type 2 enhancer, *Nucleic Acids Res.,* 15, 2719, 1987.

69. **Kovesdi, I., Reichel, R., and Nevins, J. R.,** Identification of a cellular transcription factor involved in E1A *trans*-activation, *Cell,* 45, 219, 1986.

70. **Reichel, R., Kovesdi, I., and Nevins, J. R.,** Developmental control of a promoter-specific factor that is also regulated by the E1A gene product, *Cell,* 48, 501, 1987.

71. **Kovesdi, I., Reichel, R., and Nevins, J. R.,** E1A transcription induction: enhanced binding of a factor to upstream promoter sequences, *Science,* 231, 719, 1986.

72. **Jalinot, P., Devaux, B., and Kedinger, C.,** The abundance and in vitro DNA binding of three cellular proteins interacting with the adenovirus EIIa early promoter are not modified by the Eia gene products, *Mol. Cell. Biol.,* 7, 3806, 1987.

73. **SivaRaman, L. and Thimmappaya, B.,** Two promoter-specific host factors interact with adjacent sequences in an EIA-inducible adenovirus promoter, *Proc. Natl. Acad. Sci. U.S.A.,* 84, 6112, 1987.

74. **Boeuf, H., Zajchowski, D. A., Tamura, T., Hauss, C., and Kedinger, C.,** Specific cellular proteins bind to critical promoter sequences of the adenovirus early EIIa promoter, *Nucleic Acids Res.,* 15, 509, 1987.

75. **Goding, C. R., Temperley, S. M., and Fisher, F.,** Multiple transcription factors interact with the adenovirus-2 EII-late promoter: evidence for a novel CCAAT recognition factor, *Nucleic Acids Res.,* 15, 7761, 1987.

76. **Garcia, J., Wu, F., and Gaynor, R.,** Upstream regulatory regions required to stabilize binding to the TATA sequence in an adenovirus early promoter, *Nucleic Acids Res.,* 15, 8367, 1987.

77. **Lee, K. A. W. and Green, M. R.,** A cellular transcription factor E4F1 interacts with E1a-inducible enhancer and mediates constitutive enhancer function *in vitro, EMBO J.,* 6, 1345, 1987.

78. **Montminy, M. R. and Bilezikjian, L. M.,** Binding of a nuclear protein to the cyclic-AMP response element of the somatostatin gene, *Nature (London),* 328, 175, 1987.

79. **Graves, B. J., Johnson, P. F., and McKnight, S. L.,** Homologous recognition of a promoter domain common to the MSV LTR and the HSV *tk* gene, *Cell,* 44, 565, 1986.

80. **Parslow, T. G., Jones, S. D., Bond, B., and Yamamoto, K. R.,** The immunoglobulin octanucleotide: independent activity and selective interaction with enhancers, *Science,* 235, 1498, 1987.

81. **Efstratiadis, A., Polakony, J. W., Maniatis, T., Lawn, R. M., O'Connell, C., Spritz, R. A., DeRiel, J. K., Forget, B. G., Weissman, S. M., Slightom, J. L., Blechl, A. E., Smithies, O., Baralle, F. E., Shoulders, C. C., and Proudfoot, N. J.,** The structure and evolution of the human β-globin gene family, *Cell,* 21, 653, 1980.

82. **Benoist, C., O'Hare, K., Breathnach, R., and Chambon, P.,** The ovalbumin gene-sequence of putative control regions, *Nucleic Acids Res.,* 8, 127, 1980.
83. **McKnight, S. and Tjian, R.,** Transcriptional selectivity of viral genes in mammalian cells, *Cell,* 46, 795, 1986.
84. **Myers, R. M., Tilly, K., and Maniatis, T.,** Fine structure genetic analysis of a β-globin promoter, *Science,* 232, 613, 1986.
85. **Morgan, W. D., Williams, G. T., Morimoto, R. I., Greene, J., Kingston, R. E., and Tjian, R.,** Two transcriptional activators, CCAAT-box-binding transcription factor and heat shock transcription factor, interact with a human *hsp70* gene promoter, *Mol. Cell. Biol.,* 7, 1129, 1987.
86. **Wu, B. J., Williams, G. T., and Morimoto, R. I.,** Detection of three protein binding sites in the serum-regulated promoter of the human gene encoding the 70-kDa heat shock protein, *Proc. Natl. Acad. Sci. U.S.A.,* 84, 2203, 1987.
87. **Jones, K. A., Kadonaga, J. T., Rosenfeld, P. J., Kelly, T. J., and Tjian, R.,** A cellular DNA-binding protein that activates eukaryotic transcription and DNA replication, *Cell,* 48, 79, 1987.
88. **Chodosh, L. W., Baldwin, A. S., Carthew, R. W., and Sharp, P. A.,** Human CCAAT-binding proteins have heterologous subunits, *Cell,* 53, 11, 1988.
89. **Oikarinen, J., Hatamochi, A., and de Crombrugghe, B.,** Separate binding sites for nuclear factor 1 and a CCAAT DNA binding factor in the mouse α-$_2$(I) collagen promoter, *J. Biol. Chem.,* 262, 11064, 1987.
90. **Cereghini, S., Raymondjean, M., Carranca, A. G., Herbomel, P., and Yaniv, M.,** Factors involved in control of tissue-specific expression of albumin gene, *Cell,* 50, 627, 1987.
91. **Chodosh, L. A., Olesen, J., Hahn, S., Baldwin, A. S., Guarente, L., and Sharp, P.,** A yeast and a human CCAAT-binding protein have heterologous subunits that are functionally interchangeable, *Cell,* 53, 25, 1988.
92. **Dorn, A., Bollekens, J., Staub, A., Benoist, C., and Mathis, D.,** A multiplicity of CCAAT box-binding proteins, *Cell,* 50, 863, 1987.
93. **Raymondjean, M., Cereghini, S., and Yaniv, M.,** Several distinct CCAAT box binding proteins coexist in eukaryotic cells, *Proc. Natl. Acad. Sci. U.S.A.,* 85, 757, 1988.
94. **Hooft van Huijsduijnen, R. A. M., Bollekens, J., Dorn, A., Benoist, C., and Mathis, D.,** Properties of a CCAAT box-binding protein, *Nucleic Acids Res.,* 15, 7265, 1987.
95. **Cohen, R. B., Sheffery, M., and Kim, C. G.,** Partial purification of a nuclear protein that binds to the CCAAT box of the mouse α_1-globin gene, *Mol. Cell. Biol.,* 6, 821, 1986.
96. **Barberis, A., Superti-Furga, G., and Busslinger, M.,** Mutually exclusive interaction of the CCAAT-binding factor and of a displacement protein with overlapping sequences of a histone gene promoter, *Cell,* 50, 347, 1987.
97. **Knight, G. B., Guda, J. M., and Pardee, A. B.,** Cell-cycle-specific interaction of nuclear DNA-binding proteins with a CCAAT sequence from the human thymidine kinase gene, *Proc. Natl. Acad. Sci. U.S.A.,* 84, 8350, 1987.
98. **Hatamochi, A., Paterson, B., and de Crombrugghe, B.,** Differential binding of a CCAAT DNA binding factor to the promoters of the mouse α-2(I) and 1(III) collagen genes, *J. Biol. Chem.,* 261, 11310, 1986.
99. **Bienz, M.,** A CCAAT box confers cell-type-specific regulation on the *Xenopus hsp70* gene in oocytes, *Cell,* 46, 1037, 1986.
100. **Nagata, K., Guggenheimer, R. A., Enomoto, T., Lichy, J. H., and Hurwitz, J.,** Adenovirus DNA replication *in vitro*: identification of a host factor that stimulates synthesis of the preterminal protein-dCMP complex, *Proc. Natl. Acad. Sci. U.S.A.,* 79, 6438, 1982.
101. **Nagata, K., Guggenheimer, R. A., and Hurwitz, J.,** Specific binding of a cellular DNA replication protein to the origin of replication of adenovirus DNA, *Proc. Natl. Acad. Sci. U.S.A.,* 80, 6177, 1983.
102. **Rawlins, D. R., Rosenfeld, P. J., Wides, R. J., Challberg, M. D., and Kelly, T. J., Jr.,** Structure and function of the adenovirus origin of replication, *Cell,* 37, 309, 1984.
103. **Hay, R. T.,** Origin of adenovirus DNA replication; role of the nuclear factor I binding site *in vivo, J. Mol. Biol.,* 186, 129, 1985.
104. **Nowock, J. and Sippel, A. E.,** Specific protein-DNA interaction at four sites flanking the chicken lysozyme gene, *Cell,* 30, 607, 1982.
105. **Borgmeyer, U., Nowock, J., and Sippel, A. E.,** The TGGCA-binding protein: a eukaryotic nuclear protein recognizing a symmetrical sequence on double-stranded linear DNA, *Nucleic Acids Res.,* 12, 4295, 1984.
106. **Nowock, J., Borgmeyer, U., Puschel, A. W., Rupp, R. A. W., and Sippel, A. E.,** The TGGCA protein binds to the MMTV-LTR, the adenovirus origin of replication, and the BK virus enhancer, *Nucleic Acids Res.,* 13, 2045, 1985.
107. **Leegwater, P. A. J., van der Vliet, P. C., Rupp, R. A. W., Nowock, J., and Sippel, A. E.,** Functional homology between the sequence-specific DNA-binding proteins nuclear factor I from HeLa cells and the TGGCA protein from chicken liver, *EMBO J.,* 5, 381, 1986.
108. **Miksicek, R., Borgmeyer, U., and Nowock, J.,** Interaction of the TGGCA-binding protein with upstream sequences is required for efficient transcription of mouse mammary tumor virus, *EMBO J.,* 6, 1355, 1987.

109. **Schneider, R., Gander, I., Muller, U., Mertz, R., and Winnacker, E. L.,** A sensitive and rapid gel retention assay for nuclear factor I and other DNA-binding proteins in crude nuclear extracts, *Nucleic Acids Res.*, 14, 1303, 1986.

110. **Dynan, W. S.,** A single protein that binds to enhancers, promoters and replication origins, *Trends Genet.*, 1, 269, 1985.

111. **Gronostajski, R. M., Nagata, K., and Hurwitz, J.,** Isolation of human DNA sequences that bind to nuclear factor I, a host protein involved in adenovirus DNA replication, *Proc. Natl. Acad. Sci. U.S.A.*, 81, 4013, 1984.

112. **Siebenlist, U., Hennighausen, L., Battey, J., and Leder, P.,** Chromatin structure and protein binding in the putative regulatory region of the c-*myc* gene in Burkitt lymphoma, *Cell*, 37, 381, 1984.

113. **Hennighausen, L., Siebenlist, U., Danner, D., Leder, P., Rawlins, D., Rosenfeld, P., and Kelly, T., Jr.,** High-affinity binding site for a specific nuclear protein in the human IgM gene, *Nature (London)*, 314, 289, 1985.

114. **Cordingley, M. G., Riegel, A. T., and Hager, G. L.,** Steroid-dependent interaction of transcription factors with the inducible promoter of mouse mammary tumor virus in vivo, *Cell*, 48, 261, 1987.

115. **Shaul, Y., Ben-Levy, R., and De-Medina, T.,** High affinity binding site for nuclear factor I next to the hepatitis B virus S gene promoter, *EMBO J.*, 5, 1967, 1986.

116. **Hennighausen, L. and Fleckenstein, B.,** Nuclear factor 1 interacts with five DNA elements in the promoter region of the human cytomegalovirus major immediate early gene, *EMBO J.*, 5, 1367, 1986.

117. **Jeang, K.-T., Rawlins, D. R., Rosenfeld, P. J., Shero, J. H., Kelly, T. J., and Hayward, G. S.,** Multiple tandemly repeated binding sites for cellular nuclear factor 1 that surround the major immediate-early promoters of simian and human cytomegalovirus, *J. Virol.*, 61, 1559, 1987.

118. **Ghazal, P., Lubon, H., Fleckenstein, B., and Hennighausen, L.,** Binding of transcription factors and creation of a large nucleoprotein complex on the human cytomegalovirus enhancer, *Proc. Natl. Acad. Sci. U.S.A.*, 84, 3658, 1987.

119. **Leegwater, P. A. J., van Driel, W., and van der Vliet, P. C.,** Recognition site of nuclear factor I, a sequence-specific DNA-binding protein from HeLa cells that stimulates adenovirus DNA replication, *EMBO J.*, 4, 1515, 1985.

120. **de Vries, E., van Driel, W., van den Heuvel, S. J. L., and van der Vliet, P. C.,** Contact point analysis of the HeLa nuclear factor I recognition site reveals symmetrical binding at one side of the DNA helix, *EMBO J.*, 6, 161, 1987.

121. **Rosenfeld, P. J. and Kelly, T. J.,** Purification of nuclear factor I by DNA recognition site affinity chromatography, *J. Biol. Chem.*, 261, 1398, 1986.

122. **Diffley, J. F. X. and Stillman, B.,** Purification of a cellular, double-stranded DNA-binding protein required for initiation of adenovirus DNA replication by using a rapid filter-binding assay, *Mol. Cell. Biol.*, 6, 1363, 1986.

123. **Rupp, R. A. W. and Sippel, A. E.,** Chicken liver TGGCA protein purified by preparative mobility shift electrophoresis shows a 36.8 to 29.8 Kd microheterogeneity, *Nucleic Acids Res.*, 15, 9707, 1987.

124. **Lichtsteiner, S., Wuarin, J., and Schibler, U.,** The interplay of DNA-binding proteins on the promoter of the mouse albumin gene, *Cell*, 51, 963, 1987.

125. **Buetti, E. and Kuhnel, B.,** Distinct sequence elements involved in the glucocorticoid regulation of the mouse mammary tumor virus promoter identified by linker scanning mutagenesis, *J. Mol. Biol.*, 190, 379, 1986.

126. **Cato, A. C. B., Skroch, P., Weinmann, J., Butkeraitis, P., and Ponta, H.,** DNA sequences outside the receptor-binding sites differentially modulate the responsiveness of the mouse mammary tumour virus promoter to various steroid hormones, *EMBO J.*, 7, 1403, 1988.

127. **Richard-Foy, H. and Hager, G. L.,** Sequence-specific positioning of nucleosomes over the steroid-inducible MMTV promoter, *EMBO J.*, 6, 2321, 1987.

128. **Laimins, L. A., Tsichlis, P., and Khoury, G.,** Multiple enhancer domains in the 3' terminus of the Prague strain of Rous sarcoma virus, *Nucleic Acids Res.*, 12, 6427, 1984.

129. **Cullen, B. R., Raymond, K., and Ju, G.,** Functional analysis of the transcription control region located within the avian retroviral long terminal repeat, *Mol. Cell. Biol.*, 5, 438, 1985.

130. **Sealey, L. and Chalkley, R.,** At least two nuclear proteins bind specifically to the Rous sarcoma virus long terminal repeat enhancer, *Mol. Cell. Biol.*, 7, 787, 1987.

131. **Goodwin, G. H.,** Identification of three sequence-specific DNA-binding proteins which interact with the Rous sarcoma virus enhancer and upstream promoter elements, *J. Virol.*, 62, 2186, 1988.

132. **Karnitz, L., Faber, S., and Chalkley, R.,** Specific nuclear proteins interact with the Rous sarcoma virus internal enhancer and share a common element with the enhancer located in the long terminal repeat of the virus, *Nucleic Acids Res.*, 15, 9841, 1987.

133. **Speck, N. and Baltimore, D.,** Six distinct nuclear factors interact with the 75-base-pair repeat of the Moloney murine leukemia virus enhancer, *Mol. Cell. Biol.*, 7, 1101, 1987.

134. **Flamant, F., Gurin, C. C., and Sorge, J. A.,** An embryonic DNA-binding protein specific for the promoter of the retrovirus long terminal repeat, *Mol. Cell. Biol.,* 7, 3548, 1987.

135. **Quinn, J. P., Holbrook, N., and Levens, D.,** Binding of a cellular protein to the gibbon ape leukemia virus enhancer, *Mol. Cell. Biol.,* 7, 2735, 1987.

136. **Sive, H. L. and Roeder, R. G.,** Interaction of a common factor with conserved promoter and enhancer sequences in histone H2B, immunoglobulin, and U2 small nuclear RNA (snRNA) genes, *Proc. Natl. Acad. Sci. U.S.A.,* 83, 6382, 1986.

137. **Landolfi, N. F., Capra, J. D., and Tucker, P. W.,** Interaction of cell-type-specific nuclear proteins with immunoglobulin V_H promoter region sequences, *Nature (London),* 323, 548, 1986.

138. **Singh, H., Sen, R., Baltimore, D., and Sharp, P. A.,** A nuclear factor that binds to a conserved sequence motif in transcriptional control elements of immunoglobulin genes, *Nature (London),* 319, 154, 1986.

139. **Mocikat, R., Falkner, G. F., Mertz, R., and Zachau, H. G.,** Upstream regulatory sequences of immunoglobulin genes are recognized by nuclear proteins which also bind to other gene regions, *Nucleic Acids Res.,* 14, 8830, 1986.

140. **Falkner, F. G., Mocikat, R., and Zachau, H. G.,** Sequences closely related to an immunoglobin gene promoter/enhancer element occur also upstream of other eukaryotic and of prokaryotic genes, *Nucleic Acids Res.,* 14, 8819, 1986.

141. **Ephrussi, A., Church, G. M., Tonegawa, S., and Gilbert, W.,** B lineage-specific interactions of an immunoglobulin enhancer with cellular factors in vivo, *Science,* 227, 134, 1985.

142. **Weinberger, J., Baltimore, D., and Sharp, P. A.,** Distinct factors bind to apparently homologous sequences in the immunoglobulin heavy-chain enhancer, *Nature (London),* 322, 846, 1986.

143. **Schlokat, U., Bohmann, D., Scholer, H., and Gruss, P.,** Nuclear factors binding specific sequences within the immunoglobulin enhancer interact differentially with other enhancer elements, *EMBO J.,* 5, 3251, 1986.

144. **Augereau, P. and Chambon, P.,** The mouse immunoglobulin heavy-chain enhancer: effect on transcription *in vitro* and binding of proteins present in HeLa and lymphoid B cell extracts, *EMBO J.,* 5, 1791, 1986.

145. **Gimble, J. M., Levens, D., and Max, E. E.,** B-cell nuclear proteins binding in vitro to the human immunoglobulin κ enhancer: localization by exonuclease protection, *Mol. Cell. Biol.,* 7, 1815, 1987.

146. **Peterson, C. L., Orth, K., and Calame, K. L.,** Binding in vitro of multiple cellular proteins to immunoglobulin heavy-chain enhancer DNA, *Mol. Cell. Biol.,* 6, 4168, 1986.

147. **Atchinson, M. L. and Perry, R. P.,** The role of the κ enhancer and its binding factor NF-KB in the developmental regulation of K gene transcription, *Cell,* 48, 121, 1987.

148. **Sen, R. and Baltimore, D.,** Inducibility of κ immunoglobulin enhancer-binding protein NF-KB by a posttranslational mechanism, *Cell,* 47, 921, 1986.

149. **Antoniou, M., deBoer, E., Habets, G., and Grosveld, F.,** The human β-globin gene contains multiple regulatory regions: identification of one promoter and two downstream enhancers, *EMBO J.,* 7, 377, 1988.

150. **Emerson, B. M., Lewis, C. D., and Felsenfeld, G.,** Interaction of specific nuclear factors with the nuclease-hypersensitive region of the chicken adult β-globin gene: nature of the binding domain, *Cell,* 41, 21, 1985.

151. **Plumb, M. A., Lobanenkov, V. V., Nicolas, R. H., Wright, C. A., Zavou, S., and Goodwin, G. H.,** Characterisation of chicken erythroid nuclear proteins which bind to the nuclease hypersensitive regions upstream of the β^A and β^H globin genes, *Nucleic Acids Res.,* 14, 7675, 1986.

152. **Emerson, B. M., Nickol, J. M., Jackson, P. D., and Felsenfeld, G.,** Analysis of the tissue-specific enhancer at the 3′ end of the chicken adult β-globin gene, *Proc. Natl. Acad. Sci. U.S.A.,* 84, 4786, 1987.

153. **Galson, D. L. and Housman, D. E.,** Detection of two tissue-specific DNA-binding proteins with affinity for sites in the mouse β-globin intervening sequence 2, *Mol. Cell. Biol.,* 8, 381, 1988.

154. **Kemper, B., Jackson, P. D., and Felsenfeld, G.,** Protein-binding sites within the 5′ DNase I-hypersensitive region of the chicken α^D-globin gene, *Mol. Cell. Biol.,* 7, 2059, 1987.

155. **Mantovani, R., Malgaretti, N., Giglioni, B., Comi, P., Cappellini, N., Nicolis, S., and Ottolenghi, S.,** A protein factor binding to an octamer motif in the γ-globin promoter disappears upon induction of differentiation and hemoglobin synthesis in K562 cells, *Nucleic Acids Res.,* 15, 9349, 1987.

156. **Babiss, L. E., Herbst, R. S., Bennett, A. L., and Darnell, J. E., Jr.,** Factors that interact with the rat albumin promoter are present both in hepatocytes and other cell types, *Genes Dev.,* 1, 256, 1987.

157. **Ohlsson, H. and Edlund, T.,** Sequence-specific interactions of nuclear factors with the insulin gene enhancer, *Cell,* 45, 35, 1986.

158. **Lefevre, C., Imagawa, M., Dana, S., Grindlay, J., Bodner, M., and Karin, M.,** Tissue-specific expression of the human growth hormone gene is conferred in part by the binding of a specific *trans*-acting factor, *EMBO J.,* 6, 971, 1987.

159. **Catanzaro, D. F., West, B. L., Baxter, J. D., and Reudelhuber, T. L.,** A pituitary-specific factor interacts with an upstream promoter element in the rat growth hormone gene, *Mol. Endocrinol.,* 1, 90, 1986.

160. **Glass, C. K., Franco, R., Weinberger, C., Alberts, V. R., Evans, R. M., and Rosenfeld, M. G.,** A c-*erb*-A binding site in rat growth hormone gene mediates *trans*-activation by thyroid hormone, *Nature (London),* 329, 738, 1987.
161. **Koenig, R. J., Brent, G. A., Warne, R. L., Larsen, P. R., and Moore, D. D.,** Thyroid hormone receptor binds to a site in the rate growth hormone promoter required for induction by thyroid hormone, *Proc. Natl. Acad. Sci. U.S.A.,* 84, 5670, 1987.
162. **Gutierrez-Hartmann, A., Siddiqui, S., and Loukin, S.,** Selective transcription and DNase I protection of the rat prolactin gene by GH₃ pituitary cell-free extracts, *Proc. Natl. Acad. Sci. U.S.A.,* 84, 5211, 1987.
163. **Cao, Z., Barron, E. A., Carrillo, A. J., and Sharpe, Z. D.,** Reconstitution of cell-type specific transcription of the rat prolactin gene in vitro, *Mol. Cell. Biol.,* 7, 3402, 1987.
164. **Becker, P. B., Gloss, B., Schmid, W., Strahle, U., and Schutz, G.,** *In vivo* protein-DNA interactions in a glucocorticoid response element require the presence of the hormone, *Nature (London),* 324, 686, 1986.
165. **Becker, P. B., Ruppert, S., and Schutz, G.,** Genomic footprinting reveals cell type-specific DNA binding of ubiquitous factors, *Cell,* 51, 435, 1987.
166. **Walsh, K. and Schimmel, P.,** Two nuclear factors compete for the skeletal muscle actin promoter, *J. Mol. Biol.,* 9429, 1987.
167. **Miwa, T., Boxer, L. M., and Kedes, L.,** CArG boxes in the human cardiac α-actin gene are core binding sites for positive trans-acting regulatory factors, *Proc. Natl. Acad. Sci. U.S.A.,* 84, 6702, 1987.
168. **Distel, R. J., Ro, H.-S., Rosen, B. S., Groves, D. L., and Spiegelman, B. M.,** Nucleoprotein complexes that regulate gene expression in adipocyte differentiation: direct participation of c-*fos, Cell,* 49, 835, 1987.
169. **Musti, A. M., Ursini, V. M., Avvedimento, E. V., Zimarino, V., and Di Lauro, R.,** A cell type specific factor recognizes the rat thyroglobin promoter, *Nucleic Acids Res.,* 15, 8149, 1987.
170. **Chodosh, L. A., Carthew, R. W., Morgan, J. G., Crabtree, G. R., and Sharp, P. A.,** The adenovirus major late transcription factor activates the rat γ-fibrinogen promoter, *Science,* 238, 684, 1987.
171. **Courtois, G., Morgan, J. G., Campbell, L. A., Fourel, G., and Crabtree, G. R.,** Interaction of a liver-specific nuclear factor with the fibrinogen and α₁-antitrypsin promoters, *Science,* 238, 688, 1987.
172. **van der Hoorn, F. A.,** c-*mos* upstream sequence exhibits species-specific enhancer activity and binds murine-specific nuclear proteins, *J. Mol. Biol.,* 193, 255, 1987.
173. **Royer, H. D. and Reinherz, E. L.,** Multiple nuclear proteins bind upstream sequences in the promoter region of a T-cell receptor β-chain variable-region gene: evidence for tissue specificity, *Proc. Natl. Acad. Sci. U.S.A.,* 84, 232, 1987.
174. **Dierich, A., Gaub, M.-P., Le Pennec, J.-P., Astinotti, D., and Chambon, P.,** Cell-specificity of the chicken ovalbumin and conalbumin promoters, *EMBO J.,* 6, 2305, 1987.
175. **Tsai, S. Y., Sagami, I, Wang, H., Tsai, M.-J., and O'Malley, B. W.,** Interactions between a DNA-binding transcription factor (COUP) and a non-DNA binding factor (S300-II), *Cell,* 50, 701, 1987.
176. **Hayes, T. E., Kitchen, A. M., and Cochran, B. H.,** Inducible binding of a factor to the c-*fos* regulatory region, *Proc. Natl. Acad. Sci. U.S.A.,* 84, 1272, 1987.
177. **Treisman, R.,** Identification of a protein-binding site that mediates transcriptional response of the c-*fos* gene to serum factors, *Cell,* 46, 567, 1986.
178. **Treisman, R.,** Identification and purification of a polypeptide that binds to the c-*fos* serum response element, *EMBO J.,* 6, 2711, 1987.
179. **Prywes, R. and Roeder, R. G.,** Inducible binding of a factor to the c-*fos* enhancer, *Cell,* 47, 777, 1986.
180. **Greenberg, M. E., Siegfried, Z., and Ziff, E. B.,** Mutation of the c-*fos* gene dyad symmetry element inhibits serum inducibility of transcription in vivo and the nuclear regulatory factor binding in vitro, *Mol. Cell. Biol.,* 7, 1217, 1987.
181. **Gilman, M. Z., Wilson, R. N., and Weinberg, R.,** Multiple protein-binding sites in the 5'-flanking region regulate c-*fos* expression, *Mol. Cell. Biol.,* 6, 4305, 1986.
182. **Chung, J., Sinn, E., Reed, R. R., and Leder, P.,** Trans-acting elements modulate expression of the human c-*myc* gene in Burkitt lymphoma cells, *Proc. Natl. Acad. Sci. U.S.A.,* 83, 7918, 1986.
183. **Remmers, E. F., Yang, J.-Q., and Marcu, K. B.,** A negative transcriptional control element located upstream of the murine c-*myc* gene, *EMBO J.,* 5, 899, 1986.
184. **Kakkis, E. and Calame, K.,** A plasmacytoma-specific factor binds the c-*myc* promoter region, *Proc. Natl. Acad. Sci. U.S.A.,* 84, 7031, 1987.
185. **Lobanenkov, V. V., Nicolas, R. H., Plumb, M. A., Wright, C. A., and Goodwin, G. H.,** Sequence-specific DNA-binding proteins which interact with (G + C)-rich sequences flanking the chicken c-*myc* gene, *Eur. J. Biochem.,* 159, 181, 1986.
186. **Dailey, L., Hanly, S. M., Roeder, R. G., and Heintz, N.,** Distinct transcription factors bind specifically to two regions of the human histone H4 promoter, *Proc. Natl. Acad. Sci. U.S.A.,* 83, 7241, 1986.
187. **Pauli, U., Chrysogelos, S., Stein, G., Stein, J., and Nick, H.,** Protein-DNA interactions in vivo upstream of a cell cycle-regulated human H4 histone gene, *Science,* 236, 1308, 1987.

188. **van Wijnen, A. J., Stein, J. L., and Stein, G. S.,** A nuclear protein with affinity for the 5' flanking region of a cell cycle dependent human H4 histone gene *in vitro, Nucleic Acids Res.,* 15, 1679, 1987.
189. **Kingston, R. E., Schuetz, T. J., and Larin, Z.,** Heat-inducible human factor that binds to a human *hsp70* promoter, *Mol. Cell. Biol.,* 7, 1530, 1987.
190. **Wu, C., Wilson, S., Walker, B., Dawid, I., Paisley, T., Zimarino, V., and Ueda, H.,** Purification and properties of *Drosophila* heat shock activator protein, *Science,* 238, 1247, 1987.
191. **Baldwin, A. S., Jr. and Sharp, P. A.,** Binding of a nuclear factor to a regulatory sequence in the promoter of the mouse *H-2K^b* class I major histocompatibility gene, *Mol. Cell. Biol.,* 7, 305, 1987.
192. **Israel, A., Kimura, A., Kieran, M., Yano, O., Kanellopoulous, J., Le Bail, O., and Kourilsky, P.,** A common positive trans-acting factor binds to enhancer sequences in the promoters of mouse *H-2* and β$_2$-microglobulin genes, *Proc. Natl. Acad. Sci. U.S.A.,* 84, 2653, 1987.
193. **Yano, O., Kanellopoulous, J., Kieran, M., Le Bail, O., Israel, A., and Kourilsky, P.,** Purification of KBF1, a common factor binding to both *H-2* and β$_2$-microglobulin enhancers, *EMBO J.,* 6, 3317, 1987.
194. **Miwa, K., Doyle, C., and Strominger, J. L.,** Sequence-specific interactions of nuclear factors with conserved sequences of human class II major histocompatibility complex genes, *Proc. Natl. Acad. Sci. U.S.A.,* 84, 4939, 1987.
195. **Seguin, C. and Hamer, D. H.,** Regulation in vitro of metallothionein gene binding factors, *Science,* 235, 1383, 1987.
196. **Andersen, R. D., Taplitz, S. J., Wong, S., Bristol, G., Larkin, B., and Herschman, H. R.,** Metal-dependent binding of a factor in vivo to the metal-responsive elements of the metallothionein 1 gene promoter, *Mol. Cell. Biol.,* 7, 3574, 1987.
197. **Montminy, M. R., Sevarino, K. A., Wagner, J. A., Mandel, G., and Goodman, R. H.,** Identification of a cyclic-AMP-responsive element within the rat somatostatin gene, *Proc. Natl. Acad. Sci. U.S.A.,* 83, 6682, 1986.
198. **Shot, J. M., Wynshaw-Boris, A., Short, H. P., and Hanson, R. W.,** Characterization of the phosphoenolpyruvate carboxykinase (GTP) promoter-regulatory region, *J. Biol. Chem.,* 261, 9721, 1986.
199. **Comb, M., Birnberg, N. C., SeashoItz, A., Herbert, E., and Goodman, H. M.,** A cyclic AMP-and phorbol ester-inducible DNA element, *Nature (London),* 323, 353, 1986.
200. **Delegeane, A. M., Ferland, L. H., and Mellon, P. L.,** Tissue-specific enhancer of the human glycoprotein hormone α-subunit gene: dependence on cyclic AMP-inducible elements, *Mol. Cell. Biol.,* 7, 3994, 1987.
201. **Osborne, T. F., Gil, G., Brown, M. S., Kowal, R. C., and Goldstein, J. L.,** Identification of promoter elements required for *in vitro* transcription of hamster 3-hydroxy-3-methylglutaryl coenzyme A reductase gene, *Proc. Natl. Acad. Sci. U.S.A.,* 84, 3614, 1987.
202. **Hollenberg, S. M., Weinberger, C., Ong, E. S., Cerelli, G., Oro, A., Lebo, R., Thompson, E. B., Rosenfeld, M. G., and Evans, R. M.,** Primary structure and expression of a functional human glucocorticoid receptor cDNA, *Nature (London),* 318, 635, 1985.
203. **Danielsen, M., Northrop, J. P., and Ringold, G. M.,** The mouse glucocorticoid receptor: mapping of functional domains by cloning, sequencing, and expression of wild type and mutant receptor proteins, *EMBO J.,* 5, 2513, 1986.
204. **Arriza, J. L., Weinberger, C., Cerelli, G., Glaser, T. M., Handelin, B. L., Housman, D. E., and Evans, R. M.,** Cloning of human mineralocorticoid receptor complementary DNA: structural and functional kinship with the glucocorticoid receptor, *Science,* 237, 268, 1987.
205. **Green, S., Walter, P., Kumar, V., Krust, A., Bornert, J. M., Argos, P., and Chambon, P.,** Human oestrogen receptor cDNA: sequence, expression and homology to v-*erb*-A, *Nature (London),* 320, 134, 1986.
206. **Greene, G. L., Gilna, P., Waterfield, M., Baker, A., Hort, Y., and Shine, J.,** Sequence and expression of human estrogen receptor complementary DNA, *Science,* 231, 1150, 1986.
207. **Kumar, V., Green, S., Staub, A., and Chambon, P.,** Localization of the oestradiol-binding and putative DNA-binding domains of the human oestrogen receptor, *EMBO J.,* 5, 2231, 1986.
208. **Conneely, O. M., Sullivan, W. P., Toft, D. O., Birnbaumer, M., Cook, R. G., Maxwell, B. L., Zarucki-Schulz, T., Greene, G. L., Schrader, W. T., and O'Malley, B. W.,** Molecular cloning of the chicken progesterone receptor, *Science,* 233, 767, 1986.
209. **Jeltsch, J. M., Krozowski, Z., Quirin-Stricher, C., Gronemeyer, H., Simpson, R. J., Garnier, R. J., Krust, J. M., Jacob, F., and Chambon, P.,** Cloning of the chicken progesterone receptor, *Proc. Natl. Acad. Sci. U.S.A.,* 83, 5424, 1986.
210. **Loosfelt, H., Atger, M., Misrahi, M., Guiochon-Mantel, A., Meriel, C., Logeat, F., Benarous, R., and Milgrom, E.,** Cloning and sequence analysis of rabbit progesterone-receptor complementary DNA, *Proc. Natl. Acad. Sci. U.S.A.,* 83, 9045, 1986.
211. **Sap, J., Munoz, A., Damm, K., Goldberg, Y., Ghysdael, J. L., Leutz, A., Beug, H., and Vennstrom, B.,** The c-*erb*-A protein is a high affinity receptor for thyroid hormone, *Nature (London),* 324, 635, 1986.
212. **Weinberger, C., Thompson, C., Ong, E. S., Lebo, R., Gruol, D. J., and Evans, R. M.,** The c-*erb*-A gene encodes a thyroid hormone receptor, *Nature (London),* 324, 641, 1986.

213. **Green, S. and Chambon, P.,** A superfamily of potentially oncogenic hormone receptors, *Nature (London),* 324, 615, 1986.
214. **Giguere, V., Hollenberg, S. M., Rosenfeld, M. G., and Evans, R. M.,** Functional domains of the human glucocorticoid receptor, *Cell,* 46, 645, 1986.
215. **Kumar, V., Green, S., Stack, G., Berry, M., Jin, J.-R., and Chambon, P.,** Functional domains of the human estrogen receptor, *Cell,* 51, 941, 1987.
216. **Godowski, P. J., Rusconi, S., Miesfeld, R., and Yamamoto, K. R.,** Glucocorticoid receptor mutants that are constitutive activators of transcriptional enhancement, *Nature (London),* 325, 365, 1987.
217. **Willmann, T. and Beato, M.,** Steroid-free glucocorticoid receptor binds specifically to mouse mammary tumour virus DNA, *Nature (London),* 324, 688, 1986.
218. **Joab, I., Radanyi, C., Renoir, M., Buchous, T., Catelli, M.-G., Binar, N., Mester, J., and Baulieu, E. E.,** Common non-hormone binding component in non-transformed chick oviduct receptors of four steroid hormones, *Nature (London),* 308, 850, 1984.
219. **Hollenberg, S. M., Giguere, V., Segui, P., and Evans, R. M.,** Colocalization of DNA-binding and transcriptional activation functions in the human glucocorticoid receptor, *Cell,* 49, 39, 1987.
220. **Karin, M., Haslinger, A., Holtreve, H., Richards, R. I., Krauter, P., Westphal, H. M., and Beato, M.,** Characterization of DNA sequences through which cadmium and glucocorticoid hormone induce human metallothionein-IIA gene, *Nature (London),* 308, 513, 1984.
221. **von der Ahe, D., Renoir, J.-M., Buchou, T., Baulieu, E.-E., and Beato, M.,** Receptors for glucocorticosteroid and progesterone recognize distinct features of a DNA regulatory element, *Proc. Natl. Acad. Sci. U.S.A.,* 83, 2817, 1986.
222. **Jost, J. P., Seldran, M., and Geiser, M.,** Preferential binding of estrogen-receptor complex to a region containing the estrogen-dependent hypomethylation site preceding the chicken vitellogenin II gene, *Proc. Natl. Acad. Sci. U.S.A.,* 81, 429, 1984.
223. **Maurer, R. A. and Notides, A. C.,** Identification of an estrogen-responsive element from the 5'-flanking region of the rat prolactin gene, *Mol. Cell. Biol.,* 7, 4247, 1987.
224. **Feavers, I. M., Jiricny, J., Moncharmont, B., Saluz, H. P., and Jost, J. P.,** Interaction of two nonhistone proteins with the estradiol response element in the avian vitellogenin gene modulates the binding of estradiol-receptor complex, *Proc. Natl. Acad. Sci. U.S.A.,* 84, 7453, 1987.
225. **Gordon, M. S. and Notides, A. C.,** Computer modeling of estradiol interactions with the estrogen receptor, *J. Steroid Biochem.,* 25, 177, 1986.
226. **Adler, S., Waterman, M. L., He, X., and Rosenfeld, M. G.,** Steroid receptor-mediated inhibition of rat prolactin gene expression does not require the receptor DNA-binding domain, *Cell,* 52, 685, 1988.
227. **Zenke, M., Kahn, P., Disela, C., Vennstrom, B., Leutz, A., Keegan, K., Hayman, M. J., Choi, H.-R., Yew, N., Engel, J. D., and Beug, H.,** v-erbA specifically suppresses transcription of the avian erthryocyte anion transporter (band 3) gene, *Cell,* 52, 107, 1988.
228. **Petkovich, M., Brand, N. J., Krust, A., and Chambon, P.,** A human retinoic acid receptor which belongs to the family of nuclear receptors, *Nature (London),* 330, 444, 1987.
229. **McDonnell, D. P., Mangelsdorf, D. J., Pike, J. W., Haussler, M. R., and O'Malley, B. W.,** Molecular cloning of complementary DNA encoding the avian receptor for vitamin D, *Science,* 235, 1214, 1987.
230. **Ciliberto, G., Castagnoli, L., and Cortese, R.,** Transcription by RNA polymerase III, *Curr. Top. Dev. Biol.,* 18, 59, 1983.
231. **Sharp, S. J., Shaack, J., Cooley, L., Burke, D. J., and Soll, D.,** Structure and transcription of eukaryotic tRNA genes, *CRC Crit. Rev. Biochem.,* 19, 107, 1985.
232. **Segall, J., Matsui, T., and Roeder, R. G.,** Multiple factors are required for the accurate transcription of purified genes by RNA polymerase III, *J. Biol. Chem.,* 255, 11986, 1980.
233. **Shastry, B. S., Ng, S.-Y., and Roeder, R. G.,** Multiple factors involved in the transcription of class III genes in *Xenopus laevis, J. Biol. Chem.,* 257, 12979, 1982.
234. **Fuhrman, S. A., Engelke, D. R., and Geiduschek, E. P.,** HeLa cell RNA polymerase III transcription factors, *J. Biol. Chem.,* 259, 1934, 1984.
235. **Ulla, E. and Weiner, A. M.,** Upstream sequences modulate the internal promoter of the human 7SL RNA gene, *Nature (London),* 318, 371, 1985.
236. **Murphy, S., Tripodi, M., and Melli, M.,** A sequence upstream from the coding region is required for the transcription of the 7SK RNA genes, *Nucleic Acids Res.,* 14, 9243, 1986.
237. **Ohshima, Y., Okada, N., Tani, T., Itoh, Y., and Itoh, M.,** Nucleotide sequences of mouse genomic loci including a gene or pseudogene for U6 (4.8S) nuclear RNA, *Nucleic Acids Res.,* 9, 5145, 1981.
238. **Sakonju, S. and Brown, D. D.,** The binding of a transcription factor to deletion mutants of a 5S ribosomal RNA gene, *Cell,* 23, 665, 1981.
239. **Wormington, W. M., Bogenhagen, D. F., Jordan, E., and Brown, D. D.,** A quantitative assay for *Xenopus* 5S RNA gene transcription in vitro, *Cell,* 24, 809, 1981.
240. **Korn, L. J.,** Transcription of *Xenopus* 5S ribosomal RNA genes, *Nature (London),* 295, 101, 1982.

241. **Sakonju, S. and Brown, D. D.,** Contact points between a positive transcription factor and the *Xenopus* 5S RNA gene, *Cell,* 31, 395, 1982.
242. **Sakonju, S., Bogenhagen, D. F., and Brown, D. D.,** A control region in the center of the 5S RNA gene directs specific initiation of transcription. I. The 5′ border of the region, *Cell,* 19, 13, 1980.
243. **Bogenhagen, D. F., Sakonju, S., and Brown, D. D.,** A control region in the center of the 5S RNA gene directs specific initiation of transcription. II. The 3′ border of the region, *Cell,* 19, 27, 1980.
244. **Bogenhagen, D. F.,** The intragenic control region of the *Xenopus* 5S RNA gene contains two factor A binding domains that must be aligned properly for efficient transcription initiation, *J. Biol. Chem.,* 260, 6466, 1985.
245. **Engelke, D. R., Ng, S.-T., Shastry, S., and Roeder, R. G.,** Specific interaction of a purified transcription factor with an internal control region of 5S RNA genes, *Cell,* 19, 717, 1980.
246. **Picard, B. and Wegnez, M.,** Isolation of a 7S particle from *Xenopus laevis* oocytes: a 5S RNA-protein complex, *Proc. Natl. Acad. Sci. U.S.A.,* 76, 241, 1979.
247. **Pelham, H. R. B. and Brown, D. D.,** A specific transcription factor that can bind either the 5S RNA gene or 5S RNA, *Proc. Natl. Acad. Sci. U.S.A.,* 77, 4170, 1980.
248. **Smith, D. R., Jackson, I. J., and Brown, D. D.,** Domains of the positive transcription factor specific for the *Xenopus* 5S RNA gene, *Cell,* 37, 645, 1984.
249. **Bieker, J. J. and Roeder, R. G.,** Physical properties and DNA-binding stoichiometry of a 5S gene-specific transcription factor, *J. Biol. Chem.,* 259, 6158, 1984.
250. **Wu, G.-J.,** Adenovirus DNA-directed transcription of 5.5S RNA *in vitro, Proc. Natl. Acad. Sci. U.S.A.,* 75, 2175, 1978.
251. **Weil, P. A., Segall, J., Harris, B., Ng, S.-Y., and Roeder, R. G.,** Faithful transcription of eukaryotic genes by RNA polymerase III in systems reconstituted with purified DNA templates, *J. Biol. Chem.,* 6163, 1979.
252. **Shi, L.-P., Wingender, E., Bottrich, J., and Seifart, K. H.,** Faithful transcription of ribosomal 5-S RNA *in vitro* depends on the presence of several factors, *Eur. J. Biochem.,* 131, 189, 1983.
253. **Lassar, A. B., Martin, P. L., and Roeder, R. G.,** Transcription of class III genes: formation of preinitiation complexes, *Science,* 222, 740, 1983.
254. **Bieker, J. J., Martin, P. L., and Roeder, R. G.,** Formation of a rate-limiting intermediate in 5S RNA gene transcription, *Cell,* 40, 119, 1985.
255. **Bogenhagen, D. F., Wormington, W. M., and Brown, D. D.,** Stable transcription complexes of *Xenopus* 5S RNA genes: a means to maintain the differentiated state, *Cell,* 28, 413, 1982.
256. **Setzer, D. R. and Brown, D. D.** Formation and stability of the 5S RNA transcription complex, *J. Biol. Chem.,* 260, 2483, 1984.
257. **Cozzarelli, N. R., Gerrard, S. P., Schlissel, M., Brown, D. D., and Bogenhagan, D. F.,** Purified RNA polymerase III accurately and efficiently terminates transcription of 5S RNA genes, *Cell,* 34, 829, 1983.
258. **Ginsberg, A. M., King, B. O., and Roeder, R. G.,** *Xenopus* 5S gene transcription factor, TFIIIA: characterization of a cDNA clone and measurement of RNA levels throughout development, *Cell,* 39, 479, 1984.
259. **Miller, J., McLachlan, A. D., and Klug, A.,** Repetitive zinc-binding domains in the protein transcription factor IIIA from *Xenopus* oocytes, *EMBO J.,* 4, 1609, 1985.
260. **Hanas, J. S., Hazuda, D. J., Bogenhagen, D. F., Wu., F. Y.-H., and Wu, C.-W.,** *Xenopus* transcription factor A requires zinc for binding to the 5S RNA gene, *J. Biol. Chem.,* 258, 14120, 1983.
261. **Rhodes, D. and Klug, A.,** An underlying repeat in some transcriptional control sequences corresponding to half a double helical turn of DNA, *Cell,* 46, 123, 1986.
262. **Fairall, L., Rhodes, D., and Klug, A.,** Mapping of the sites of protection on a 5S RNA gene by the *Xenopus* transcription factor IIIA. A model for the interaction, *J. Mol. Biol.,* 192, 577, 1986.
263. **Vrana, K. E., Churchill, M. E. A., Tullius, T. D., and Brown, D. D.,** Mapping functional regions of transcription factor TFIIIA, *Mol. Cell. Biol.,* 8, 1684, 1988.
264. **Pieler, T., Oei, S.-L., Hamm, J., Engelke, U., and Erdmann, V. A.,** Functional domains of the *Xenopus laevis* 5S gene promoter, *EMBO J.,* 4, 3751, 1985.
265. **Pieler, T., Hamm, J., and Roeder, R. G.,** The 5S gene internal control region is composed of three distinct sequence elements, organized as two functional domains with variable spacing, *Cell,* 48, 91, 1987.
266. **Brown, D. D.,** How a simple animal gene works, *Harvey Lect.,* 76, 27, 1982.
267. **Brown, D. D.,** The role of stable complexes that repress and activate eucaryotic genes, *Cell,* 37, 359, 1984.
268. **Peck, L. J., Millstein, L., Eversole-Cire, P., Gottesfeld, J. M., and Varshavsky, A.,** Transcriptionally inactive oocyte-type 5S RNA genes of *Xenopus laevis* are complexed with TFIIIA *in vitro, Mol. Cell. Biol.,* 7, 3503, 1987.
269. **Millstein, L., Eversole-Cire, P., Blanco, J., and Gottesfeld, J. M.,** Differential transcription of *Xenopus* oocyte and somatic-type 5S genes in a *Xenopus* oocyte extract, *J. Biol. Chem.,* 262, 17100, 1987.

270. **Wolfe, A. P.,** Transcription fraction TFIIIC can regulate differential *Xenopus* 5S RNA gene transcription *in vitro, EMBO J.,* 7, 1071, 1988.
271. **McConkey, G. A. and Bogenhagen, D. F.,** TFIIIA binds with equal affinity to somatic and major oocyte 5S RNA genes, *Genes Dev.,* 2, 205, 1987.
272. **Hoeffler, W. K. and Roeder, R. G.,** Enhancement of RNA polymerase III transcription by the EIA gene product of adenovirus, *Cell,* 41, 955, 1985.
273. **Gaynor, R. B., Feldman, L. T., and Berk, A. J.,** Transcription of class III genes activated by viral immediate early proteins, *Science,* 230, 447, 1985.
274. **Yoshinaga, S., Dean, N., Han, M., and Berk, A. J.,** Adenovirus stimulation of transcription by RNA polymerase III: evidence for an EIA-dependent increase in transcription factor IIIC concentration, *EMBO J.,* 5, 343, 1986.
275. **Fradkin, L. G., Yoshinaga, S. K., Berk, A. J., and Dasgupta, A.,** Inhibition of host cell RNA polymerase III-mediated transcription by poliovirus: inactivation of specific transcription factors, *Mol. Cell. Biol.,* 7, 3880, 1987.
276. **Carey, M. F., Gerrard, S. P., and Cozzarelli, N. R.,** Analysis of RNA polymerase III transcription complexes by gel filtration, *J. Biol. Chem.,* 261, 4309, 1986.
277. **Majowski, K., Mentzel, H., and Pieler, T.,** A split binding site for TFIIIC on the *Xenopus* 5S gene, *EMBO J.,* 6, 3057, 1987.
278. **Yoshinaga, S. K., Boulanger, P. A., and Berk, A. J.,** Resolution of human transcription factor TFIIIC into two functional components, *Proc. Natl. Acad. Sci. U.S.A.,* 84, 3585, 1987.
279. **Dean, N. and Berk, A. J.,** Separation of TFIIIC into two functional components by sequence specific DNA affinity chromatography, *Nucleic Acids Res.,* 15, 9895, 1987.
280. **Sharp, S. J. and Garcia, A. D.,** Transcription of the *Drosophila melanogaster* 5S RNA gene requires an upstream promoter and four intragenic sequence elements, *Mol. Cell. Biol.,* 8, 1266, 1988.
281. **Klekamp, M. S. and Weil, P. A.,** Specific transcription of homologous class III genes in yeast-soluble cell-free extracts, *J. Biol. Chem.,* 257, 8432, 1982.
282. **Burke, D. J., Schaack, J., Sharp, S., and Soll, D.,** Partial purification of *Drosophila* Kc cell RNA polymerase III transcription components, *J. Biol. Chem.,* 258, 15224, 1983.
283. **Stillman, D. J. and Geiduschek, E. P.,** Differential binding of a *S. cerevisiae* RNA polymerase III transcription factor to two promoter segments of a tRNA gene, *EMBO J.,* 3, 847, 1984.
284. **Baker, R. E., Gabrielsen, O., and Hall, B. D.,** Effects of tRNA point mutations on the binding of yeast RNA polymerase III transcription factor C, *J. Biol. Chem.,* 261, 5275, 1986.
285. **Camier, S., Gabrielsen, O., Baker, R., and Sentenac, A.,** A split binding site for transcription factor τ on the tRNA$_3^{Glu}$ gene, *EMBO J.,* 4, 491, 1985.
286. **Klekamp, M. S. and Weil, P. A.,** Partial purification and characterization of the *Saccharomyces cerevisiae* transcription factor TFIIIB, *J. Biol. Chem.,* 261, 2819, 1986.
287. **Ruet, A., Camier, S., Smagowicz, W., Sentenac, A., and Fromageot, P.,** Isolation of a class C transcription factor which forms a stable complex with tRNA genes, *EMBO J.,* 3, 343, 1984.
288. **Weinmann, R., Jaehning, J. A., Raskas, H. J., and Roeder, R. G.,** Viral RNA synthesis and levels of DNA-dependent RNA polymerases during replication of adenovirus 2, *J. Virol.,* 17, 114, 1976.
289. **Soderlund, H., Pettersson, U., Vennstrom, B., Philipson, L., and Mathews, M. B.,** A new species of virus-coded low molecular weight RNA from cells infected with adenovirus type 2, *Cell,* 7, 585, 1976.
290. **Weinmann, R., Raskas, H. J., and Roeder, R. G.,** Role of DNA-dependent RNA polymerases II and III in transcription of the adenovirus genome late in productive infection, *Proc. Natl. Acad. Sci. U.S.A.,* 71, 3426, 1974.
291. **Fowlkes, D. M. and Shenk, T.,** Transcriptional control regions of the adenovirus VA1 RNA gene, *Cell,* 22, 405, 1980.
292. **Guilfoyle, R. and Weinmann, R.,** Control region for adenovirus VA RNA transcription, *Proc. Natl. Acad. Sci. U.S.A.,* 78, 3378, 1981.
293. **Bhat, R. A., Metz, B., and Thimmappaya, B.,** Organization of the non-contiguous promoter components of adenovirus VA1 gene is strikingly similar to that of eucaryotic tRNA genes, *Mol. Cell. Biol.,* 3, 1996, 1983.
294. **Van Dyke, M. W. and Roeder, R. G.,** Multiple proteins bind to VA RNA genes of adenovirus type 2, *Mol. Cell. Biol.,* 7, 1021, 1987.
295. **Hoffman, M. L., Korf, G. M., McNamara, K. J., and Stumph, W. E.,** Structural and functional analysis of chicken U4 small nuclear RNA genes, *Mol. Cell. Biol.,* 6, 3901, 1986.
296. **Kunkel, G. R., Maser, R. L., Calvet, J. P., and Pederson, T.,** U6 small nuclear RNA is transcribed by RNA polymerase III, *Proc. Natl. Acad. Sci. U.S.A.,* 83, 8575, 1986.
297. **Reddy, R., Henning, D., Das, G., Harless, M., and Wright, D.,** The capped U6 small nuclear RNA is transcribed by RNA polymerase III, *J. Biol. Chem.,* 262, 75, 1987.
298. **Bark, C., Weller, P., Zabielski, J., Janson, L., and Petterson, U.,** A distant enhancer element is required for polymerase III transcription of a U6 RNA gene, *Nature (London),* 328, 356, 1987.

299. **Krol, A., Carbon, P., Ebel, J.-P., and Appel, B.,** *Xenopus tropicalis* U6 snRNA genes transcribed by Pol III contain the upstream promoter elements used by Pol II dependent U2 snRNA genes, *Nucleic Acids Res.,* 15, 2463, 1987.

300. **Carbon, P., Murgo, S., Ebel, J.-P., Krol, A., Tebb, G., and Mattaj, I. W.,** A common octamer motif binding protein is involved in the transcription of U6 snRNA by polymerase III and U2 snRNA by RNA polymerase II, *Cell,* 51, 71, 1987.

301. **Das, G., Henning, D., Wright, D., and Reddy, R.,** Upstream regulatory elements are necessary and sufficient for transcription of a U6 RNA gene by RNA polymerase III, *EMBO J.,* 7, 503, 1988.

302. **Kunkel, G. R. and Pederson, T.,** Upstream elements required for efficient transcription of a human U6 RNA gene resemble those of U1 and U2 genes even though a different polymerase is used, *Genes Dev.,* 2, 196, 1988.

303. **Reinberg, D. and Roeder, R. G.,** Factors involved in specific transcription by mammalian RNA polymerase II. Purification and functional analysis of initiation factors IIB and IIE, *J. Biol. Chem.,* 262, 3310, 1987.

304. **Reinberg, D., Horikoshi, M., and Roeder, R. G.,** Factors involved in specific transcription in mammalian RNA polymerase II. Functional analysis of initiation factors IIA and IID and identification of a new factor operating at sequences downstream of the initiation site, *J. Biol. Chem.,* 262, 3322, 1987.

305. **Reinberg, D. and Roeder, R. G.,** Factors involved in specific transcription by mammalian RNA polymerase II. Transcription factor IIS stimulates elongation of RNA chains, *J. Biol. Chem.,* 262, 3331, 1987.

306. **Conaway, J. W., Bond, M. W., and Conaway, R. C.,** An RNA polymerase II transcription system from rat liver, *J. Biol. Chem.,* 262, 8293, 1987.

307. **Perkins, N., Nicolas, R. H., Plumb, M., Goodwin, G. H.,** *Nucleic Acids Res.,* 17, 1299, 1989.

Chapter 3

PHYSICAL AND TOPOLOGICAL PROPERTIES OF CLOSED CIRCULAR DNA

William R. Bauer and Robert Gallo

TABLE OF CONTENTS

I. INTRODUCTION

A. HISTORICAL PERSPECTIVE

Closed circular duplex DNA, once regarded as an interesting special case, is clearly an integral part of the general chemistry and biology of DNA. The first suggestion that a circular structure could be important for DNA came from the work of Sinsheimer[1] who showed that the single-stranded viral DNA from ΦX174 behaves like a ring, based upon hydrodynamic and electron microscopic evidence. This circular structure is clearly fragile, a single-chain scission being sufficient to remove it. The concept that the conversion of a linear structure to a ring can account for altered hydrodynamic properties was naturally applied to the superficially similar situation that arose with polyoma viral DNA.[2,3] Here the sedimentation pattern consists of three distinct components, the slowest a linear molecule, the intermediate a simple ring, and the leading component having a more compact (twisted) structure.[4] The relative compactness of the most rapidly sedimenting component, which brought about its discovery by Vinograd[4], is one of the many consequences of the covalent continuity of the two strands.

It was soon recognized that the changes in DNA physical properties upon covalent closure are associated with a fundamental topological conservation condition.[5-7] This condition states the requirement that an integer quantity, known as the linking number, is constant in closed DNA in the absence of a chain scission. Both the physical and chemical properties of closed circular DNA readjust as required to maintain the constancy of the linking number, and the effects upon structure and reactivity can be highly significant. Among the physical properties affected by covalent closure are sedimentation, diffusion, electrophoresis, and others associated with hydrodynamic motion; the radius of gyration; and the resistance to shear degradation. Among the chemical properties affected are the denaturation profile, whether produced thermally, by pH change, or otherwise; the binding constant, both to proteins and to low molecular weight ligands; the interaction of reagents that bind covalently to the bases; the ease of local structural transitions such as cruciform extrusion and Z DNA formation; and integrative recombination.

B. SCOPE OF CHAPTER

We will principally describe the physical and topological properties of closed circular DNA as they have been determined during the past decade. The period from 1963 through 1977 has been the subject of previous general reviews[8-11] and will be dealt with only briefly here. During this earlier period, many basic properties of superhelical DNA were extensively documented, including the hydrodynamic behavior, the buoyant density under various conditions, and the drug-binding isotherms. The superhelix densities of DNAs from a variety of sources were determined, and attempts were made to examine the structure of the superhelix by electron microscopy. A new class of enzymes, the topoisomerases and gyrases, was discovered and characterized; and gel electrophoresis was applied to the resolution and analysis of individual DNA topoisomers.

Although toposisomerases are of central importance in the study of the properties of closed circular DNA, these enzymes have been the subject of several recent reviews[12-17] and they will not be dealt with in detail here. The application of topological analysis to DNA replication and recombination, under conditions in which either knots or catenanes are formed, has proved to be of great utility. This area has also been extensively reviewed.[18-20] Finally, the *in vivo* aspects of DNA supercoiling have also been the subject of much recent investigation. These may currently be divided into two general areas: (1) the genetics of topoisomerases and gyrases and (2) the effects of supercoiling upon biological processes. Among the latter, the best studied have been transcription and the control of gene expression in prokaryotes. This general area, which has been the subject of recent reviews,[21,22] will be dealt with only incidentally in the present article.

II. TOPOLOGICAL AND GEOMETRIC RELATIONSHIPS

A. GENERAL GEOMETRIC RELATIONSHIPS

The technical description of closed circular DNA involves the use of several specialized terms and relationships, both topological and geometric. The most generally important terms will be presented and defined in this section, and others will be introduced as needed. A more complete discussion of closed circular DNA topology and geometry has recently appeared.[23] The relationships discussed in the remainder of this section require the use of certain geometric terms and concepts, which are illustrated in Figure 1.

Description of the geometric properties of closed circular DNA requires, in general, specification of two curves and of a correspondence between them.[24] It is often convenient to choose the backbone strands, here denoted C_1 and C_2 as the two curves. Alternatively, either C_1 or C_2 may be selected, with the axis curve A being used as the second curve. Figure 1 illustrates the set of vectors that is used to specify the trajectory of the curves and their correspondence. The normal vector $\mathbf{N}_{12}$ is equal to the vector cross product $\mathbf{T}_1 \times \mathbf{v}_{12}$, where $\mathbf{T}_1$ is the unit tangent vector to C_1 at any point and $\mathbf{v}_{12}$ is the component of $\mathbf{z}_{12}$ that is perpendicular to $\mathbf{T}_{12}$. The correspondence vector $\mathbf{z}_{12}$ connects corresponding base pairs on the strands C_1 and C_2.

B. TOPOLOGICAL QUANTITY: THE LINKING NUMBER, LK

The linking number between two closed backbone chains C_1 and C_2 is topological (nonmetrical) and is given by the Gauss integral over curves C_1 and C_2:

$$Lk(C_1, C_2) = \left(\frac{1}{4\pi}\right) \iint\limits_{C_1 \times C_2} \frac{\hat{e} \cdot (\mathbf{T}_2 \times \mathbf{T}_1)}{r^2} \, ds_1 ds_2 \qquad (1)$$

Here, $\hat{e}$ is the unit vector from any point x on C_1 to any point y on C_2, and r is the linear distance between x and y. The linking number is an integer and is positive for right-handed duplex DNA. In principle, the most straightforward way to determine the value of Lk is to project the strands onto a plane and to count the number of cross-overs. Then Lk is one half the sum of the number of signed cross-overs. The convention for determining the sign of each cross-over point is illustrated in Figure 2.

C. THE GEOMETRIC QUANTITIES: TWIST AND WRITHE
1. The Twist, Tw

The twist of either strand C (C_1 or C_2) about the axis A is given by the line integral over A:

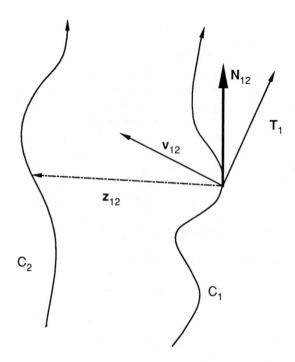

FIGURE 1.   Definition of vectors required for the description of twist and writhe in closed circular DNA. The vector z_{12} connects corresponding points on curves C_1 and C_2. The vector v_{12} is a unit vector in the plane spanned by z_{12} and the tangent vector T_1. The normal vector $N_1 = T_1 \times v_{12}$ is perpendicular to both T_1 and v_{12} and is used in Equation 2 to calculate $Tw(C_1, C_2)$. For calculation of the Tw (C,A), the subscripts 1,2 are replaced by C,A.

$$Tw(C, A) = \left(\frac{1}{2\pi}\right) \int_A N_{CA} \cdot dv_{CA} \tag{2}$$

The twist is a real number, and its value depends upon the choice of curves and upon the relationship between them. For example, the above definition refers to the twist of either backbone chain about the axis. Alternatively, Tw may be defined and calculated as the twist of either backbone chain about the other.[24] In general, the function $Tw(C_2, C_1)$ designates the twist of curve C_2 about curve C_1. For any three curves, the twists are related by permutations of the equation:[25]

$$Tw(C_3, C_1) = Tw(C_2, C_1) + \Phi(C_1) \tag{3a}$$

Here, $\Phi(C_1)$ is the winding number of curve C_1 and is given by

$$\Phi(C_1) = \left(\frac{1}{2\pi}\right) \int_{C_1} d\phi \tag{3b}$$

and ϕ is the angle between the vector pair v_{12} and N_{12} in Figure 1 and the corresponding pair, v_{13} and N_{13}. The twist of DNA is often confused with the duplex winding number, Φ. These two quantities are equal only in the special case that STw, the surface twist, vanishes (see Section II.E below).

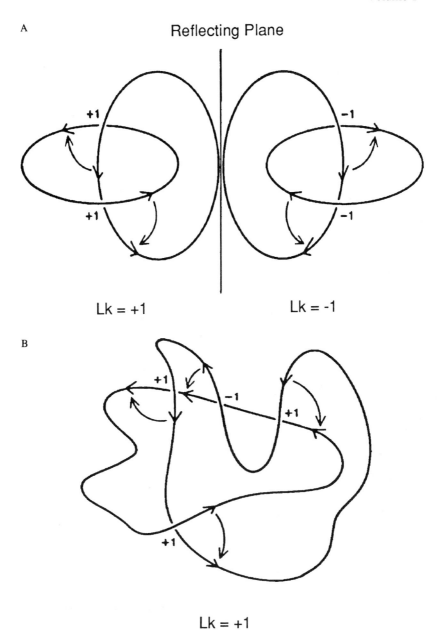

FIGURE 2.   Convention for the calculation of the linking number in closed circular DNA. Each crossover contributes $+1$ to Lk if the tangent to the upper curve coincides with the tangent vector to the lower curve following a counterclockwise rotation. The contribution is -1 if the corresponding rotation is clockwise. In (A), the sign of Lk changes if the orientation of either of the curves is changed or if either of the curves is reflected through a mirror. In (B), the value of Lk remains unchanged regardless of how the two curves are deformed in space.

2. The Writhe, Wr

The writhe is a property of a single curve, C, only. It is given formally by the Gauss integral, with points x and y on the same curve (cf. Equation 1).

$$\mathrm{Wr(C)} = \left(\frac{1}{4\pi}\right) \iint_{C \times C} \frac{\hat{e} \cdot (T_2 \times T_1)}{r^2} \, ds_1 ds_2 \qquad (4)$$

The direct calculation of the writhe from Equation 4 presents severe problems of numerical integration, consequently, Wr(C) for a closed DNA is usually obtained from the second fundamental relationship, Equation 6a, b. Several detailed examples of the direct calculation of Wr(C) for special cases have appeared in the literature.[26-28]

D. RELATIONSHIPS AMONG TOPOLOGICAL AND GEOMETRIC QUANTITIES

If C_1 and C_2 are the backbone chains of a DNA, and A is the axis, then the first fundamental relationship[5] is:

$$Lk(C_1, C_2) = Lk(C_1, A) = Lk(C_2, A) = constant \qquad (5)$$

The validity of the above relationship for A requires that the axis be continuous and well-defined. This may not be the case for certain local structural transitions, such as cruciform extrusion.[29]

The second fundamental relationship[5] may also be expressed either in terms of the two backbone chains or in terms of either one and the axis curve:

$$Lk(C_1, C_2) = Tw(C_1, C_2) + Wr(C_2) \qquad (6a)$$

$$Lk(C, A) = Tw(C, A) + Wr(A) \qquad (6b)$$

Several points must be emphasized: (1) the value of Tw depends upon the choice of curves and upon the ordering; that is, $Tw(C_1, C_2) \neq Tw(C_2, C_1)$ and $Tw(C, A) \neq Tw(A, C)$, etc.[24] In particular, only in a few special cases is any of these twist expressions equal to the duplex winding number, Φ. Failure to realize this has led to misconceptions and misinterpretations of experiments involving closed circular DNA.[30]

E. SURFACE-WRAPPED DNA

The relations described in this section apply, for example, to the wrapping of DNA about the histone octamer in a nucleosome.[30] The twist of either backbone curve, C, about the DNA axis, A, is given by the sum of the surface twist, STw, and the winding number, Φ:

$$Tw(C, A) = STw + \Phi \qquad (7)$$

The winding number is equal to one half the number of times that C cuts the wrapping surface on which A lies as the surface is traversed. The surface twist, STw is calculated from knowledge of the surface geometry. For example, if the DNA axis A wraps n times, with a pitch $2\pi p$, on a cylindrical surface of radius r, then STw is simply the reference frame twist, or $-np/(p^2 + r^2)^{1/2}$, for left handed wrapping.

The surface linking number, SLk, is related to the surface twist and the writhe by:[30]

$$SLk = STw + Wr \qquad (8)$$

Combining Equations 6b, 7, and 8, the linking number for surface-wrapped DNA is then given by:

$$Lk = SLk + \Phi \qquad (9)$$

Equation 9 states that the linking number of closed DNA wrapped on a surface can be written as the sum of two integers: SLk and Φ. For the nucleosome, Φ is measured by determining the periodicity of nuclease digestion and is 14.36 turns.[31] SLk is determined from straightforward considerations of geometry. For example, SLk = 0 for any spheroidal surface. SLk = n, the number of superhelical turns, for a toroidal surface. For the nucleosome, where the wrapping is left-handed, SLk = -1.8.[32]

III. DETERMINATION OF THE SUPERHELIX DENSITY

A. DEFINITION OF THE SUPERHELIX DENSITY
The superhelix density, σ, is defined as:

$$\sigma = (Lk - Lk_0)/Lk_0 = \Delta Lk/Lk_0 \qquad (10)$$

Here, Lk_0 is the center of the Gaussian distribution of topoisomers under conditions that relax the DNA, and Lk is the corresponding center of the Gaussian distribution of topoisomers for the DNA sample in question. For all known naturally occurring DNAs $\sigma < 0$. The superhelix density is essentially a measure of the extent of underwinding of a DNA, normalized by the DNA length. If the helical repeat of the relaxed closed DNA, h_0, is known, then $Lk_0 = N/h_0$, where N is the number of base pairs in the DNA. For the sodium salt of random sequence DNA, $h_0 = 10.6 \pm 0.1$.[33,34] For the magnesium salt, $h_0 = 10.5 \pm 0.1$.[35]

Since Lk_0 is a function of environmental conditions, including temperature, ionic strength, and counterion,[35-41] the superhelix density σ also varies with these factors. Little variability is associated with the anion at low to moderate concentrations[35] or with the pH in nonmelting regions.[41] For purposes of comparison, the superhelix densities of naturally occurring closed circular DNAs should be reported under standard conditions: 37°, 0.2 M NaCl, neutral pH.[11] The corresponding value of σ is designated $\sigma°$ and is denoted the standard superhelix density.

B. CORRECTION TO STANDARD CONDITIONS
A more complete summary of this data has been given previously.[11] The temperature coefficient of σ is given by $\Delta\sigma/\Delta T = 3.1 \times 10^{-4}$ deg^{-1}, and $\Delta\sigma = +0.0053$, for an increase in temperature from 20° to 37°. Over the ionic strength range 0.05 to 0.3 M, $\Delta\sigma/\Delta pX^+$ is 4.47×10^{-3} for the ions Na$^+$, K$^+$, Li$^+$, and NH$_4^+$ and is 6.70×10^{-3} for Rb$^+$, Cs$^+$, and Mg^{2+}. No effect upon σ results from an increase in [Cs$^+$] > 1.0 M.

C. METHODS OF MEASURING THE SUPERHELIX DENSITY
A more complete description of the various methods for the determination of $\sigma°$ is given in Reference 11. An abbreviated description, along with references containing technical details, is given here.

1. Sedimentation velocity/dye titration (Dye/s)[9,42] — This method requires determination of the critical binding ratio, ν_c, associated with the principal minimum in the sedimentation velocity profile (refer to Figure 8). Provided that the dye unwinding angle ϕ, expressed in degrees is known, the superhelix density may be calculated from the relationship:

$$\sigma = -(h_0\phi/180)\nu_c = -1.53_1\nu_c \qquad (11)$$

2. Viscometric dye titration (Dye/η)[43-45] — This method is also based upon determination of ν_c. Here ν_c is determined by locating the maximum in the relative viscosity as a function of the concentration of added dye. This method has the advantage that knowledge of neither the dye binding constant nor of the maximum binding ratio is required. Viscometric titrations are performed as a function of the nucleotide concentration, N_T, and the data are plotted according to Equation 12,

$$c_T = \nu_c(N_T)_c + (c)_c \tag{12}$$

where c_T is the total dye concentration. A plot of c_T versus $(N_T)_c$ gives a straight line, the slope of which is ν_c.

3. Gel electrophoresis dye titration (Dye/GE)[46-48] — Here, ν_c is determined by location of the minimum in a plot of distance migrated versus dye concentration.

4. Dye buoyant density (Dye/θ)[49-52] — In these experiments, a mixture of DNAs I and II (or of I and III) is placed in a solution of approximately 5.8 M CsCl containing ethidium bromide (EtdBr) at a concentration of 50 to 100 μg/ml. A reference mixture is similarly prepared for a DNA of known σ°. Following the attainment of equilibrium, the locations of the average distance of each pair from the center of rotation, $\bar{r}$, and the separation between bands, Δr, are measured for each pair. The standard superhelix density of the unknown closed DNA is then given in terms of that of the reference DNA by:[53]

$$\sigma^0 = \sigma^0_{ref} + (\sigma^0_{ref} + 0.306)(1/f)[(\bar{r}\Delta r)/(\bar{r}\Delta r)_{ref} - 1] \tag{13}$$

The factor f is given by $f = (\theta_o - 1)^2/(\theta_o - 1)^2_{ref}$, where θ_o is the buoyant density of the dye-free DNA.

5. Band counting in gel electrophoresis (BC/GE)[54,55] — This method is based on the ability of agarose gels to separate closed DNAs based solely upon differences in the gel-dependent frictional coefficient and, indirectly, in ΔLk. In particular, a population of closed DNAs characterized by a Gaussian or other distribution with mean linking number Lk_m will resolve into separate bands centered around Lk_m. If two populations differing only in Lk_m migrate on the same gel, bands of equal Lk nearly overlap. In this manner, it is possible to calculate the corresponding differences in Lk_m by actual band counting. Then $\Delta Lk = (\Delta Lk)_{ref} + \Delta Lk_m$. Expression of the result in terms of the superhelix density requires independent determination of Lk_o. Since the resolution of topoisomer bands becomes increasingly difficult as the molecular weight of the DNA increases, this method is most useful for relatively small DNAs (below 10,000 bp). A more complete analysis has been presented previously.[11]

6. Early hyperchromicity, gel electrophoresis (H/GE)[56] — In this method, closed DNA samples are progressively heat denatured in low ionic strength buffer, then treated with carbodiimide to stabilize denatured regions. The extent of denaturation needed to remove all superhelical turns is ascertained from the minimum in electrophoretic mobility. The superhelix density is then assessed by comparing the change in absorbance at the equivalence point to that of the untitrated DNA. The method has not been widely used.

7. Methods involving electron microscopy (EM)[57] — Attempts are sometimes made to count the number of crossovers in a superhelical DNA by direct examination following spreading and electron microscopy. This is unreliable in principle because surface spreading forces are expected to cause an unknown amount of conversion of writhe into twist. The observed number of crossovers is therefore not a reliable measure of ΔLk. For a more extensive critique, see Reference 33. In a second electron microscopic approach, the length of the complementary single-stranded fragment necessary to remove all superhelical turns is estimated by length measurements on the relaxed complexes. This method determines the titratable superhelix density.[11] The method, which is tedious, has not been widely employed.

8. Buoyant density in denaturing solvents (Alk θ)[58,59] — These methods have the advantage that they do not depend upon knowledge of a drug unwinding angle. The alkaline buoyant titration was first used with polyoma virion DNA[58] and was later refined and applied to virion PM-2 DNAs of various superhelix density.[59] The relative difficulty of the method renders it unsuitable for routine applications.

An alternative buoyant method involves the chaotropic solvent neutral aqueous Rb trichloroacetate (RbTCA).[60] This solvent brings about large separations between DNAs I and II, and the magnitude of the separations increases linearly with $\sigma°$. Although the method is highly precise, the unavailability of the solvent on a commercial basis renders it difficult to apply.

D. SUPERHELIX DENSITIES OF NATIVE DNAs

Naturally occurring DNAs consist of population distributions of topoisomers, as is the case with closed DNA samples that have been incubated with topoisomerase. No systematic study has yet appeared of the extent of variation of the linking number in DNA populations from different sources. Extensive measurements are, however, available on the average value of $\sigma°$ for DNA from various organisms. Table 1 lists the mean standard superhelix densities of selected naturally occurring closed circular DNAs from both prokaryotic and eukaryotic sources.

Since the superhelix density is often subject to cell-cycle and other variation, compilation of an inclusive list is no longer practical or meaningful. The examples listed here have been selected to encompass the range of superhelix densities known to occur. The entries in Table 1 are listed in order of increasingly negative $\sigma°$. The extent of the range in values of $\sigma°$ is surprisingly large, varying from -0.03 to -0.09. No naturally occurring superhelical DNA has yet been found for which $\sigma°$ is more negative than -0.1, the largest value that has yet been shown to be introduced by the DNA gyrase.[61]

IV. SEDIMENTATION PROPERTIES OF CLOSED CIRCULAR DNA

A. GENERAL DESCRIPTION

Sedimentation velocity methods are generally incapable of resolving individual topoisomers, due to diffusional spreading. Sedimentation of a sample that is paucidisperse in ΔLk yields results that are applicable to the Gaussian mean of the topoisomer distribution, corresponding to the average value of σ for the DNA sample in question. Most of the work in this area was reviewed previously[11] and the description here will be brief. The current hydrodynamic method of choice is gel electrophoresis, in which diffusion is relatively greatly reduced. As a consequence, the migration of individual topoisomers can also be resolved under suitable gel conditions. This topic is covered separately in Section V.

B. SEDIMENTATION AT NEUTRAL pH

The dependence of the sedimentation coefficient of closed circular SV40 DNA at neutral pH upon σ, ionic strength, counterion, and temperature has been determined.[40] The general shape of the curve describing the variation of $s_{20,w}$ with σ is shown in Figure 3.

The results obtained in 2.83 M CsCl, 20°C can be summarized in terms of four regions. In Region A, only a slight change occurs in the superhelix density and the corresponding change in R is slight. The conformational change here might be associated with a small amount of toroidal supercoiling. In Region B, the superhelix density changes from 0 to -0.042 and the ratio of sedimentation coefficients of DNAs I and II (R) increases nearly linearly from 1.00 to 1.33. In Region C, σ changes from -0.042 to -0.085 and R decreases from 1.33 to 1.23. In Region D, σ changes from -0.085 to -0.20, accompanied by a nearly linear increase in R from 1.23 to 1.62. A local maximum is found at $\sigma = -0.042$, and a local minimum is found at $\sigma = -0.085$. The tertiary structural changes responsible for the shape of the sedimentation velocity curve are not clear. The general shape of the curve of R vs. $-\sigma$ is similar to that found in the sedimentation velocity/dye titration, provided that the starting closed-circular DNA is sufficiently supercoiled (see below).

It was originally suggested[40] that Region B is associated with a progressive reduction in the radius of gyration as toroidal supercoiling increases. The maximum between Regions

TABLE 1
Superhelix Densities of Naturally Occurring DNAs

Host	DNA	Method	σ°	Ref.
Escherichia coli	$\lambda b_2 b_5 c$	Dye/s	−0.032	41
Mouse L cell	mDNA	Dye/θ	−0.035	128
E. coli	lcI857[a]	Dye/s	−0.036	41
Monkey	SV40, IC[b]	Dye/θ	−0.042	129
E. coli	lcI857[c]	Dye/s	−0.044	41
Lytechinus pictus	mDNA	Dye/θ	−0.047	130
Human lymphoid liver cell	EB, IC	Dye/θ	−0.048	131
Mouse	Polyoma virion	(d)	−0.048 ± 0.008	11
E. coli	F factor	Dye/s	−0.053	51
Bovine	Papilloma vir.	Dye/s	−0.053	132
E. coli	pBRbG[e]	H/GE	−0.0545	133
E. coli	pBRbG[e]	BC/GE	−0.0549	133
Chicken liver	mDNA	Dye/s	−0.055	134
E. coli	FX RF I	Dye/s	−0.057	41
E. coli	ColE1 amp plasmid	BC/GE	−0.057 ± 0.004	95
Monkey	SV40 virion	(d)	−0.0572 ± 0.010	11
E. coli	15 plasmid	Dye/s	−0.061	41
HeLa	mDNA	Dye/θ	−0.061	130
HeLa	spcDNA	Dye/θ	−0.061	135
HeLa	mDNA	Dye/θ	−0.061	135
E. coli	minicol	BC/GE	−0.0651	55
E. coli	pSM4	Dye/θ	−0.068	65
Rabbit liver	mDNA	Dye/θ	−0.069	130
E. coli	M13 RFI	Dye/GE	−0.071	47
Rat liver	mDNA	Dye/θ	−0.071	130
HeLa	spcDNA[f]	Dye/θ	−0.074	135
E. coli	pSM2	Dye/θ	−0.075	65
BAL-31	pM2, IC	Dye/θ	−0.076	136
E. coli	pSM1	Dye/θ	−0.078	65
E. coli	Col E1	Dye/GE	−0.078	46
E. coli	pSC101	Dye/GE	−0.085	46
E. coli	pSM6	Dye/θ	−0.088	65
E. coli	pSM7	Dye/θ	−0.088	65
E. coli	pSM8	Dye/θ	−0.088	65
E. coli	pSM9	Dye/θ	−0.088	65
BAL-31	pM-2 virion	(d)	−0.089	11
BAL-31	PM-2 virion	Dye/η	−0.0896	44
E. coli	pSM5	Dye/θ	−0.092	65

[a]   From sensitive cells.
[b]   IC refers to the intracellular form of the DNA.
[c]   From superinfected immune cells.
[d]   Average of tabulated values.
[e]   Rabbit β-globin cDNA insert.
[f]   Cycloheximide-treated cells.

B and C would mark the conversion from toroidal to interwound superhelical structures. The decrease in s with $-\sigma$ in Region C would be due to the reduction in mass per unit length along the plectonemic superhelix axis as ΔLk increases negatively. The minimum in s between Regions C and D would be due to the onset of branch formation in the plectonemic superhelix. Finally, the nearly linear increase of s with $-\sigma$ in Region D would arise from steadily increasing branch formation. This interpretation now appears to be less likely since toroidal superhelical structures do not appear to form at superhelix densities as great as those found in Region B.[61a] In view of the theoretical calculations of Le Bret,[61b] it now appears

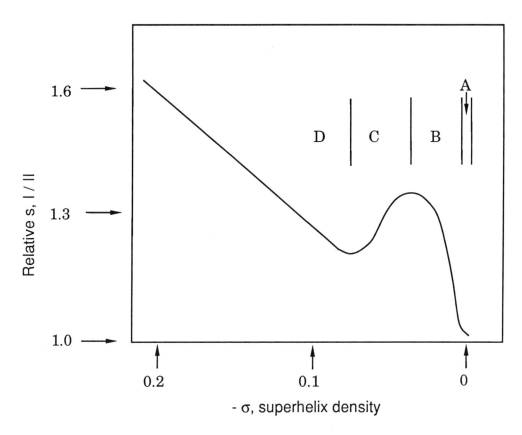

FIGURE 3. Schematic diagram of the variation of the sedimentation coefficient of closed circular DNA (I), relative to that of nicked circular DNA (II), with the superhelix density, σ. The delineation of Regions B, C, and D follows the discussion in Reference 40. Region A is possibly due to the initial introduction of a very few toroidal superhelical turns, as discussed in the text.

more likely that Region A is produced by a few (1 to 3) toroidal superhelical turns. Region B is associated with progressive shape changes of a plectonemic superhelix, Region C is dominated by the effects of branch formation, and Region D is dominated by progressive superhelix-associated denaturation. No conclusive evidence has yet appeared to provide a completely convincing explanation of the curve shape shown in Figure 3.

The results obtained in 1.0 M NaCl are nearly identical to those reported above, except that the location of the local minimum is shifted to a slightly more negative superhelix density. Similar results were obtained with $\lambda b_2 b_5 c$ DNA.[41] In the latter case, R increases from 1.0 to 1.6 in Regions A and B. Partial results were obtained for Region C, the extent of the decrease in R being to 1.5. No data are available for Region D. The exact values of R in each region depend upon ionic strength. In Regions A and B, there is a linear increase from R = 1.0 at 25 mM to R = 1.1 at 1 M. In Region C, R increases rapidly to 1.43 at 10 mM, then decreases slightly to 1.32 at 60 mM, remaining nearly constant at higher salt concentrations. In Region D, R increases rapidly to 1.37 at 10 mM, increases rather less rapidly to 1.48 at 60 mM, and increases slowly to 1.62 at 1 M. As a function of temperature, R is approximately constant in Regions A, B, and C over the range of 5 to 30°. In Region D, R decreases from 1.2 to 1.0 as the temperature increases from 5 to 40° (all in 2.85 M CsCl).

C. SEDIMENTATION AT ACID AND ALKALINE pH

For polyoma DNA, $s_{20,w}$ is independent of pH over the range of $3.0 \leqslant pH \leqslant 11.5$.[8] For PM-2 DNA, similar behavior in $s_{20,w}$ was found over the pH range 7.0 to 11.1.[59]

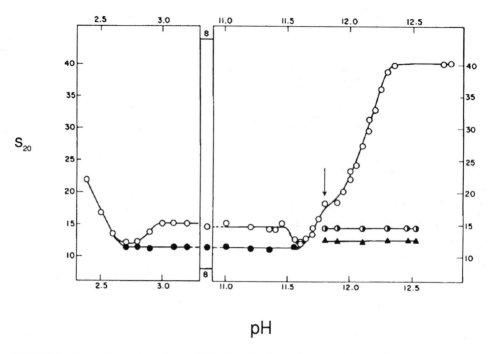

pH

FIGURE 4.   The sedimentation velocity-pH titration of polyoma DNA at 20°C in aqueous CsCl, ρ = 1.35 g/ml. (O) Closed duplex DNA; (●) nicked duplex DNA; (◑) circular single strands; (▲) linear single strands. Sedimentation coefficients are not corrected for solvent viscosity or buoyant effects. The arrow indicates the pH at which strand separation occurs in the nicked molecule. (From Vinograd, J. and Lebowitz, J., *J. Gen. Physiol.*, 49, 103, 1966. With permission.)

Reduction in pH decreases $s_{20,w}$ until R = 1, at pH = 2.70.[8] Further reduction in pH results in an increase in R, to 2.2 at pH = 2.4. Increase in pH above 11.5 results in a reduction in R to 1.0 at pH = 11.7, followed by an increase in R to 4.0 at pH = 12.35.[8] At pH > 12.5, R decreases by approximately 5% for a 10% decrease in $|\sigma|$.[62]

The most comprehensive study of the variation of R with pH was done for polyoma DNA[8] and is shown in Figure 4.

D. SEDIMENTATION IN CHAOTROPIC SOLVENTS

The addition of high concentrations of chaotropic salts to aqueous solutions containing DNA lowers the helix-coil transition temperature by an amount proportional to the salt concentration.[63] The sedimentation profile of closed circular DNA is generally altered under the same conditions in a direction that mimics unwinding by an intercalating drug (Section VII). Data for three solvents are available, and all experiments were conducted with PM-2 viral DNA. For aqueous $NaClO_4$,[37] Region D is spanned by a change in salt concentration from 0 to 2.0 *M*; Region C is spanned from 2.0 to 6.0 *M*; and Regions A and B are spanned from 6.0 to 7.2 *M*. For aqueous $RbCCl_3CO_2$,[64] the corresponding salt concentration ranges are 0 to 1.5 *M*, 1.5 to 2.4 *M*, and 2.4 to 2.6 *M*. The helix-coil transition temperature is reached in Region III in this case, and the effects of melting dominate those of regular duplex unwinding. Finally, for aqueous $NaCCl_3CO_2$,[64] the corresponding salt concentration ranges are 0 to 1.25 *M*, 1.25 to 2.60 *M*, and 2.60 to 3.28 *M*. In this solvent a fourth region is apparent, corresponding to the introduction of positive superhelical turns. The value of R increases from 1.0 to 1.6 as the salt concentration is increased from 3.28 to 3.78 *M*.

E. BUOYANT EQUILIBRIUM UNDER DENATURING CONDITIONS

In the absence of other reagents and at neutral pH and near room temperature, the

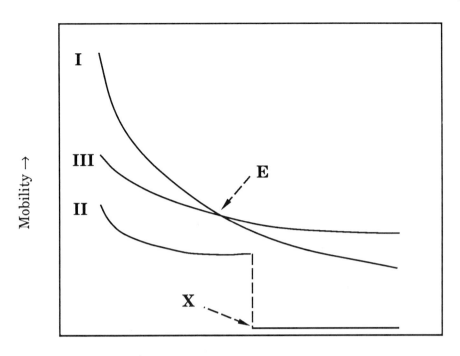

FIGURE 5.   Dependence of the electrophoretic mobility of closed circular (I), nicked circular (II), and linear (III) DNAs upon the molecular length, N, usually measured in base pairs. The point E represents the mobility and length at which DNAs I and III have equal mobility. The point X represents the molecular length above which DNA II is excluded from the gel.

buoyant density of closed circular DNA is usually indistinguishable from that of nicked circular or linear DNA. PM-2 DNA, having a very large negative superhelix density ($\sigma^\circ =$ -0.095), exhibits a shift in buoyant density, $\Delta\theta$, relative to nicked circular DNA of 2 mg/ml.[59] Experiments at acid pH are generally not practical because of the attendant depurination reactions over the time scale of the buoyant experiments.

The buoyant density of closed circular DNA increases with $-\sigma$ in buoyant $RbCCl_3CO_2$, a chaotropic solvent.[60] The increase in buoyant density is linear in $-\sigma$, according to the relationship $\Delta\theta(mg/ml) = -2.565 - 177.18\ \sigma$. Under these conditions the extent of denaturation is linearly related to the buoyant shift.

V. GEL ELECTROPHORESIS OF CLOSED CIRCULAR DNA

A. AGAROSE GELS

In the present section, we confine the treatment to a single dimension gel electrophoresis. Two dimension gels are described in Section VI.F. For the molecular length range 5.42 to 83.0 kbp (M_r range of 3.59 to 55.0), the general gel electrophoresis behavior of closed circular (I), nicked circular (II), and linear (III) DNAs has been determined[65] and is drawn diagrammatically in Figure 5, where the point labeled X designates the molecular weight at which a nicked (or relaxed closed) circular DNA is excluded from the gel. The point labeled E designates the molecular weight at which a closed circular DNA comigrates with its linear cognate. The relative electrophoretic mobilities and the molecular weights characteristic of X and E depend upon the detailed experimental conditions.

At E, the point of equality of mobility between DNAs I and III, the DNA length L_E is related to the agarose concentration, $A_\%$, by:

$$\log(L_E) = 2.1970 - 1.1143 \, A_\% , \quad A_\% \leq 1.27 \tag{14a}$$

$$L_E = 6.92 \text{ kb}, \quad A_\% > 1.27 \tag{14b}$$

The relative mobility, $u_{r,E}$, at E depends upon $A_\%$, according to:

$$u_{r,E} = 0.76075 + 0.47695 \, A_\% , \quad A_\% \leq 1.4 \tag{15}$$

The molecular length at which exclusion of nicked circles occurs, L_X, is 23.5 kb, independent of gel percent agarose.

Reduction in the voltage gradient from 8 to 4 V/cm results in an increase in length L_E by 3.1 kb, with no change in $u_{r,E}$. The value of L_X increases by 6.3 kb. At a voltage gradient of 4 V/cm, an increase in the salt concentration in the electrophoresis buffer from 0.05 to 0.08 M sodium acetate increases L_E by 1.1 kb, with no change in $u_{r,E}$. The value of L_X is increased by 17.4 kb.

B. PULSED-FIELD AGAROSE GELS

This technique, consisting of cycles of forward and reverse electric field pulses, allows open (relaxed closed-circular and nicked circular) DNAs to enter agarose gels, over the molecular length range of 2.9 to 56 kb.[66] In a subsequent study, the relative mobilities of supercoiled plasmids ranging in size from 4 to 16 kb were investigated.[67] The extent of reduction in ΔLk has a pronounced effect on the gel electrophoretic mobility. In general the mobility of supercoiled DNA $>>$ mobility of relaxed closed circular DNA $>$ mobility of nicked circular DNA.

C. SMALL CIRCULAR DNAs

In the very small size range, 230 to 370 bp, resolution of individual topoisomers can be attained by electrophoresis at high voltage gradients (approximately 30 V/cm) in 40-μm thick gels of 5% polyacrylamide.[68] Alternatively, topoisomers in this size range can be resolved by electrophoresis in thicker polyacrylamide gels containing ethidium bromide.[69]

For small closed-circular DNAs in the size range of 350 to approximately 750 bp, the shape of the trough in the above figure becomes V-shaped in 4% polyacrylamide gels.[70] The more usual U-shaped behavior is obtained at higher gel concentrations and with longer DNAs. The V-shaped behavior was found for circular DNAs as large as 1374 bp in the relatively high concentration of 2.5% agarose, for the first 4 to 5 topoisomers at either side of the minimum.

VI. ENERGETICS OF CLOSED CIRCULAR DNA

A. GENERAL CONSIDERATIONS

The free energy of superhelix formation, $\Delta G(\Delta Lk)$, is found empirically to satisfy the expression

$$\Delta G(\Delta Lk) = K(\Delta Lk)^2 \tag{16}$$

where the constant $K = a/N$, N is the number of base pairs in the DNA, and a is an experimentally determined constant. The constant a was first estimated by comparison of the binding isotherm of ethidium bromide with closed DNA to that with nicked circular DNA.[42] The constant was later estimated from the width of a Boltzman distribution of topoisomers.[38,71] Little information is available concerning the other thermodynamic-state functions. Two estimates have been reported for the enthalpy of superhelix formation and only one estimate has been made of the entropy of superhelix formation, as discussed below. Equation 16 is central to any application that requires inclusion of the free energy contribution due to a change in the extent of supercoiling. The assumption is usually made that Equation 16 applies without regard to nucleotide sequence or to the prior occurrence of local structural transitions in the DNA. The generally good agreement between the EtdBr binding method and the topoisomer distribution method supports these assumptions.

B. THEORETICAL CALCULATIONS

Energy calculations on superhelical DNA have been generally directed toward two different objectives. The first of these is prediction of the most stable structure; that is, of the most probable superhelix geometry at any given value of ΔLk. This is readily extended to include calculation of the population distribution functions of Lk, Tw, and Wr. The second is the assessment of the influence of supercoiling upon chemical reactions and structural transitions. We will treat these topics separately in the present section. We then review the empirical definition and determination of $\Delta G(\Delta Lk)$, the quantity commonly termed the free energy of superhelix formation and given in Equation 16.

1. Calculation of the Elastic Energy of Supercoiling

The general approach in energy calculations is to model the DNA as a thin elastic wire or rod, having characteristic bending (B) and torsional (T) stiffness. The energy equation is then set up as if for a macroscopic flexible rod, and the calculations employ the best available estimates of the torsional and bending force constants as applied to DNA. The assumption is also made that fluctuations in the twist and the writhe are uncoupled, meaning that contributions to the total energy are supposed to arise independently from Tw and Wr. The energy in question is actually a Helmholtz free energy, for which we will use the symbol U. The total energy is thus the sum.

$$U = U_B + U_T \tag{17}$$

The bending energy for an isotropic rod is given by an expression of the following type[72]

$$U_B = (A/2)\kappa^2 L \tag{18}$$

where κ is the curvature, A is a force constant, and L is the contour length of the DNA. In general, both κ and A are expected to vary from point to point along a DNA molecule: κ due to the flexibility of the chain and A due to the effects of base sequence. The constant A can also be written in terms of the persistence length, P, of DNA. For a single DNA molecule, P (also called $1/2\lambda$) and A are related by[73] $A = k_b TP$, where k_b is the Boltzmann constant (1.381×10^{-23} J/degree). (The related quantity $2P = 1/\lambda$ is known as the Kuhn statistical segment length.) The persistence length is a measure of the stiffness of a chain polymer and is the average of the sum of the projections of all bonds upon the first bond of the chain. For an average DNA, $P = 65 \pm 6$ nm, based upon the combined results of light scattering and hydrodynamic measurements.[74,77] The value of the bending constant is then $A = 2.67 \times 10^{-28}$ J-m (2.67×10^{-19} erg-cm). On a molar basis, the corresponding constant is $A' = 38.5$ kcal-nm/mol.

Equation 18 would be more realistically written with the curvature vector, $\kappa(s)$, where s measures distance along the DNA axis. Then κ^2 is given by $\kappa(s) \cdot \kappa(s)$. This requires, in turn, specification of both the position of each chain element and of the chain curvature at all locations along the DNA axis. It is therefore necessary to know the DNA geometry before the bending energy can be calculated.

The torsional energy of deformation for an isotropic rod is given by[78]

$$U_T = (C/2)(\Delta Tw)^2 L \qquad (19)$$

where ΔTw is the deviation from the twist of the fully relaxed DNA; i.e., $\Delta Tw = Tw - Tw_o$. As with bending, the torsional rigidity C and ΔTw are expected to exhibit variation along the contour length of the DNA: C due to the effects of base sequence and ΔTw due to chain flexibility.

2. Application to a Distribution of Topoisomers

Equation 19 may be applied to a distribution of topoisomers, either Gaussian or discrete, provided at least two assumptions are satisfied: (1) the variance of Wr over the distribution is negligible compared to the variance of Tw[79] and (2) the thermodynamic work function is the sole contributor to $\Delta G(\Delta Lk)$. In these experiments, a distribution of topoisomers is obtained by thermal equilibration of a DNA sample under conditions such that free duplex rotation is allowed, followed by strand closure under conditions in which the thermal equilibrium is maintained.

The relationship between the variances in Lk and in the geometric quantities may be written beginning with the fundamental relationship:[5]

$$\Delta Lk = \Delta Tw + Wr \qquad (20)$$

The complete expression for the variance is then:

$$<(\Delta Lk)^2> = <(\Delta Tw)^2> + 2 <\Delta Tw \cdot \Delta Wr> + <Wr^2> \qquad (21)$$

Experimentally, it is $<(\Delta Lk)^2>$ that is determined. For the Gaussian distribution,[38,71] the variance $<(\Delta Lk)^2> = RT/2K$. In order to compare with the elasticity theory, however, $<(\Delta Tw)^2>$ is the quantity that is required. From Equation 21, it is seen that both the variance in Wr and the covariance in (Tw,Wr) must also be taken into account. This latter term, $<\Delta Tw \cdot \Delta Wr>$, is not necessarily equal to zero. For example, negative fluctuations in Tw (duplex unwinding) might be less (or more) strongly correlated with long-range chain fluctuations (variance in Wr) than positive fluctuations. Since no information is currently available concerning the magnitude of this term, it is generally neglected. It has been suggested[79] that the assumption is reasonable for nicked circular DNA, but even this is uncertain.

Given the assumption that the covariance may be neglected, it follows from Equation 21 that $<(\Delta Tw)^2>$ may be obtained from the experimental value of $<(\Delta Lk)^2>$, either by estimating $<Wr^2>$ and subtracting or by selecting conditions such that $<Wr^2>$ is expected to be small. The calculation of Wr for a given Lk is difficult (refer to Section I). For small circular (sc) DNA, it has been argued that the second assumption is satisfied.[68,69] Given the correctness of these assumptions, it follows that $<(\Delta Lk)^2> = <(\Delta Tw)^2>_{sc}$. Under such conditions, it is expected that $<(\Delta Tw)^2>_{sc} = RT/2K$.

Employing elasticity theory, an alternative expression for $<(\Delta Tw)^2>$ is[26]

$$<(\Delta Tw)^2> = k_b TL/(4\pi^2 C) \qquad (22)$$

Employing Equation 21, with the covariance neglected and substituting for $<(\Delta Lk)^2>$, the result is:[28]

$$RT/NK = k_b Tl/2\pi^2 C + <Wr^2>/N \qquad (23)$$

where $l = L/N$.

If the variance of Wr vanishes, Equation 23 immediately reduces to the expression for the torsional rigidity.[80]

$$C = (NK/RT)(k_b Tl/2\pi^2) \qquad (24)$$

3. Effect of Supercoiling Upon Chemical Reactions and Structural Transitions

Because of the constraint imposed by the requirement of constant Lk, the state of minimum energy for a closed circular DNA is often different from that for a nicked or linear DNA. In the simplest case, the total energy is simply the sum of the chemical energy due to weak interactions along the chain (especially those that maintain the duplex) plus the distortion energy due to superhelix formation. The sum of these two is always greater than the energy of a relaxed circular DNA, which consists of the first term only.

The most feasible alternative structure for a closed circular DNA of negative ΔLk is one in which the extent of supercoiling is reduced at the expense of the denaturation of a portion of the duplex. In studies of the temperature dependence of the gel electrophoretic mobility of plasmid pSM1 DNA,[81] N = 5420, no indication of ΔLk-induced denaturation was found at any temperature below 37°. At 37°, a portion of the duplex appeared to melt, beginning at a superhelix density of $\sigma = -0.06$. As the temperature increased, denaturation appeared to be initiated at progressively lower values of $|\sigma|$, finally occurring at $\sigma = 0$ at 72°. Thus, for most naturally occurring DNAs at physiological temperatures and below, the superhelix may best be thought of as supercoiling, with little if any duplex disruption. The exceptions to the above statement are DNAs that contain certain special local structures, such as inverted repeats and alternating purine-pyrimidine sequences.

The influence of supercoiling upon denaturation has been examined both by thermodynamics[82] and by statistical mechanics.[83] The thermodynamic theory proceeds by employing ΔTw as a measure of deformation in the elasticity theory, then calculating the fraction of closed circular DNA molecules expected to be at least partly denatured as a function of the ratio ΔTw/Lk$_o$. Independent estimates of Tw are, however, not available. The statistical mechanics treatment extends the range of the thermodynamic treatment to allow predictions of the denaturation profile for any assumed (small) sequence. The general prediction from both approaches, in agreement with experiment, is that temperature and ΔLk changes are complementary over some ranges of the variables.

Various theoretical (actually, semiempirical) treatments have been made of B-Z[84,87] transition, the cruciform[88,89] transition, and competitions between them.[90,91] The usual approach employs a variation of models previously developed for the DNA helix-coil transition, writing the partition function as the sum of terms in ν, the equilibrium constant for initiation (nucleation) and of an altered region, and s, the equilibrium constant for extension of the region. (The symbol σ is normally used for the nucleation equilibrium constant. We substitute ν here to avoid confusion with the superhelix density.) For a closed circular DNA, additional terms must be included to take into account the free energy of superhelix formation, as well as the change in ΔG(ΔLk) as the transition proceeds. The theoretical treatments predict the average ΔTw (actually the intersection number,[29] In) for a given ΔLk.

The results of two dimension gel electrophoresis experiments indicate the values of ΔLk that are associated with the structural transition in question. From this, ΔG(ΔLk) can be calculated and the undetermined energy and length parameters may be estimated by nonlinear least squares curve fitting. For example, the free energy of extending a stretch of Z DNA by one base pair is given by $\Delta G_{BZ} = -(RT/2)\ln s$. The free energy of initiating a B-Z transition is given by $\Delta G_J = -(RT/2)\ln \nu$. The curve fitting results suggest that ΔG_J is approximately 21 kJ/mol and that ΔG_{BZ} is 1.25 to 2.5 kJ/mol, depending upon ionic conditions.[83,92] Further discussion of the experimental data for both B-Z and cruciform transitions are given in Section VI.F below. The effect of covalent closure on the binding of unwinding drugs is discussed separately in Section VII.

C. EXPERIMENTAL DETERMINATIONS OF THE FREE ENERGY
1. Free Energy of Superhelix Formation

Numerous determinations of the free energy of superhelix formation have now appeared in the literature, over the molecular length range of approximately 200 to 10,000 bp. Although

TABLE 2
Experimental Determinations of $\Delta G(\Delta Lk)$

DNA	N, bp	NK/RT	NK, kJ	T, °C	Solvent[a]	Method[b]	Ref.
pBR322 Cf[h]	210	3910	9890	37	A	a	69
ΦX-174 RF Cf	237—254	3420	8337 (Tw)	20	B	b	94
pBR322 Cf	551	3230	8330	37	A	a	69
ΦX-174 RF Cf	245—880	4190	9964 (Av)	2—17	B	c	68
pBR322 Cf	1182	1870	4220	37	A	a	69
ΦX-174 RF Cf	1361	2860	6730	2—17	B	c	68
pNT7 monomer	2025	1296	3160	0—38	C	d	96
Escherichia coli 15 plasmid	2200	1414	3360 (Av)	0—26	D	d	38
ΦX-174 RF Cf	2302	1750	4265	20	B	c	68
pBR322 Cf	2434	1220	3150	37	A	a	69
Minicol	3300	1060	2730	37	D	e	71
pNT7 Dimer	4050	1198	2920	0—38	C	d	96
ΦX-174 RF Cf	4362	1610	3924	20	B	c	68
pBR322	4362	1152	2809	0—38	C	d	96
pBR322 Cf	4362	1130	2915	37	A	a	69
SV40	5243	1100	2625 (Av)	0—26	D	d	38
SV40	5243	890	2205	25	E	f	42
fd RF	5750	975	2330 (Av)	10—20	D	d	38
ColE1	6500	1105	2850	37	D	e	71
PM-2	9850	985	2540	37	D	e	71
PM-2	9850	998	2430 (Av)	14—29	D	d	38
PM-2	9850	1200	2975	25	E	f	38
PM-2	9850	525	1300	3—50	E	g	93

[a] (A) 10 mM MgCl$_2$, 1 mM ATP, 2.5 mM potassium phosphate, 20 mM BME, 50 mM Tris, pH 7.8 (at 25°), (B) 10 mM MgCl$_2$, 50 mM NaCl, 10 mM Tris, pH 7.5, 5 mM DTT, (C) 6.6 mM MgCl$_2$, 1 mM KCl, 66 mM Tris, pH 7.6, 1 mM DTT, (D) 200 mM NaCl, 10 mM Tris, pH 7.4, and (E) concentrated CsCl. 5.8 M (buoyant) in Reference 42, with EtdBr; 7.3 M (buoyant) in Reference 38, lacking EtdBr; 3 M in Reference 93.

[b] (a) Analysis was by polyacrylamide gel electrophoresis. Topoisomers were obtained by ligation with T4 DNA ligase. (b) Analysis was by electrophoresis in thin polyacrylamide gels. The probability of cyclization of short restriction fragments was related to the twist energy. The ligation temperature was varied to produce band shifts. NK = NK$_{Tw}$. (c) Data were obtained from analysis of the topoisomer distribution in polyacrylamide gel electrophoresis (thin gels). The distributions were obtained alternatively by varying temperature or EtdBr concentration during ligation. (d) Analysis was by two-dimensional agarose gel electrophoresis. Topoisomers were obtained by ligation of whole DNA restriction enzyme digestion products with T4 DNA ligase. (e) Analysis was by agarose gel electrophoresis. Topoisomers were obtained with calf thymus type I topoisomerase. (f) Data were obtained from determination of the EtdBr binding isotherms to closed and nicked DNAs, based upon the shift in buoyant density produced by EtdBr binding. (Values have been corrected to EtdBr unwinding angle of 26°.) (g) Data were obtained from spectrophotometric determination of EtdBr binding isotherms to closed and nicked DNAs. (Values have been corrected to EtdBr unwinding angle of 26°.) (h) Cyclized restriction fragments.

the value of K is strongly dependent upon N, it is relatively insensitive both to temperature and to the ionic composition of the solution. All available estimates of K are listed in Table 2, with NK reported in units of kJ.

2. Length Dependence of Free Energy

The length dependence of the free energy coefficient is best illustrated by a plot of the product NK versus N, the DNA length in bp. This relationship is shown in Figure 6. This plot includes all the published values of NK. All data taken together are reasonably well fit by a logarithmic expression, as indicated by the dashed line, with the result (in kJ):

$$NK = 1.0015 \times 10^5 \, N^{-0.42115} \tag{25a}$$

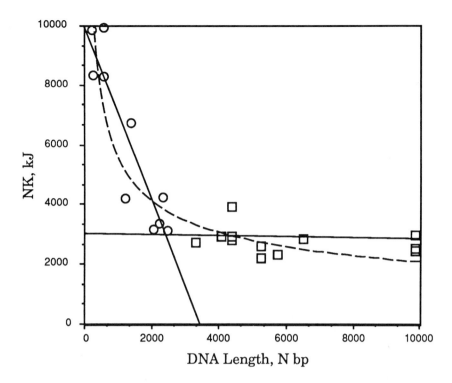

FIGURE 6. Plot of the available experimental estimates of NK, the product of N, the number of base pairs in a DNA, and K, the free energy coefficient, versus N. The dashed line represents the logarithmic fit to all the data. The solid lines represent the best linear fits over the regions to either side of the intersection point at 2400 base pairs.

Alternatively, as indicated by the two solid lines in Figure 6, the curve may be fit by two separate straight lines, one a linear relationship and the other a constant value. These are (in kJ):

$$NK = 9950 - 2.913 \, N, \quad N \leqslant 2400 \qquad (25b)$$

$$NK = 2820, \quad N > 2400 \qquad (25c)$$

(Here, the data point from Hsieh and Wang,[93] being significantly outside the range of the remainder, has been omitted.) In terms of ΔLk, the free energy is $\Delta G(\Delta Lk) = K(\Delta Lk)^2$ in each case. In terms of σ, the superhelix density, $\Delta Lk = \sigma Lk_o$, and $Lk_o = N/h_o$, where $h_o = 10.6$ base pairs/turn for a relaxed DNA in solution at room temperature. Then the free energy is (in kJ):

$$\Delta G(\sigma) = 890 \, N^{0.579} \sigma^2, \quad \text{all data} \qquad (26a)$$

$$\Delta G(\sigma) = [88.5 - 0.026 \, N]N\sigma^2, \quad N \leqslant 2400 \qquad (26b)$$

$$\Delta G(\sigma) = 25.12 \, N\sigma^2, \quad N > 2400 \qquad (26c)$$

Relative to the thermal energy at 25°, where $RT = 2.478$ kJ, the expressions for the free energy become:

$$\Delta G(\sigma)/RT = 359 \, N^{0.579} \sigma^2, \quad \text{all data} \qquad (26d)$$

$$\Delta G(\sigma)/RT = [35.7 - 0.010 \text{ N}]N\sigma^2, \quad N \leqslant 2400 \tag{26e}$$

$$\Delta G(\sigma)/RT = 10.14 \text{ N}\sigma^2, \quad N > 2400 \tag{26f}$$

The value of NK/RT for N > 2400, which is often used for calculations, is:

$$NK/RT = 1138, \quad N > 2400 \tag{26g}$$

D. OTHER THERMODYNAMIC AND ELASTIC PROPERTIES

1. The Torsional Rigidity

The measure of the torsional contribution to the elastic stiffness of DNA is defined in Equation 24. As discussed above, calculations of the torsional rigidity of DNA that employ the K derived from the variance of Lk are in principle incorrect since K includes, in general, contributions from both writhe and twist. It seems reasonable that only twist contributions should be included in the calculation of C. Results employing very small DNAs[69,94] would be expected to meet this criterion since sufficiently small chains are expected to have little if any writhe.

Two estimates of C are available based upon relative topoisomer populations employing DNAs of length 210 to 245 bp. Horowitz and Wang[69] obtained C = 2.9 × 10⁻¹⁹ erg-cm for pBR322 restriction fragments at 37°, in solvent A using method a (refer to Table 2). The same number was obtained by Shore and Baldwin[68] for restriction fragments of ΦX174 DNA at 20°, in solvent B using method c. A somewhat lower estimate, C = 2.4 × 10⁻¹⁹ erg-cm was obtained by Shore and Baldwin[94] at 37°, using solvent B and method b. A yet lower estimate of C = 1.79 × 10⁻¹⁹ erg-cm was obtained from the ΔH of superhelix formation,[95] using ColE1 amp plasmid DNA of N = 11,000. (The experimental conditions were 37°, 2 mM MgCl$_2$, 1 mM EDTGA, 10 mM Tris, pH 8.0.) The calculation in this case neglects bending contributions, as pointed out by the authors.

2. The Enthalpy of Superhelix Formation

The two published estimates of ΔH itself agree within about 70%, even though the methods of determination are quite different. The temperature dependence of ΔG was originally reported to be small and not suitable for a van't Hoff plot to determine ΔH and ΔS.[38,71] A later report, containing additional data, nonetheless presented a van't Hoff analysis and reported ΔH and ΔS values.[96] These numbers should, at present, be regarded as tentative. The second approach to the determination of ΔH employed microcalorimetry.[95] In particular, the heat of relaxation of a closed circular ColE1 DNA sample was measured in the presence of the *Escherichia coli* ω protein.[97] This heat was then identified with the enthalpy of superhelix formation. In this study, the ΔS of superhelix formation was calculated, but only by difference from the published values of ΔG. As can readily be seen from the data in Table 2, the free energy data are sufficiently imprecise to render this procedure questionable. For these reasons, the ΔS values are not reported here.

E. THE HELIX-COIL TRANSITION

1. Effect of DNA Closure on DNA Melting

The transition from helical duplex to denatured coil is profoundly affected by the constancy of the linking number and by the free energy of superhelix formation. Negatively supercoiled circular DNAs are underwound with respect to their linear counterparts and the resulting torsional strain destabilizes the secondary structure relative to the corresponding relaxed DNA. This leads to an early melting region in which (negatively supercoiled) closed DNA initiates melting before linear or nicked circular DNA. The early melting continues until the closed DNA is completely relaxed, at which point all preferential melting ceases.

Further denaturation, such as by an increase in the temperature or by the addition of alkali, initially causes the formation of positive supercoils and later results in the formation of a highly compact coiled structure. This late melting region is characterized by loss of cooperativeness in melting and by a highly asymmetric denaturation profile relative to open DNA.

The thermal helix-coil transition may be monitored by measuring the percent hyperchromicity, %h, as a function of temperature. Thus, at 260 nm the absorbance typically increases by a factor of 1.4 upon denaturation, while the %h goes from 0 to 100 No hyperchromicity is associated with superhelicity per se. Thermal melting in closed DNA is most conveniently carried out in a chaotropic solvent, such as 7.2 M NaClO$_4$, that reduces the melting region to accessible temperatures.[58] Only the late melting transition is seen in these experiments, since chaotropic solvents themselves unwind duplex DNA by an amount sufficient to remove all native superhelical turns.[36] For polyoma DNA, which has $\sigma° = -0.075$, the temperature at the midpoint of the transition (T_m) in 0.15 M NaCl is 89° for the nicked DNA and is >100° for the closed DNA. In 7.2 M NaClO$_4$, however, T_m is 72.8° for the closed DNA compared to 47.8° for the nicked molecule. In this solvent, the central 80% of the transition encompasses 30° for the closed DNA, but only 2.4° for the nicked DNA. In contrast to open DNA, the width of the thermal melting transition for closed DNA does not appear to vary with the base composition.[98]

Alternatively, the helix-coil transition can be monitored by determining the buoyant density, θ, as a function of pH.[58,59] For example,[59] PM-2 DNA I of very high superhelix density (-0.228) evidences a buoyant shift that is 10% of the maximum at pH 11.2, while the same DNA of low superhelix density (-0.013) requires pH 11.7 for the same increment in θ. As is the case with the thermal denaturation, the buoyant shift curves are greatly broadened for closed DNA compared to nicked.

2. Single Strand Specific Reagents and Chemical Modification

Reagents that bind specifically to single-stranded DNA promote the early melting of DNA by binding to and stabilizing denatured regions. The affinity of these probes for single-stranded regions in a closed circular duplex DNA is generally determined by the initial superhelix density. The binding of single-strand-specific reagents causes a shift in the sedimentation velocity and buoyant density of the closed circular form, but not in the nicked counterpart. The carbodiimides (CMC), for example, are water-soluble reagents that bind to guanine and thymine bases in unpaired regions of DNA.[99] The effect of this binding when CMC is present in excess is to increase $s_{20,w}$ from 21.0 S to 22.5 S for native SV40 DNA,[100] which is moderately supercoiled (-0.06). For PM-2 DNA, which is highly supercoiled (-0.093), an immediate shift from 28.5 S to 33.0 S occurs at very low CMC levels.[101] If $s_{20,w}$ is measured as a function of the extent of the CMC reaction, a curve similar to that shown in Figure 8 is generated.[101,102] Similar results are obtained with other reagents such as CH$_3$HgOH[103,104] and formaldehyde.[105,106]

F. LOCAL STRUCTURAL TRANSITIONS IN SUPERHELICAL DNA
1. Description of Transitions

The free energy of superhelix formation has been shown to stabilize certain types of altered secondary structures in closed circular DNA (a more complete list of references is provided in the recent review by Wells[107]). Sequences of alternating G and C residues, for example, can be made to adopt a left-handed helical structure (Z DNA) having 12 base pairs per turn.[108] The transition from a right- to a left-handed helix results in relaxation of negative supercoils by changing the local duplex winding from a positive to a negative value.[109,110] Palindromic sequences and other inverted repeats may respond to torsional strain by extrusion to a cruciform. (Recent reviews have appeared.[111,112]) Base pairs located in the arms of the cruciform are unwound in terms of the original double-stranded duplex, allowing a decrease in the negative superhelical tension.

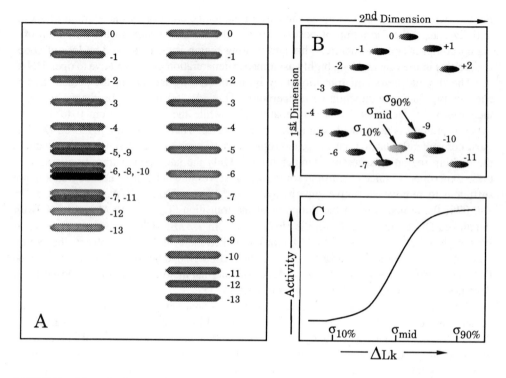

FIGURE 7.   Schematic diagram of results from local structural transitions by agarose gel electrophoresis. (A) Agarose gel electrophoresis in one dimension, visualized by postelectrophoresis staining with ethidium bromide. Both lanes contain plasmids of the same molecular weight and average superhelical density. The plasmid on the left contains, in addition, a sequence that can undergo transition to an altered conformation, bringing about a defined amount of local duplex unwinding. The plasmid on the right lacks the sequence. (B) Agarose gel electrophoresis run in two dimensions. The first dimension is run exactly as in A. The second dimension is obtained by removing the left-hand ladder in A and placing it on top of a second gel containing chloroquine. The bands that overlapped in the first dimension are now resolved. (C) Representative results from various nuclease assay techniques. Activity is defined as the degree of advancement of the transition. $\sigma_{10\%}$ is the superhelix density required to attain 10% of the transition, σ_{mid} signifies the 50% conversion point, and $\sigma_{90\%}$ is the superhelix density required to achieve 90% transition.

2. Analysis by Gel Electrophoresis

Both types of transitions result in characteristic changes in the electrophoretic mobility. This is illustrated in cartoon form in Figure 7. Figure 7A illustrates the typical ladder of topoisomers observed[54] with closed circular DNA (right-hand ladder). Local structural transitions generally require a threshold $-\sigma$ for initiation, are characterized by a midpoint $-\sigma_{mid}$, and are completed over a fairly narrow region. Figure 7A, left-hand ladder, illustrates the complex appearance of the topoisomer distribution upon the occurrence of such a transition. The energy considerations that pertain to transitions of these types were discussed briefly in Section VI.B above.

Resolution of topoisomers has been greatly improved with the recent use of two-dimensional gel electrophoresis.[83,96] In these experiments, the ladder obtained in the normal electrophoresis experiment is rotated 90° relative to the field and electrophoresed again, this time in the presence of the intercalating drug chloroquine,[113] which reduces ΔLk by essentially the same amount for all topoisomers and, in particular, which reverses the structural transition in question. Topoisomers that had the same mobility in the first dimension, due to the transition, are now separated. This is illustrated by the cartoon in Figure 7B, which represents the two-dimensional gel analysis of the first dimension shown in Figure 7A.

3. Relationship Between Superhelix Density and Structural Transitions

Although individual local structural transitions may be brought about by other means, we restrict the present discussion to the production of a local structural transition by supercoiling. In general, any particular structural transition occurs at a threshold superhelix density, taken to be $\sigma_{10\%}$, is half-way complete at a median value σ_{mid}, and is essentially complete by $\sigma_{90\%}$ (refer to Figures 7B, C). In this section, we review the corresponding superhelical densities associated with representative B—Z and cruciform extrusions. Examination of the available experimental data makes it clear that the three σ-transition parameters vary considerably with the sequence of the structural element in question and, in the case of certain cruciform transitions, with the sequences of the surrounding regions.

Tables 3 and 4 contain representative results for the critical σ values associated with B-Z (Table 3) and cruciform (Table 4) transitions. The critical σ values are seen to depend upon the length, composition, and detailed structure of the element in question. Local structural elements are ordered in each table in order of decreasing requirement for torsional strain for attainment of 50% of the transition (σ_{mid}). All superhelix densities are calculated for a helical repeat of 10.6 base pairs per turn[34] and an ethidium unwinding angle[59] of 26°. The superhelix densities are not corrected for salt and temperature conditions[11] since the torsional stress is expected to vary with the physical superhelix density. The various assay methods used to detect the transitions, in addition to two-dimensional gel electrophoresis, are briefly described as footnotes to the tables.

It is apparent from the tables that various structural transitions take place over a wide range of superhelix densities. For Z DNA, the range in σ_{mid} is at least from -0.035 to -0.093, a change in $\Delta G(\sigma)/N$ of 0.2 kJ/mol base pair (nearly 1000 kJ/mol DNA for a molecule the size of pBR322). The corresponding energy range in σ_{mid} for the cruciform extrusion is -0.028 to -0.066, or a range in $\Delta G(\sigma)/N$ of about 0.1 kJ/mol base pairs (500 kJ/mol pBR322 DNA). Although not indicated in Table 3, the assays used to detect supercoil-specific transition take place at NaCl concentrations ranging from 40 to 150 mM and at pH varying from 4.6 to 8.3. Reference should be made to the original articles for the experimental details.

In the case of cruciform extrusion, kinetic considerations are also of primary importance. It has been reported that extrusion is typically a very slow process at room temperature, but that extrusion is kinetically favored at elevated temperatures.[114,115] More recently, Lilley and co-workers[116-118] have shown that supercoil-driven cruciform transitions are of two general types, based upon their kinetic characteristics. The C-type transition takes place at low ionic strength ($\leqslant$ 30 mM NaCl), has a large activation energy (750 kJ/mol), and has AT-rich sequences flanking the inverted repeat. The S-type transition has an optimum salt concentration of 30 to 70 mM NaCl, a lower activation energy (125 to 200 kJ/mol), and no particular flanking sequence composition requirements.

VII. LIGAND BINDING TO CLOSED CIRCULAR DNA

A. INTRODUCTION

The binding of drugs and other ligands to closed circular DNA often results in major changes in DNA structure and in other properties. Certain types of structural alteration follow immediately from the requirement that Lk be constant, since superhelical winding can compensate duplex unwinding or overwinding. Other types of structural alteration, such as occur following the binding of platinum antitumor drugs, have a less straightforward structural basis. It is convenient to classify ligand/closed DNA interactions according to four general categories:

Intercalation — These reagents unwind the duplex by insertion between adjoining base pairs, hence, they induce large changes in the superhelical structure. No denatured regions

TABLE 3
Superhelix Density-Dependent B-Z Transitions

Plasmid[a]	Z-DNA Sequence[b]	$\sigma_{10\%}$	$\sigma_{50\%}$	$\sigma_{90\%}$	Assay[c]	Ref.
pBR322	CACGGTGCGCATG	−.086	−.093	−.101	E	137
pGF3	(−218:−304) Ref. 138, Fig. 2A.	−.069	−.077	−.085	F	138
SV40	See Reference 139, Fig. 6	−.057	−.075	−.091	E	139
pLP014	d(CG)$_{14}$	−.066	−.071	−.079	E	137
pGF3—meth	d(^{m}CG)$_4$	−.055	−.059	−.073	F	138
pRW757	d(CG)$_5$	−.042	−.058	−.070	A	140
pRW155, tr2	d(TG)$_{20}$-[*lac* 95]-d(CA)$_{20}$		−.057		C	141
pAN064	d(CA)$_{32}$	−.045	−.056	−.066	E	142
pRW751	d(CG)$_{16}$-[*lac* 95]-d(CG)$_{13}$	−.040	−.053	−.086	G	140
pDHf14	d(TG)$_{30}$	−.046	−.052	−.057	C	143
pRW1004	d(CG)$_5$-TTTT-d(CG)$_5$	−.042	−.049	−.058	C	144
pRW751	d(CG)$_{16}$-[*lac* 95]-d(CG)$_{13}$	−.035	−.049	−.064	D	145
pDHf2	d(TG)$_{11}$	−.043	−.047	−.053	C	146
pLP32	d(CG)$_{32}$	−.036	−.047	−.055	E	137
pRW1101	d(CG)$_4$-TAC-d(CG)$_3$		−.047		C	144
pRW1003	(CG)$_5$-TT-(CG)$_5$		−.046		C	144
pRW1002	d(CG)$_4$		−.045		C	144
pRW1007	d(CG)$_5$-TA-d(CG)$_5$		−.045		C	144
pRW1009	d(CG)$_5$-d(TA)$_3$-d(CG)$_4$	−.040	−.045	−.047	C	144
pRW1008	d(CG)$_5$-d(TA)$_2$-d(CG)$_5$	−.042	−.044	−.047	C	144
pRW777	d(TG)$_{32}$	−.036	−.044	−.053	A	147
pRW751	d(CG)$_{16}$-[*lac* 95]-d(CG)$_{13}$	−.030	−.044	−.068	A	145
pRW155, tr1	d(TG)$_{20}$-[*lac* 95]-d(CA)$_{20}$		−.042		C	141
pRW756	d(CG)$_{16}$	−.036	−.038	−.041	C	141
pRW756	d(CG)$_{16}$	−.027	−.038	−.052	A	140
pRW1001	d(CG)$_9$		−.037		C	144
pDHg16	d(GC)$_{11}$	−.033	−.035	−.038	C	146
pRW755	d(CG)$_{13}$	−.023	−.035	−.046	A	140

[a] meth denotes HhaI methylation; tr 1 denotes first transition in (TG) block; tr 2 denotes second transition in (CA) block.

[b] Z-DNA sequence (xxx:yyy) denotes location of potential Z-DNA relative to the start site of transcription.

[c] (A)*S1 Nuclease*: Topoisomers were generated by the addition of topoisomerase I to purified plasmid at various ethidium bromide concentrations. Individual topoisomers were isolated and treated with S1 nuclease. The cleavage activity is observed in one-dimension agarose gels in the presence of ethidium bromide and is measured by the fraction of the total sample cleaved.[145] (Refer to Figure 7C.) (B) *Psoralen cross-linking activity*: Topoisomers were generated as in A above. Trimethylpsoralen cross-links the arms of the cruciform upon irradiation, and *Eco*RI endonuclease excises the cruciform arms. The fragments of the plasmid and cruciform arms are resolved in single-dimension agarose gels.[149] In Figure 7C, the activity is measured by the amount of cruciform-arm-specific band that is assayed by gel electrophoresis. (C) *2D Gels*: Partial relaxation of supercoiled DNA by topoisomerase I, followed by agarose gel electrophoresis in two dimensions, with chloroquine present in the second dimension. Structural transitions are indicated by a discontinuity in the topoisomer distribution, as described in the text.[156] (See Figure 7C.) (D) *1D Gels*: Partial relaxation of supercoiled DNA as in C above, followed by agarose gel electrophoresis in one dimension with ethidium staining.[145] (See Figure 7C.) (E) *Anti-Z-DNA antibody*: Individual topoisomers of radioactive DNA were generated as in A above. Anti-Z-DNA antibody recognizes and binds to left-handed helical regions. Extent of antibody binding is assayed by retention of labeled DNA on nitrocellulose filters.[137] (In Figure 7C, activity is indicated by the extent of binding of antibody.) (F) *DEP-HR*: Individual topoisomers were isolated as in A and treated with the alkylating agent diethyl pyrocarbonate (DEP). DEP preferentially carbethoxylates the N7 atom of purines when in the Z-form. Piperidine is used to cleave the destabilized ring system, and fragments are labeled and resolved by autoradiography.[147a] (G) *Inhibition of Bam HI cleavage*: Individual topoisomers were generated as in A above. The inhibition of *Bam*HI cleavage activity is observed on agarose gels run in one dimension and stained with ethidium bromide.[140] In Figure 7C, activity is given by the % inhibition.

TABLE 4
Superhelix Density-Dependent Cruciform Transitions

Plasmid	Inverted Repeat[a]	Type[b]	$\sigma_{10\%}$	σ_{mid}	$\sigma_{90\%}$	Assay[c]	Ref.
pRW810	7bp-(5bp)-7bp	S		$< -.152$		a	148
pRW809	10bp-(5bp)-10bp	S	-.057	-.066	-.071	a	148
pRW808	13bp-(5bp)-13bp	S	-.059	-.065	-.072	a	148
pOCE12	66bp*	S	-.055	-.064	-.077	b	149
pVH51 HT[d]	13bp-(5bp)-13bp ×2	S	-.054	-.062	-.066	a	150
pAO3	31bp*	C	-.057	-.060	-.063	c	151
pColIR515	13bp-(5bp)-13bp	C	-.045	-.057	-.065	a	152
pSA1B.56A	322bp (8bp)[e]	S	-.048	-.051	-.053	c	153
pLNc40	d(CATG)$_{10}$*	S	-.045	-.051	-.056	c	154
pUC7	48bp*	S	-.043	-.047	-.052	c	115
pAT34	d(AT)$_{17}$*	S	-.035	-.039	-.043	c	155
pAC103	61bp*	C			-.033	c	114
pXG540	d(AT)$_{17}$*	S	-.025	-.028	-.030	c	156

[a] Palindromes are specified by an asterisk following the total length of the repeat. For other inverted repeats, the number of unmatched base pairs is given in parentheses.

[b] Cruciform extrusion have been classified into two groups (C and S), based upon the kinetic characteristics of the extrusion reactions.[116] See text for a more detailed description.

[c] Refer to Footnote c of Table 3.

[d] HT denotes a head-to-tail dimer.

[e] See Reference 153, Figure 2.

are formed in the binding process. Intercalative drug binding to closed circular DNA has two general applications: (1) assessment of the superhelix density, σ, with a reagent of known unwinding angle and (2) determination of the unwinding angle, ϕ, of a previously uncharacterized drug by comparison with an already characterized drug. Example: ethidium bromide (EtdBr).

Nonintercalative unwinding — This class includes ligands that cannot, for structural reasons, intercalate. No denaturation of the duplex is involved. The mechanism is often unknown. Example: irehidiamine A. (Proteins and polypeptide chains are strictly members of this class, but are discussed separately in Section VIII below.)

Covalent or strong coordination bonds — These ligands usually exert their differential effects upon closed circular DNA by stabilizing local denatured regions. Large alterations in superhelical winding can result since the locally denatured region can wind left-handed in addition to the direct duplex unwinding caused by the denaturation.[81] Example: *cis*-dichlorodiammine Pt(II).

Strand scission — This reaction results in the loss of the topological constraint and in conversion of closed circular DNA to a species that behaves like the nicked circular species. Example: hedamycin.

B. METHODS FOR DETERMINATION OF DNA-LIGAND INTERACTIONS
1. Definitions

The binding ratio ν is the mole ratio of bound drug to nucleotide; the critical binding ratio, ν_c, is the value of ν at which there is equal binding to closed and nicked circular DNA. We note in particular that these definitions of ν and of ν_c specifically include only that fraction of bound drug that participates in unwinding. For more complex cases, in which there is more than one unwinding mode, the present formulations should be employed with caution. In the event that a fraction, α, of the bound drug participates in duplex unwinding, ν and ν_c should be replaced by $\alpha\nu$ and $\alpha\nu_c$, respectively.

2. Quantitative Relationships
a. Relations Involving the Superhelix Density, σ

$$\Delta Lk = (2N)(\phi/360)(\nu - \nu_c) \tag{27a}$$

$$Lk_0 = N/h_0 \tag{27b}$$

$$\sigma(\nu) = (h_0\phi/180)(\nu - \nu_c) \tag{27c}$$

$$\sigma = -(h_0\phi/180)\nu_c = -1.53_1\nu_c \tag{27d}$$

b. Relations Involving the Drug Binding Constant, $K(\sigma)$

$$K(\sigma) = K\exp[-\Delta G'(\sigma)\delta\sigma/RT] \tag{28}$$

where $\Delta G(\sigma)$ is given in Section V, above, $\Delta G'(\sigma) = d\Delta G/d\sigma$, and $\delta\sigma = h_0\phi/360N$.

The above expression for $K(\sigma)$ holds, provided that no σ-dependent interactions occur between bound drug molecules. Even with such interactions, it is often the case that $K(\sigma) = K$ at $\sigma = 0$. An alternative form of this relationship[119] is useful in the assessment of the unwinding angle for a ligand whose binding constant to closed DNA, $K(\sigma)$, is known as a function of σ,

$$\ln[K(\sigma)/K_\Pi] = -0.591\sigma\phi \tag{29}$$

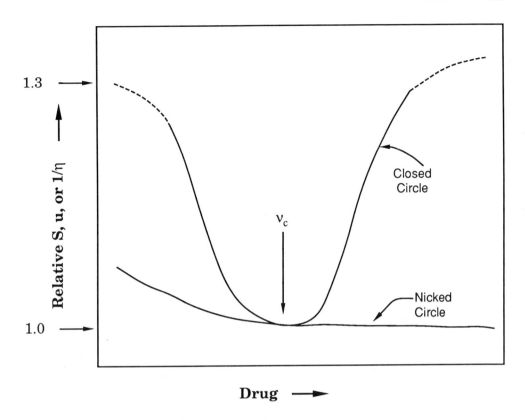

FIGURE 8. Schematic plot of hydrodynamic data for the determination of ν_c for the binding of a drug to closed circular DNA. The curves for the relative sedimentation coefficient of DNAs I and II, for the relative electrophoretic mobility (u), or for the reciprocal of the relative viscosity $(1/\eta)$ are of similar appearance. The binding ratio, ν, is the ratio of moles of drug bound to moles of nucleotide. The critical binding ratio, ν_c, is that value of ν required to bring about equality in the appropriate hydrodynamic ratio.

where σ is given by Equation 27d immediately above, and ϕ is measured in degrees. (Equation 29 is valid for N > 2400. For smaller N, refer to Section VI.)

3. Hydrodynamic Methods

The form of the experimental data obtained by hydrodynamic experiments is shown schematically in Figure 8.

The drugs for which data are given in this section bind to DNA by any mechanism that produces the type of curve illustrated in Figure 8. The quantity determined in these experiments is the apparent ϕ_{rel}, the ratio of the helix unwinding angle of the drug in question to that of EtdBr. The various hydrodynamic methods require determination of two kinds of data: (1) the hydrodynamic location at which the closed and nicked DNAs comigrate and (2) ν_c, the drug binding ratio at this location. Then $\phi_{rel} = (\nu_c)_{EtdBr}/(\nu_c)_{Drug}$. The critical binding ratio is normally expressed as moles drug per mole nucleotide. Three principle hydrodynamic methods are used for this purpose. The first of these is sedimentation velocity (S),[120] in which a minimum constant occurs in the plot (Figure 8) at ν_c. Calculation of ϕ_{rel} requires the binding constant, K, to relaxed DNA. Topoisomers are not resolved. The second method is viscometric (V),[43] requiring measurement of the flow time, usually with a flow viscometer. The hydrodynamic plot here shows a maximum at $\nu = \nu_c$, but the binding constant K is not needed. Finally, gel electrophoresis (GE) can be used in a way analogous to sedimentation,[47] except that here individual topoisomers are resolved.

TABLE 5
Experimental Determinations of the EtdBr Unwinding
Angle

ϕ Degrees	Method	Ref.
26 ± 2.6	Buoyant titration of PM2 DNA in alkaline CsCl	59
24—36	Comparison with carbodiimide-induced unwinding	102
28.5	EM length measurements of PM-2 DNA I hybridized with known restriction fragments	57
28	Comparison with band counting, gel electrophoresis, and SV40 DNA	54
23	Comparison with band counting, gel electrophoresis, SV40, and minicol DNAs	55
26	X-ray crystallography of iodinated dinucleotide-monophosphate complexes with EtdBr	157—159
23	X-ray diffraction of complex between CpG, EtdBr	160
27.2 ± 0.5	Buoyant shift in rubidium trichloroacetate	60

4. Nonhydrodynamic Methods

Two other, nonhydrodynamic, methods have also been developed for the determination of ϕ. In the first method (Topo GE), the extent of unwinding is estimated by mixing drug and DNA under conditions of duplex-free rotation, closing the DNA before removal of the ligand, and analysis of the resulting topoisomer distribution. This method originally employed incubation of nicked circular DNA with ligand, sealing the nick with DNA ligase, and analysis by differential sedimentation.[121] A later modification also employed DNA ligase, but analyzed the product by gel electrophoresis.[122] The current method of choice (designated TGE) is incubation of closed DNA with ligand in the presence of topoisomerase, and analysis by agarose gel electrophoresis.[123] This method actually measures the ΔLk of the center of the Gaussian distribution of topoisomers due to ligand binding. ΔLk is not, however, generally equal to the product $N\phi\nu_c$ (see above). The utility of this approach requires that SLk = 0 under binding conditions, a condition that is often not satisfied (histone octamer binding, for example). We therefore refer to the "unwinding angles" determined by this procedure as ϕ_a, or apparent angles. The final method (KEq) employs the measurement of the binding constant as a function of σ, $K(\sigma)$, relative to K_{II}.[119] (Refer to Equation 29 above). This is often the method of choice for ligands that have very large binding constants, in which case the binding constants may be estimated from the ratio of rate constants.[124] This method also yields the apparent unwinding angle, ϕ_a.

C. ESTIMATES OF THE ETHIDIUM BROMIDE UNWINDING ANGLE

The original estimate of ϕ for ethidium bromide was 12°, based upon model-building studies.[125] This estimate was presented as a minimum value and it is now generally regarded to be a serious underestimate. Table 5 lists the currently available experimental determinations of ϕ. The consensus value of ϕ is 26°.

D. UNWINDING ANGLES OF REPRESENTATIVE DNA-UNWINDING DRUGS

Table 6 lists representative drugs and their unwinding angles, as determined by the indicated method. The purpose of the table is to provide useful and specific literature references of applications of the above methods, from the point of view of the properties and applications of closed circular DNA. No attempt has been made to strive for completeness in terms of a more general catalog of the molecular characteristics of drug binding to DNA.

TABLE 6
Ligand Unwinding Angles Determined with Closed Circular DNAs

Ligand	ϕ Deg[a]	Method[b]	Ref.
Acridine orange, chloroquine, quinine, chlorpromazine, methylene blue, primaquine	+[c]	V	161
Acridine orange, trimeric derivative	55	V	162
Antitumor antibiotics BBM-928A and echinomycin	43	V	163
Triacridine derivatives[d]	22—37	V	164
Diacridine derivatives[d]	28—35	V	164
Binuclear Pt(II)—terpyridine complexes	23—36[e]	V	165
Acridine orange, dimeric derivative	32	V	162
Coralyne	23	V	166
Amsacrine, anilino ring-substitutes	18—20	V	167
Acridine orange, monomer	17	V	162
2,7-Dialkyl-substituted proflavines	16—17	V	168
Butane-1-thiolato-Pt(II)-terpyridine, monomer	17	V	165
Anthracyclines, various derivatives	10—13	V	169
Stereoisometric androstane-3,17-bisquaternary ammonium salts	5.9—13[f]	V, S	170
Diacridines, separation < 4 residues	15—16	V, S	171
Diacridines, separation 5 residues	22	V, S	171
Diacridines, separation > 5 residues	26—35	V, S	171
Chloroquine	+[c]	S	120
Quinomycin and triostin antibiotics	52	S	172
Echinomycin	31—47	S	120
Propidium diiodide	26	S	120
Actinomycin D	26	S[g]	173
Actinomycin D	25	S	120
2-Hydroxyethanethiolato(2,2′,2″-terpyridine)-Pt(II)	18—21	S	174
Proflavine	19	S	120
Nogalomycin	18	S	120
Isomethamidium	17.7	S	175
Ellipticine	17	S	176
Hycanthone	15	S	120
3-Nitrobenzothiazolo(3,2-a) quinolinium perchlorate	13	S	177
Prothidium dibromide	12.2	S	175
Bleomycin (bithiazole ring intercalation)	12	S	178
Daunomycin	11	S	120
Proflavine derivative	10[c]	S	179
Desaminoisometamidium	6.3	S	175
Cationic thiazole amides, bleomycin A2 derivatives	<0.03	TGE	180
Peptide antibiotics: tyrocidin, gramicidin (low conc.)	+[c]	TGE	180a
Spermine, Streptomycin, berenil	−	S	120
Irehidiamine A	±[h]	S	120

[a] Here, ϕ for EtdBr is taken to be 26°.
[b] V = viscometry, S = sedimentation, TGE = topoisomerase gel electrophoresis.
[c] Qualitative results only. Evidence was obtained for intercalative binding.
[d] Values are given for several derivatives. The exact value of ϕ depends upon the specific structure of the derivative.
[e] The unwinding angle varies in a nonlinear way with the length of the hydrocarbon chain that separates the two intercalating ligands.
[f] Although the hydrodynamic curves of these drugs resemble those of intercalators, other evidence suggests a nonintercalative mode of binding.

TABLE 6 (continued)
**Ligand Unwinding Angles Determined with Closed Circular
DNAs**

g Nicked circular DNA was closed with polynucleotide ligase in the presence of various
concentrations of drug, then titrated with EtdBr, as monitored by sedimentation.

h The sedimentation velocity profile closely resembles that of an intercalator, yet
structural considerations apparently rule out this binding mode. No binding data
were available.

TABLE 7
Drugs That Bind Covalently or Cause Strand Scissions

Drug/Reaction	Method/Comments	Ref.
cis-, *trans*-dichlorodiammine Pt(II)	GE. A minimum in mobility occurs, and the ladder of topoisomer bands blurs	181—183
Distamycin A, Berenil, terephthalanilides, Bis(guanylhydrazones)	V. No apparent unwinding of the DNA is observed	184
UV radiation (thymidine dimer production)	S. ϕ_a = 12° per dimer	185
Benzy[a]pyrene diol epoxides	GE. The value of ϕ_a varies from 30 to 330°	186
Bleomycin	EM. Cross-links correlate with σ	187
Ellipticine derivatives	GE. DNA moves through reagent; obtain min. in mobility, broadened bands	188
Benzo[a]pyrene	Reversible binding, ϕ_a = 13°	189
Gold coordination complexes	Similar to Pt(II) compounds	190, 191
Anthracyclines (various)	GE. Diffuse bands	192
Hedamycin	GE. Breakage, preferentially breaks I	193
O,O,S-trimethyl phosphorthiolate	GE. Diffuse bands, no minimum occurs in mobility plot	194
S-(2-haloethyl)-mercapturic acid analogs	GE (and EM). Strand scissions	195

a GE = gel electrophoresis, V = viscometry, S = sedimentation, EM = electron microscopy.

E. LIGANDS THAT BIND COVALENTLY OR THAT PRODUCE CHAIN SCISSIONS

Gel electrophoresis (GE) is commonly used to detect covalent binding that results in
unwinding of the duplex. In these experiments, native superhelical DNA (containing a
mixture of topoisomers) is typically incubated with drug for varying times, then subjected
to electrophoresis in the absence of drug. The mobility of the family of topoisomers decreases
with incubation, becoming indistinguishable from that of nicked circular DNA and often
increasing thereafter. Unlike noncovalent unwinding, the family of topoisomers migrates as
an unresolved smear at intermediate mobilities.[126] Alternatively, the extent of strand scission
due to drug binding can readily be detected by monitoring the amount of DNA remaining
in the closed circular band in gel electrophoresis as a function of drug dose. Results obtained
with a variety of ligands that bind either covalently or by very strong coordination are listed
in Table 7. Since these results are not readily reduced to an unwinding angle, qualitative
descriptions are generally given.

VIII. BINDING OF PROTEINS TO CLOSED CIRCULAR DNA

The binding of superhelical DNA to proteins is characterized by both local and global
effects. Local DNA structural alterations can be of a type similar to those discussed in
Section VI.F or they can result in the induction of bends or kinks. Global DNA structural
alterations can arise from the surface wrapping of the DNA on the protein. Closed circular

DNA is often employed in protein binding experiments because of the constancy of the linking number. Typically, the DNA is incubated with the putative binding protein, followed by incubation with topoisomerase and removal of all proteins. The quantity measured is then the ΔLk that is associated with the protein binding. Two methods have generally been employed for the determination of ΔLk, those designated TGE and KEq in Section VII. We emphasize that both methods determine only the total change in Lk (ΔLk) that is associated with the binding of protein, followed by a nicking-closing cycle, under the appropriate experimental conditions. No estimate of ϕ_{EtdBr} is, however, required in either method. Calculation of ΔLk per protein molecule requires, in addition, that an independent estimate be made of the number of protein molecules bound. Table 8 lists values of ΔLk associated with the binding of proteins to closed circular DNA.

A more complete description of the topological and geometric relationships that apply to these cases is presented in Section II. The linking number deficiency upon protein binding is easy to measure, but its interpretation is not always straightforward. As an example, we assume that each protein bound alters the winding of the DNA by ϕ degrees. Then the change in winding per bound protein is $\phi = 360 (\Phi - \Phi_o)$, where Φ is the winding number of the DNA in association with the protein, and Φ_o is the winding number of an equivalent length of unassociated DNA. If m proteins are bound per DNA, and topoisomeric equilibrium has been established, the linking number is:

$$Lk = SLk + \Phi_0 + m\phi/360 \qquad (30)$$

For the corresponding relaxed circular DNA with m = 0, $Lk_o = \Phi_o$. Combining,

$$\Delta Lk = SLk + m\phi/360 \qquad (31)$$

Rearranging, the result is

$$\phi = 360[(\Delta Lk - SLk)/m] \qquad (32)$$

For specificity, consider minichromosome for which ΔLk = -26[127] and m = 26.[55] The apparent unwinding angle, ϕ_a, would be calculated naively from ΔLk by neglecting SLk, giving $\phi_a = 360\Delta Lk/m$. The result is $\phi_a = -360°$.

The *actual* duplex unwinding angle, ϕ, requires the inclusion of the surface linking number, SLk (refer to Section II.B). For toroidal wrapping, as in a minichrososome, SLk = $-1.8m$.[24] Combining, the correct result is $\phi = +290°$.

In view of the large disparity in these values, it is clear that additional knowledge of the structure of the complex is required to interpret linking number changes for protein binding. Not only the magnitude, but even the direction of the winding change is otherwise subject to uncertainty. In particular, structural information is required to determine if the binding mode involves surface wrapping of the DNA about the protein surface or duplex unwinding. The use of closed circular DNA for the measurement of the ΔLk associated with protein binding and a nicking-closing cycle does, however, provide a valuable test for the validity of structural models.

TABLE 8
Binding of Proteins to Closed Circular DNA[a]

Protein	DNA	$-\Delta Lk$/protein	Solvent	T, °C	Ref.
Escherichia coli RNA polymerase	λ	0.77—1.53[b]	0.12 KCl	37	121
	SV40	1.61 ± 0.08	0.18 KCl	37	196
	SV40 chromosomes	0.6—1.2	0.15 KCl	37	123
	fd RF	0.7	0.05 KCl/0.01 $MgCl_2$	37	122
		0.39	0.2 KCl	37	122
		0.17—0.22	0.2 KCl	31	122
		0	0.2 KCl	0	122
	pBR322/tet promoter	1.19	0.1 KCl	37	197
	pBR322/all promoters	1.31	0.1 KCl	37	197
	pMI4/gal promoter	1.5	0.1 KCl	37	198
TFIIIA (nonspecific)[c]	pMB9	1.6/DNA	0.07 KCl	37	199
	pKA2	0.5/DNA	0.07 NH_4Cl	22	200
TFIIIA (specific)	pMB9/HBV	0.2—0.4	0.07 KCl	37	199
	pXbs/5S (4 sites)	0.1	0.07 NH_4Cl	22	200
	pXbs/5S (1 site)	1.0	0.1 KCl	37	201
DNA gyrase (−ATP)	PM-2 DNA	$\Delta Lk>0$			202
Minichromosome	SV40	1.01 ± 0.08	0.2 NaCl	37	127
lac repressor	*lac* operator	0.07	0.01 Mg/KCl	37	124
Cyclic AMP receptor		0.17	0.2 KCl	37	203
		0.1 ± 0.06	0.1 KCl	35	204

[a] Results with the *lac* operator/repressor were obtained using the KEq method, and all others used the TGE method.

[b] These experiments employed differential sedimentation to obtain the temperature dependence of the distance sedimented. Combining with the known dependence of σ upon T gave the change in ΔLk as a function of distance sedimented. The quantitative results depend upon φ$_{EtdBr}$ and have been corrected to 26°. The range of a factor of two in the value of ΔLk is due to uncertainty in the number of polymerase molecules bound per DNA.

[c] The extent of nonspecific unwinding varies with the protein/DNA ratio.

REFERENCES

1. **Fiers, W. and Sinsheimer, R. L.,** The structure of the DNA of bacteriophage ΦX174. III. Ultracentrifugal evidence for a ring structure, *J. Mol. Biol.,* 5, 424, 1962.
2. **Dulbecco, R. and Vogt, M.,** Evidence for a ring structure of polyoma virus DNA, *Proc. Natl. Acad. Sci. U.S.A.,* 50, 236, 1963.
3. **Weil, R. and Vinograd, J.,** The cyclic helix and cyclic coil forms of polyoma viral DNA, *Proc. Natl. Acad. Sci. U.S.A.,* 50, 730, 1963.
4. **Vinograd, J., Lebowitz, J., Radloff, R., Watson, R., and Laipis, P.,** The twisted circular form of polyoma viral DNA, *Proc. Natl. Acad. Sci. U.S.A.,* 53, 1104, 1965.
5. **White, J. H.,** Self-linking and the Gauss integral in higher dimensions, *Am. J. Math.,* 91, 693, 1969.
6. **Fuller, F. B.,** Decomposition of the linking number of a closed ribbon, *Proc. Natl. Acad. Sci. U.S.A.,* 75, 3557, 1978.
7. **Crick, F. H. C.,** Linking numbers and nucleosomes, *Proc. Natl. Acad. Sci. U.S.A.,* 73, 2639, 1976.
8. **Vinograd, J. and Lebowitz, J.,** Physical and topological properties of circular DNA, *J. Gen. Physiol.,* 49, 103, 1966.
9. **Bauer, W. and Vinograd, J.,** The use of intercalative dyes in the study of closed circular DNA, in *Progress in Molecular and Subcellular Biology,* Vol. 2, Hahn, F. E., Ed., Springer-Verlag, New York, 1971, 181.
10. **Bauer, W. and Vinograd, J.,** Circular DNA, in *Basic Principles in Nucleic Acid Chemistry,* Vol. 2, Ts'O, P. O. P., Ed., Academic Press, New York, 1974, 265.
11. **Bauer, W.,** Structure and reactions of closed duplex DNA, *Annu. Rev. Biophys. Bioeng.,* 7, 287, 1978.
12. **Gellert, M.,** DNA topoisomerases, *Annu. Rev. Biochem.,* 50, 879, 1981.
13. **Drlica, K.,** Biology of bacterial deoxyribonucleic acid topoisomerases, *Microbiol. Rev.,* 48, 273, 1984.
14. **Wang, J. C.,** DNA topoisomerases, *Annu. Rev. Biochem.,* 54, 665, 1985.
15. **Liu, L.,** DNA topoisomerases — enzymes that catalyze the breaking and rejoining of DNA, *CRC Crit. Rev. Biochem.,* 15, 1, 1983.
16. **Vosberg, H.-P.,** DNA topoisomerases: enzymes that control DNA conformation, *Curr. Top. Microbiol. Immunol.,* 114, 19, 1985.
17. **Wang, J. C.,** Recent studies of DNA topoisomerases, *Biochim. Biophys. Acta,* 909, 1, 1987.
18. **Cozzarelli, N. R., Krasnow, M. A., Gerrard, S. P., and White, J. H.,** A topological treatment of recombination and topoisomerases, *Cold Spring Harbor Symp. Quant. Biol.,* 49, 383, 1984.
19. **Wasserman, S. A. and Cozzarelli, N. R.,** Biochemical topology: applications to DNA recombination and replication, *Science,* 232, 951, 1986.
20. **White, J. H., Millett, K. C., and Cozzarelli, N. R.,** Description of the topological entanglement of DNA catenanes and knots by a powerful method involving strand passage and recombination, *J. Mol. Biol.,* 197, 585, 1987.
21. **Lilley, D. M. J.,** The genetic control of DNA supercoiling, and vice versa, in *Regulation of Gene Expression,* Booth, I. and Higgins, C., Eds., Cambridge University Press, Cambridge, 1986, 105.
22. **Biaever, G. N., Lnyder, L., and Wang, J. C.,** DNA supercoiling in vivo, *Biophys. Chem.,* 29, 7, 1988.
23. **White, J. H.,** An introduction to the topology and geometry of DNA structure, in *Mathematical Models for DNA Sequences,* Waterman, M. S., Ed., CRC Press, Boca Raton, FL, 1989, 225.
24. **White, J. H. and Bauer, W. R.,** Calculation of the twist and the writhe for representative models of DNA, *J. Mol. Biol.,* 189, 329, 1986.
25. **White, J. H. and Bauer, W. R.,** Applications of the twist difference to DNA structural analysis, *Proc. Natl. Acad. Sci. U.S.A.,* 85, 772, 1988.
26. **Shimada, J. and Yamakawa, H.,** Statistical mechanics of DNA topoisomers: the helical worm-like chain, *J. Mol. Biol.,* 184, 319, 1985.
27. **Le Bret, M.,** Computation of the helical twist of nucleosomal DNA, *J. Mol. Biol.,* 200, 285, 1988.
28. **Shimada, J. and Yamakawa, H.,** Moments for DNA topoisomers: the helical wormlike chain, *Biopolymers,* 27, 657, 1988.
29. **White, J. H. and Bauer, W. R.,** Superhelical DNA with local substructures: a generalization of the topological constraint in terms of the intersection number and the ladder-like correspondence surface, *J. Mol. Biol.,* 195, 205, 1987.
30. **White, J. H., Cozzarelli, N. R., and Bauer, W. R.,** The helical repeat and linking number of surface wrapped DNA, *Science,* 241, 323 1988.
31. **Drew, H. R. and Travers, A. A.,** DNA bending and its relation to nucleosome positioning, *J. Mol. Biol.,* 186, 773, 1985.
32. **Finch, J. T., Brown, R. S., Rhodes, D., Richmond, R., Rushton, B., Lutter, L. C., and Klug, A.,** X-ray diffraction study of a new crystal form of the nucleosome core showing higher resolution, *J. Mol. Biol.,* 145, 757, 1981.
33. **Wang, J. C.,** Variation of the average rotation angle of the DNA helix and the superhelical turns of covalently closed cyclic λ DNA, *J. Mol. Biol.,* 43, 25, 1969.

34. **Peck, L. J. and Wang, J. C.**, Sequence dependence of the helical repeat of DNA in solution, *Nature (London)*, 292, 375, 1981.

35. **Anderson, P. and Bauer, W.**, Supercoiling in closed circular DNA: dependence upon ion type and concentration, *Biochemistry*, 17, 594, 1978.

36. **Bauer, W. R.**, Structure of DNA in denaturing solvents. I. Bacteriophage PM2 DNA in aqueous sodium perchlorate, *J. Mol. Biol.*, 67, 183, 1972.

37. **Bauer, W. R.**, Premelting unwinding of the deoxyribonucleic acid duplex by aqueous magnesium perchlorate, *Biochemistry*, 11, 2915, 1972.

38. **Depew, R. E. and Wang, J. C.**, Conformation fluctuations of DNA helix, *Proc. Natl. Acad. Sci. U.S.A.*, 72, 4275, 1975.

39. **Hinton, D. M. and Bode, V. C.**, Purification of closed circular lambda deoxyribonucleic acid and its sedimentation properties as a function of sodium chloride concentration and ethidium binding, *J. Biol. Chem.*, 250, 1071, 1975.

40. **Upholt, W. B., Gray, H. B., Jr., and Vinograd, J.**, Sedimentation velocity behavior of closed circular SV40 DNA as a function of superhelix density, ionic strength, counterion and temperature, *J. Mol. Biol.*, 62, 21, 1971.

41. **Wang, J. C.**, Degree of superhelicity of covalently closed cyclic DNAs from *Escherichia coli*, *J. Mol. Biol.*, 43, 263, 1969.

42. **Bauer, W. and Vinograd, J.**, Interaction of closed circular DNA with intercalative dyes. II. The free energy of superhelix formation in SV40 DNA, *J. Mol. Biol.*, 47, 419, 1970.

43. **Revet, B. M., Schmir, M., and Vinograd, J.**, Direct determination of the superhelix density of closed circular DNA by viscometric titration, *Nature New Biol.*, 229, 10, 1971.

44. **Smit, E. M. and Borst, P.**, The superhelix density of bacteriophage PM2 DNA, determined by a viscometric method, *FEBS Lett.*, 14, 125, 1971.

45. **Watson, R. and Bauer, W. R.**, The viscometric behavior of native and relaxed closed circular PM2 DNAs at intermediate and high ethidium bromide concentrations, *Biopolymers*, 16, 1343, 1977.

46. **DeLeys, R. A. and Jackson, D. A.**, Dye titrations of covalently closed supercoiled DNA analyzed by agarose gel electrophoresis, *Biochem. Biophys. Res. Commun.*, 69, 446, 1976.

47. **Espejo, R. T. and Lebowitz, J.**, A simple electrophoretic method for the determination of superhelix density of closed circular DNAs and for observation of their superhelix density heterogeneity, *Anal. Biochem.*, 72, 95, 1976.

48. **Johnson, P. H. and Grossman, L. I.**, Electrophoresis of DNA in agarose gels. Optimizing separations of conformational isomers of double- and single-stranded DNAs, *Biochemistry*, 16, 4217, 1977.

49. **Radloff, R., Bauer, W., and Vinograd, J.**, A dye-buoyant-density method for the detection and isolation of closed circular duplex DNA: the closed circular DNA in HeLa cells, *Proc. Natl. Acad. Sci. U.S.A.*, 57, 1514, 1967.

50. **Upholt, W. B.**, Superhelix densities of circular DNAs: a generalized equation for their determination by the buoyant method, *Science*, 195, 891, 1977.

51. **Gray, H. B., Jr., Upholt, W. B., and Vinograd, J.**, A buoyant method for the determination of the superhelix density of closed circular DNA, *J. Mol. Biol.*, 62, 1, 1971.

52. **Hudson, B., Dawson, J. H., Desiderio, R., and Mosher, C. W.**, Ethidium analogues with improved resolution in the dye-buoyant density procedure, *Nucleic Acids Res.*, 4, 1349, 1977.

53. **Burke, R. L. and Bauer, W.**, Measurement of superhelix densities in buoyant dye/CsCl. The use of a standard other than native SV40 DNA, *J. Biol. Chem.*, 252, 291, 1977.

54. **Keller, W.**, Determination of the number of superhelical turns in simian virus 40 DNA by gel electrophoresis, *Proc. Natl. Acad. Sci. U.S.A.*, 72, 4876, 1975.

55. **Shure, M. and Vinograd, J.**, The number of superhelical turns in native virion SV40 DNA and minicol DNA determined by the band counting method, *Cell*, 8, 215, 1976.

56. **Dougherty, G. and Koller, T.**, Determination of the number of superhelical turns in the hyperchromicity of partially denatured covalently-closed DNA molecules, *Nucleic Acids Res.*, 10, 525, 1982.

57. **Liu, L. F. and Wang, J. C.**, On the degree of unwinding of the DNA helix by ethidium. II. Studies by electron microscopy, *Biochim. Biophys. Acta*, 395, 401, 1975.

58. **Vinograd, J., Lebowitz, J., and Watson, R.**, Early and late helix-coil transitions in closed circular DNA: the number of superhelical turns in polyoma DNA, *J. Mol. Biol.*, 33, 173, 1968.

59. **Wang, J. C.**, The degree of unwinding of the DNA helix by ethidium. I. Titration of twisted PM2 DNA molecules in alkaline cesium chloride density gradients, *J. Mol. Biol.*, 89, 783, 1974.

60. **Burke, R. L. and Bauer, W. R.**, The early melting of closed duplex DNA: analysis by banding buoyant neutral rubidium trichloroacetate, *Nucleic Acids Res.*, 8, 1145, 1980.

61. **Garner, M. M., Felsenfeld, G., O'Dea, M. H., and Gellert, M.**, Effects of DNA supercoiling on the topological properties of nucleosomes, *Proc. Natl. Acad. Sci. U.S.A.*, 84, 2620, 1987.

61a. **Spengler, S. J., Stasiak, A., and Cozzarelli, N. R.**, The stereostructure of knots and catananes produced by phage λ integrative recombination: implications for mechanism and DNA structure, *Cell*, 42, 325, 1985.

61b. **Le Bret, M.,** Catastrophic variation of twist and writhing of circular DNAs with constraint, *Biopolymers,* 18, 1709, 1979.

62. **Schmir, M., Revet, B. M., and Vinograd, J.,** Dependence of the sedimentation coefficient of denatured closed circular DNA in alkali on the degree of strand interwinding. The absolute sense of supercoils, *J. Mol. Biol.,* 83, 35, 1974.

63. **Hamaguchi, K. and Geiduschek, E. P.,** The effect of electrolytes on the stability of the deoxyribonucleate helix, *J. Am. Chem. Soc.,* 84, 1329, 1962.

64. **Burke, R. L. and Bauer, W. R.,** The helix-coil transition of closed and nicked DNAs in aqueous neutral trichloroacetate solutions, *Nucleic Acids Res.,* 5, 4819, 1978.

65. **Mickel, S., Arena, V., and Bauer, W.,** Physical properties and gel electrophoresis behavior of R12-derived plasmid DNAs, *Nucleic Acids Res.,* 4, 1465, 1977.

66. **Levene, S. D. and Zimm, B. H.,** Separations of open-circular DNA using pulsed-field electrophoresis, *Proc. Natl. Acad. Sci. U.S.A.,* 84, 4054, 1987.

67. **Hightower, R. C., Metge, D. W., and Santi, D. V.,** Plasmid migration using orthogonal-field-alternation gel electrophoresis, *Nucleic Acids Res.,* 15, 8387, 1987.

68. **Shore, D. and Baldwin, R. L.,** Energetics of DNA twisting. II. Topoisomer analysis, *J. Mol. Biol.,* 170, 983, 1983.

69. **Horowitz, D. S. and Wang, J. C.,** Torsional rigidity of DNA and length dependence of the free energy of DNA supercoiling, *J. Mol. Biol.,* 173, 75, 1984.

70. **Zivanovic, Y., Goulet, I., and Prunell, A.,** Properties of supercoiled DNA in gel electrophoresis. The V-like dependence of mobility on topological constraint. DNA-matrix interactions, *J. Mol. Biol.,* 192, 645, 1986.

71. **Pulleyblank, D. E., Shure, M., Tang, D., Vinograd, J., and Vosberg, H. P.,** Action of nicking-closing enzyme on supercoiled and nonsupercoiled closed circular DNA: formation of a Boltzmann distribution of topological isomers, *Proc. Natl. Acad. Sci. U.S.A.,* 72, 4280, 1975.

72. **Laudau, L. D. and Lifshitz, E. M.,** *Statistical Physics,* Addison-Wesley, Reading, MA, 1958, 478.

73. **Gray, H. B., Jr. and Hearst, J. E.,** Flexibility of native DNA from the sedimentation behavior as a function of molecular weight and temperature, *J. Mol. Biol.,* 35, 111, 1968.

74. **Schmid, C. W., Rinehart, F. P., and Hearst, J. E.,** Statistical length of DNA from light scattering, *Biopolymers,* 10, 883, 1971.

75. **Record, M. T., Jr. and Woodbury, C. P.,** Characterization of rodlike DNA fragments, *Biopolymers,* 14, 393, 1975.

76. **Jolly, D. and Eisenberg, H.,** Photon correlation spectroscopy, total intensity light scattering with laser radiation, and hydrodynamic studies of a well fractionated DNA sample, *Biopolymers,* 15, 61, 1976.

77. **Kovacic, R. T. and van Holde, K. E.,** Sedimentation of homogeneous double-strand DNA molecules, *Biochemistry,* 16, 1470, 1977.

78. **Landau, L. D. and Lifshitz, E. M.,** *Theory of Elasticity,* 2nd ed., Pergamon Press, New York, 1970, 72.

79. **Vologodskii, A. V., Anshelevich, V. V., Lukashin, A. V., and Frank-Kamenetskii, M. D.,** Statistical mechanics of supercoils and the torsional stiffness of the DNA double helix, *Nature (London),* 280, 294, 1979.

80. **Barkley, M. D. and Zimm, B. H.,** Theory of twisting and bending of chain macromolecules; analysis of the fluorescence depolarization of DNA, *J. Chem. Phys.,* 70, 2991, 1979.

81. **Lee, F. S. and Bauer, W. R.,** Temperature dependence of the gel electrophoretic mobility of superhelical DNA, *Nucleic Acids Res.,* 13, 1665, 1985.

82. **Benham, C. J.,** Torsional stress and local denaturation in supercoiled DNA, *Proc. Natl. Acad. Sci. U.S.A.,* 76, 3870, 1979.

83. **Benham, C. J.,** The equilibrium statistical mechanics of the helix-coil transition in torsionally stressed DNA, *J. Chem. Phys.,* 72, 3633, 1980.

84. **Peck, L. J. and Wang, J. C.,** Energetics of B-to-Z transition in DNA, *Proc. Natl. Acad. Sci. U.S.A.,* 80, 6206, 1983.

85. **Benham, C. J.,** Theoretical analysis of transitions between B- and Z-conformations in torsionally stressed DNA, *Nature (London),* 286, 637, 1980.

86. **Mirkin, S. M., Lyamichev, V. I., Kumarev, V. P., Kobzev, V. F., Nosikov, V. V., and Vologodskii, A. V.,** The energetics of the B-Z transition in DNA, *J. Biomol. Struct. Dynam.,* 5, 79, 1987.

87. **Sen, S. and Majumdar, R.,** Thermodynamics of B-Z transition in supercoiled DNA, *Nucleic Acids Res.,* 15, 5863, 1987.

88. **Benham, C.,** Stable cruciform formation at inverted repeat sequences in supercoiled DNA, *Biopolymers,* 21, 679, 1982.

89. **Vologodskii, A. F. and Frank-Kamenetskii, M. D.,** Theoretical study of cruciform states in superhelical DNAs, *FEBS. Lett.,* 143, 257, 1982.

90. **Kelleher, R. J., III, Ellison, M. J. U., Ho, P. S., and Rich, A.,** Competitive behavior of multiple, discrete B-Z transitions in supercoiled DNA, *Proc. Natl. Acad. Sci. U.S.A.,* 83, 6342, 1986.

91. **Benham, C. J.,** Statistical mechanical analysis of competing conformational transitions in superhelical DNA, *Cold Spring Harbor Symp. Quant. Biol.,* 47, 219, 1982.

92. **Vologodskii, A. V. and Frank-Kamenetskii, M. D.,** Left-handed Z form in superhelical DNA: a theoretical study, *J. Biomol. Struct. Dyn.,* 1, 1325, 1984.

93. **Hsieh, T. S. and Wang, J. C.,** Thermodynamic properties of superhelical DNAs, *Biochemistry,* 14, 527, 1975.

94. **Shore, D. and Baldwin, R. L.,** Energetics of DNA twisting. I. Relation between twist and cyclization probability, *J. Mol. Biol.,* 170, 957, 1983.

95. **Seidl, A. and Hinz, H.-J.,** The free energy of DNA supercoiling is enthalpy-determined, *Proc. Natl. Acad. Sci. U.S.A.,* 81, 1312, 1984.

96. **Lee, C. H., Mizusawa, H., and Kakefuda, T.,** Unwinding of double-stranded DNA helix by dehydration, *Proc. Natl. Acad. Sci. U.S.A.,* 78, 2838, 1981.

97. **Wang, J. C.,** Interaction between DNA and an *Escherichia coli* protein omega, *J. Mol. Biol.,* 55, 523, 1971.

98. **Gagus, A. V., Belintsev, V. N., and Lyubchenko, Yu. L.,** Effect of base-pair stability on the melting of superhelical DNA, *Nature (London),* 294, 662, 1981.

99. **Gilham, P. T.,** An addition reaction specific for uridine and guanosine nucleotides and its application to the modification of ribonuclease action, *J. Am. Chem. Soc.,* 84, 687, 1962.

100. **Lebowitz, J., Garon, C. G., Chen, M. C. Y., and Salzman, N. P.,** Chemical modification of simian virus 40 DNA by reaction with a water-soluble carbodiimide, *J. Virol.,* 18, 205, 1976.

101. **Lebowitz, J., Chaudhuri, A. K., Gonenne, A., and Kitos, G.,** Carbodiimide modification of superhelical PM2 DNA: considerations regarding reaction at unpaired bases and the unwinding of superhelical DNA with chemical probes, *Nucleic Acids Res.,* 4, 1695, 1977.

102. **Pulleyblank, D. E. and Morgan, A. R.,** The sense of naturally occurring superhelices and the unwinding angle of intercalated ethidium, *J. Mol. Biol.,* 91, 1, 1975.

103. **Beerman, T. A. and Lebowitz, J.,** Further analysis of the altered secondary structure of superhelical DNA. Sensitivity to methylmercuric hydroxide, a chemical probe for unpaired bases, *J. Mol. Biol.,* 79, 451, 1973.

104. **Woodworth-Gutai, M. and Lebowitz, J.,** Introduction of interrupted secondary structure in supercoiled DNA as a function of superhelix density: consideration of hairpin structures in superhelical DNA, *J. Virol.,* 18, 195, 1976.

105. **Dean, W. W. and Lebowitz, J.,** Partial alteration of secondary structure in native superhelical DNA, *Nature New Biol.,* 231, 5, 1971.

106. **Jacob, R. J., Lebowitz, J., and Kleinschmidt, A. K.,** Locating interrupted hydrogen bonding in the secondary structure of PM2 circular DNA by comparative denaturation mapping, *J. Virol.,* 13, 1176, 1974.

107. **Wells, R. D.,** Unusual DNA structures, *J. Biol. Chem.,* 263, 1095, 1988.

108. **Rich, A., Nordheim, A., and Wang, A. H.-J.,** The chemistry and biology of left-handed Z-DNA, *Annu. Rev. Biochem.,* 53, 791, 1984.

109. **Peck, L. J., Nordheim, A., Rich, A., and Wang, J. C.,** Flipping of cloned $d(pCpG)_n$ DNA sequences from right- to left-handed helical structure by salt, Co(III), or negative supercoiling, *Proc. Natl. Acad. Sci. U.S.A.,* 79, 4560, 1982.

110. **Stirdivant, S. M., Klysik, J., and Wells, R. D.,** Energetic and structural inter-relationship between DNA supercoiling and the right- to left-handed Z helix transitions in recombinant plasmids, *J. Biol. Chem.,* 257, 10159, 1982.

111. **Lilley, D. M. J., Gough, G. W., Hallam, L. R., and Sullivan, K. M.,** The physical chemistry of cruciform structures in supercoiled DNA molecules, *Biochimie,* 67, 697, 1985.

112. **Lilley, D. M. J., Sullivan, K. M., Murchie, A. I. H., and Furlong, J. C.,** Cruciform extrusion in supercoiled DNA — mechanisms and contextual influence, in *Unusual DNA Structures,* Wells, R. D. and Harvey, S. C., Eds., Springer-Verlag, New York, 1988, 55.

113. **Shure, M., Pulleyblank, D. E., and Vinograd, J.,** The problems of eucaryotic and procaryotic DNA packaging and in vivo conformation posed by superhelix density heterogeneity, *Nucleic Acids Res.,* 4, 1183, 1977.

114. **Courey, A. J. and Wang, J. C.,** Cruciform formation in a negatively supercoiled DNA may be kinetically forbidden under physiological conditions, *Cell,* 33, 817, 1983.

115. **Gellert, M., O'Dea, M. H., and Mizuuchi, K.,** Slow cruciform transitions in palindromic DNA, *Proc. Natl. Acad. Sci. U.S.A.,* 80, 5545, 1983.

116. **Sullivan, K. M. and Lilley, D. M. J.,** A dominant influence of flanking sequences on a local structural transition in DNA, *Cell,* 47, 817, 1986.

117. **Murchie, A. I. H. and Lilley, D. M. J.,** The mechanism of cruciform formation in supercoiled DNA: initial opening of central base pairs in salt-dependent extrusion, *Nucleic Acids Res.,* 15, 9641, 1987.

118. **Sullivan, K. M. and Lilley, D. M. J.,** Helix stability and the mechanism of cruciform extrusion in supercoiled DNA molecules, *Nucleic Acids Res.,* 16, 1079, 1988.

119. **Davidson, N.,** Effect of DNA length on the free energy of binding of an unwinding ligand to a supercoiled DNA, *J. Mol. Biol.,* 66, 307, 1972.

120. **Waring, M.,** Variation of the supercoils in closed circular DNA by binding of antibiotics and drugs: evidence for molecular models involving intercalation, *J. Mol. Biol.,* 54, 247, 1970.

121. **Saucier, J. M. and Wang, J. C.,** Angular alteration of the DNA helix by *E. coli* RNA polymerase, *Nature New Biol.,* 239, 167, 1972.

122. **Wang, J. C., Jacobsen, J. H., and Saucier, J. M.,** Physicochemical studies on interactions between DNA and RNA polymerase. Unwinding of the DNA helix by *Escherichia coli* RNA polymerase, *Nucleic Acids Res.,* 4, 1225, 1977.

123. **Beard, P., Hughes, M., Nyfeler, D., and Hoey, M.,** Unwinding of the DNA helix in simian virus 40 chromosome templates by RNA polymerase, *Eur. J. Biochem.,* 143, 39, 1984.

124. **Wang, J. C., Barkley, M. D., and Bourgeois, S.,** Measurements of unwinding of lac operator by repressor, *Nature (London),* 251, 247, 1974.

125. **Fuller, W. and Waring, M. J.,** A molecular model for the interacting of ethidium bromide with deoxyribonucleic acid, *Ber. Bunsenges. Phys. Chem.,* 68, 805, 1964.

126. **Cohen, G. L., Bauer, W. R., Barton, J. K., and Lippard, S. J.,** Binding of cis- and trans-dichlorodiammineplatinum(II) to DNA: evidence for unwinding and shortening of the double helix, *Science,* 203, 1014, 1979.

127. **Simpson, R. T., Thoma, F., and Brubaker, J. M.,** Chromatin reconstituted from tandemly repeated cloned DNA fragments and core histones: a model system for study of higher order structure, *Cell,* 42, 799, 1985.

128. **Kasmatsu, H., Robberson, D. L., and Vinograd, J.,** A novel closed-circular mitochondrial DNA with properties of a replicating intermediate, *Proc. Natl. Acad. Sci. U.S.A.,* 68, 2252, 1971.

129. **Eason, R. and Vinograd, J.,** Superhelix density heterogeneity of intracellular simian virus 40 deoxyribonucleic acid, *J. Virol.,* 7, 1, 1971.

130. **Hudson, B., Upholt, W. B., Devinny, J., and Vinograd, J.,** The use of an ethidium analogue in the dye-buoyant density procedure for the isolation of closed circular DNA: the variation of the superhelix density of mitochondrial DNA, *Proc. Natl. Acad. Sci. U.S.A.,* 62, 813, 1969.

131. **Lindahl, T., Adams, A., Bjursell, G., Bornkamm, G. W., Kaschka Dierich, C., and Jehn, U.,** Covalently closed circular duplex DNA of Epstein-Barr virus in a human lymphoid cell line, *J. Mol. Biol.,* 102, 511, 1976.

132. **Bujard, H.,** Studies on circular DNA. II. Number of tertiary turns in papilloma DNA, *J. Mol. Biol.,* 33, 503, 1968.

133. **Dougherty, G. and Koller, T.,** Determination of the number of superhelical turns by the hyperchromicity of partially denatured covalently-closed DNA molecules, *Nucleic Acids Res.,* 10, 525, 1982.

134. **Ruttenberg, G. J., Smit, E. M., Borst, P., and van Bruggen, E. G.,** The number of superhelical turns in mitochondrial DNA, *Biochim. Biophys. Acta,* 157, 429, 1968.

135. **Smith, C. A. and Vinograd, J.,** Small polydisperse circular DNA of HeLa cells, *J. Mol. Biol.,* 69, 163, 1972.

136. **Espejo, R., Espejo-Canelo, E., and Sinsheimer, R. L.,** A difference between intracellular and viral supercoiled PM2 DNA, *J. Mol. Biol.,* 56, 623, 1971.

137. **Nordheim, A., Lafer, E. M., Peck, L. J., Wang, J. C., Stollar, B. D., and Rich, A.,** Negatively supercoiled plasmids contain left-handed Z-DNA segments as detected by specific antibody binding, *Cell,* 31, 309, 1982.

138. **Runkel, L. and Nordeim, A.,** Conformational DNA transition in the *in vitro* torsionally strained chicken β-globin 5' region, *Nucleic Acids Res.,* 14, 7143, 1986.

139. **Nordheim, A., Peck, L. J., Lafer, E. M., Stollar, B. D., Wang, J. C., and Rich, A.,** Supercoiling and left-handed Z-DNA, *Cold Spring Harbor Symp. Quant. Biol.,* 47, 93, 1983.

140. **Singleton, C. K., Klysik, J., and Wells, R. D.,** Conformational flexibility of junctions between contiguous B- and Z-DNAs in supercoiled plasmids, *Proc. Natl. Acad. Sci. U.S.A.,* 80, 2447, 1983.

141. **O'Connor, T. R., Kang, D. S., and Wells, R. D.,** Thermodynamic parameters are sequence-dependent for the supercoil-induced B to Z transition in recombinant plasmids, *J. Biol. Chem.,* 261, 13302, 1986.

142. **Nordheim, F. and Rich, A.,** The sequence $(dC-dA)_n \cdot (dG-dT)_n$ forms left-handed Z-DNA in negatively supercoiled plasmids, *Proc. Natl. Acad. Sci. U.S.A.,* 80, 1821, 1983.

143. **Haniford, D. B. and Pulleyblank, D. E.,** Facile transition of $poly[d(TG) \times d(CA)]$ into a left-handed helix in physiological conditions, *Nature (London),* 302, 632, 1983.

144. **McLean, M. J., Blaho, J. A., Kilpatrick M. W., and Wells, R. D.,** Consecutive A·T pairs can adopt a left-handed DNA structure, *Proc. Natl. Acad. Sci. U.S.A.,* 83, 5884, 1986.

145. **Singleton, C. K., Klysik, J., Stirdivant, S. M., and Wells, R. D.,** Left-handed Z-DNA is induced by supercoiling in physiological ionic conditions, *Nature (London),* 299, 312, 1982.

146. **Haniford, D. B. and Pulleyblank, D. E.,** The *in vivo* occurrence of Z DNA, *J. Biomol. Struct. Dyn.*, 1, 593, 1983.

147. **Singleton, C. K., Kilpatrick, M. W., and Wells, R. D.,** S1 nuclease recognized DNA conformational junctions between left-handed helical $(dT-dG)_n \cdot (dC-dA)_n$ and contiguous right-handed sequences, *J. Biol. Chem.*, 259, 1963, 1984.

147a. **Herr, W.,** Diethylpyrocarbonate: a chemical probe for secondary structure in negatively supercoiled DNA, *Proc. Natl. Acad. Sci. U.S.A.*, 82, 8009, 1985.

148. **Singleton, C. K.,** Effects of salts, temperature, and stem length on supercoil-induced formation of cruciforms, *J. Biol. Chem.*, 258, 7661, 1983.

149. **Sinden, R. R., Broyles, S. S., and Pettijohn, D. E.,** Perfect palindromic lac operator DNA sequence exists as a stable cruciform structure in supercoiled DNA *in vitro* but not *in vivo*, *Proc. Natl. Acad. Sci. U.S.A.*, 80, 1797, 1983.

150. **Singleton, D. R. and Wells, R. W.,** Relationship between superhelical density and cruciform formation in plasmid pVH51, *J. Biol. Chem.*, 257, 6292, 1982.

151. **Lymichev, V. I., Panyutin, I. G., and Frank-Kamenetskii, M. D.,** Evidence of cruciform structures in superhelical DNA provided by two-dimensional gel electrophoresis, *FEBS Lett.*, 153, 298, 1983.

152. **Lilley, D. M. and Hallam, L. R.,** Thermodynamics of the ColE1 cruciform. Comparisons between probing and topological experiments using single topoisomers, *J. Mol. Biol.*, 180, 179, 1984.

153. **Dickie, P., Morgan, A. R., and McFadden, G.,** Cruciform extrusion in plasmids bearing the replicative intermediate configuration of a poxvirus telomere, *J. Mol. Biol.*, 196, 541, 1987.

154. **Naylor, L. H., Lilley, D. M. J., and van de Sande, J. H.,** Stress-induced cruciform formation in a cloned $d(CATG)_{10}$ sequence, *EMBO J.*, 5, 2407, 1986.

155. **Haniford, D. B. and Pulleyblank, D. E.,** Transition of a cloned $d(AT)_n$-$d(AT)_n$ tract to a cruciform *in vivo*, *Nucleic Acids Res.*, 13, 4343, 1985.

156. **Greaves, D. R., Patient, R. K., and Lilley, D. M. J.,** Facile cruciform formation by an $(A-T)_{34}$ sequence from a *Xenopus* globin gene, *J. Mol. Biol.*, 185, 461, 1985.

157. **Tsai, C.-C., Jain, S. C., and Sobell, H. M.,** Visualization of drug-nucleic acid interactions at atomic resolution. Structure of an ethidium/dinucleoside monophosphate crystalline complex, ethidium:5-iodouridylyl (3'-5') adenosine, *J. Mol. Biol.*, 114, 301, 1977.

158. **Jain, S. C., Tsai, C.-C., and Gilbert, S. G.,** Visualization of drug-nucleic acid interactions at atomic resolution. Structure of an ethidium/dinucleoside monophosphate crystalline complex, ethidium:5-iodouridylyl (3'-5') quanosine, *J. Mol. Biol.*, 114, 317, 1977.

159. **Sobell, H. M., Tsai, C. C., Jain, S. C., and Gilbert, S. G.,** Visualization of drug-nucleic acid interactions at atomic resolution. III. Unifying structural concepts in understanding drug-DNA interactions and their broader implications in understanding protein-DNA interactions, *J. Mol. Biol.*, 114, 333, 1977.

160. **Wang, H.-J., Quigley, G. J., and Rich, A.,** Atomic resolution analysis of a 2:1 complex of CpC and acridine orange, *Nucleic Acids Res.*, 6, 3879, 1978.

161. **Allison, R. G. and Hahn, F. E.,** Changes in superhelical density of closed circular deoxyribonucleic acid by intercalation of anti-R-plasmid drugs and primaquine, *Antimicrob. Agents Chemother.*, 11, 251, 1977.

162. **Gaugain, B., Markovits, J., LePecq, J. B., and Roques, B. P.,** DNA polyintercalation: comparison of DNA binding properties of an acridine dimer and trimer, *FEBS Lett.*, 169, 123, 1984.

163. **Huang, C. H., Prestayko, A. W., and Crooke, S. T.,** Bifunctional intercalation of antitumor antibiotics BBM-928A and echinomycin with deoxyribonucleic acid. Effects of intercalation on deoxyribonucleic acid degradative activity of bleomycin and phleomycin, *Biochemistry*, 21, 3704, 1982.

164. **Denny, W. A., Atwell, G. J., Willmott, G. A., and Wakelin, L. P.,** Interaction of paired homologous series of diacridines and triacridines with deoxyribonucleic acid, *Biophys. Chem.*, 22, 17, 1985.

165. **McFadyen, W. D., Wakelin, L. P., Roos, I. A., and Hillcoat, B. L.,** Binuclear platinum (II)-terpyridine complexes. A new class of bifunctional DNA-intercalating agent, *Biochem. J.*, 238, 757, 1986.

166. **Wakelin, S. P. and Waring, M. J.,** The binding of echinomycin to deoxyribonucleic acid, *Biochem. J.*, 157, 721, 1976.

167. **Denny, W. A. and Wakelin, L. P.,** Kinetic and equilibrium studies of the interaction of amsacrine and anilino ring-substituted analogues with DNA, *Cancer Res.*, 46, 1717, 1986.

168. **Baguley, B. C., Ferguson, L. R., and Denny, W. A.,** DNA binding and growth inhibitory properties of a series of 2,7-di-alkyl-substituted derivatives of proflavine, *Chem. Biol. Interact.*, 42, 97, 1982.

169. **Pachter, J. A., Huang, C. H., DuVernay, V. H., Jr., Prestayko, A. W., and Crooke, S. T.,** Viscometric and fluorometric studies of deoxyribonucleic acid interactions of several new anthracyclines, *Biochemistry*, 21, 1541, 1982.

170. **Waring, M. J. and Henley, S. M.,** Stereochemical aspects of the interaction between steroidal diamines and DNA, *Nucleic Acids Res.*, 2, 567, 1975.

171. **Wakelin, L. P. G., Romanos, E. S., Canellakis, E. S., and Waring, M. J.,** Diacridines as bifunctional DNA-intercalating agents, *Stud. Biophys.*, 60, 111, 1976.

172. **Lee, J. S. and Waring, M. J.,** Bifunctional intercalation and sequence specificity in the binding of quinomycin and triostin antibiotics to deoxyribonucleic acid, *Biochem. J.,* 173, 115, 1978.

173. **Wang, J. C.,** Unwinding of DNA by actinomycin D binding, *Biochim. Biophys. Acta,* 232, 246, 1971.

174. **Jennette, K. W., Lippard, S. J., Vassiliades, G. A., and Bauer, W. R.,** Metallointercalation reagents. 2-hydroxyethanethiolato(2,2′,2″-terpyridine)-platinum(II) monocation binds strongly to DNA by intercalation, *Proc. Natl. Acad. Sci. U.S.A.,* 71, 3839, 1974.

175. **Dougherty, G. and Waring, M. J.,** The interaction between prothidium dibromide and DNA at the molecular level, *Biophys. Chem.,* 15, 27, 1982.

176. **Kohn, K. W., Waring, M. J., Glaubiger, D., and Friedman, C. A.,** Intercalative binding of ellipticine to DNA, *Cancer Res.,* 35, 71, 1975.

177. **Baez, A., Gonzalez, F. A., Vazquez, D., and Waring, M. J.,** Interaction between a 3-nitrobenzothiazolo (3,2-a) quinolinium antitumour drug and deoxyribonucleic acid, *Biochem. Pharmacol.,* 32, 2089, 1983.

178. **Povirk, L. F., Hogan, M., and Dattagupta, N.,** Binding of bleomycin to DNA: intercalation of the bithiazole rings, *Biochemistry,* 18, 96, 1979.

179. **Muller, W., Crothers, D. M., and Waring, M. J.,** A non-intercalating proflavine derivative, *Eur. J. Biochem.,* 39, 223, 1973.

180. **Fisher, L. M., Kuroda, R., and Sakai, T. T.,** Interaction of bleomycin A2 with deoxyribonucleic acid: DNA unwinding and inhibition of bleomycin-induced DNA breakage by cationic thiazole amides related to bleomycin A2, *Biochemistry,* 24, 3199, 1985.

180a. **Hansen, J., Pschorn, W., and Ristow, H.,** Functions of the peptide antibiotics tyrocidine and gramicidin. Induction of conformational and structural changes of superhelical DNA, *Eur. J. Biochem.,* 126, 297, 1982.

181. **Cohen, G. L., Bauer, W. R., Barton, J. K., and Lippard, S. J.,** Binding of cis- and trans-dichloro-diammineplatinum(II) to DNA: evidence for unwinding and shortening of the double helix, *Science,* 203, 1014, 1979.

182. **Mong, S., Huang, C. H., Prestayko, A. W., and Crooke, S. T.,** Interaction of cis-diamminedichloro-platinum(II) with PM-2 DNA, *Cancer Res.,* 40, 3313, 1980.

183. **McFadyen, W. D., Wakelin, L. P., Roos, I. A., and Hillcoat, B. L.,** Binuclear platinum (II)-terpyridine complexes. A new class of bifunctional DNA-intercalating agent, *Biochem. J.,* 238, 757, 1986.

184. **Braithwaite, A. W. and Baguley, B. C.,** Existence of an extended series of antitumor compounds which bind to deoxyribonucleic acid by nonintercalative means, *Biochemistry,* 19, 1101, 1980.

185. **Denhardt, D. T. and Kato, A. C.,** Comparison of the effect of ultraviolet radiation and ethidium bromide intercalation on the conformation of superhelical phiX174 replicative form DNA, *J. Mol. Biol.,* 77, 479, 1973.

186. **Gamper, H. B., Straub, K., Calvin, M., and Bartholomew, J. C.,** DNA alkylation and unwinding induced by benzo[a]pyrene diol epoxide: modulation by ionic strength and superhelicity, *Proc. Natl. Acad. Sci. U.S.A.,* 77, 2000, 1980.

187. **Lloyd, R. S., Robberson, D. L., and Haidle, C. W.,** Bleomycin-mediated DNA cross-links are dependent on closed-circular molecules with superhelical turns, *Chem. Biol. Interact.,* 34, 39, 1981.

188. **Malvy, C. and Cros, S.,** Interaction between ellipticine derivatives and circular supercoiled DNA as revealed by gel electrophoresis. Possible relationship with the mechanisms of cytotoxicity, *Biochem. Pharmacol.,* 35, 2264, 1986.

189. **Meehan, T., Gamper, H., and Becker, J. F.,** Characterization of reversible, physical binding of benzo[a]pyrene derivatives to DNA, *J. Biol. Chem.,* 257, 10479, 1982.

190. **Mirabelli, C. K., Sung, C. M., Zimmerman, J. P., Hill, D. T., Mong, S., and Crooke, S. T.,** Interactions of gold coordination complexes with DNA, *Biochem. Pharmacol.,* 35, 1427, 1986.

191. **Mirabelli, C. K., Zimmerman, J. P., Bartus, H. R., Sung, C. M., and Crooke, S. T.,** Inter-strand cross-links and single-strand breaks produced by gold(I) and gold(III) coordination complexes, *Biochem. Pharmacol.,* 35, 1435, 1986.

192. **Mong, S., DuVernay, V. H., Strong, J. E., and Crooke, S. T.,** Interaction of anthracyclines with covalently closed circular DNA, *Mol. Pharmacol.,* 17, 100, 1980.

193. **Mong, S., Strong, J. E., and Crooke, S. T.,** Interaction of covalently closed circular PM-2 DNA and hedamycin, *Biochem. Biophys. Res. Commun.,* 88, 237, 1979.

194. **Richardson, R. J., II and Imamura, T.,** Interaction of O,O,S-trimethyl phosphorothioate and O,S,S-trimethyl phosphorodithioate, the impurities of malathion with supercoiled PM2 DNA, *Biochem. Biophys. Res. Commun.,* 126, 1251, 1985.

195. **Vadi, H. V., Schasteen, C. S., and Reed, D. J.,** Interactions of S-(2-haloethyl)-mercapturic acid analogs with plasmid DNA, *Toxicol. Appl. Pharmacol.,* 80, 386, 1985.

196. **Gamper, H. B. and Hearst, J. E.,** A topological model for transcription based on unwinding angle analysis of *E. coli* RNA polymerase binary, initiation and ternary complexes, *Cell,* 29, 81, 1982.

197. **Bertrand-Burggraf, E., Schnarr, M., Lefevre, J. F., and Daune, M.,** Effect of superhelicity on the transcription from the tet promoter of pBR322. Abortive initiation and unwinding experiments, *Nucleic Acids Res.,* 12, 7741, 1984.

198. **Amouyal, M. and Buc, H.,** Topological unwinding of strong and weak promoters by RNA polymerase. A comparison between the lac wild-type and the UV5 sites of *Escherichia coli, J. Mol. Biol.,* 195, 795, 1987.

199. **Reynolds, W. F. and Gottesfeld, J. M.,** 5S RNA gene transcription factor IIIA alters the helical configuration of DNA, *Proc. Natl. Acad. Sci. U.S.A.,* 80, 1862, 1983.

200. **Hanas, J. S., Bogenhagen, D. F., and Wu, C.-W.,** DNA unwinding ability of *Xenopus* transcription factor A, *Nucleic Acids Res.,* 12, 1265, 1984.

201. **Shastry, B. S.,** 5S RNA gene transcription factor (TFIIIA) changes the linking number of the DNA, *Biochem. Biophys. Res. Commun.,* 134, 1086, 1986.

202. **Liu, L. F. and Wang, J. C.,** *Micrococcus luteus,* DNA gyrase: active components and a model for its supercoiling of DNA, *Proc. Natl. Acad. Sci. U.S.A.,* 75, 2098, 1978.

203. **Whitson, P. A., Hsieh, W. T., Wells, R. D., and Matthews, K. S.,** Supercoiling facilitates lac operator-repressor-pseudooperator interactions, *J. Biol. Chem.,* 262, 4943, 1987.

204. **Kolb, A. and Buc, H.,** Is DNA unwound by the cyclic AMP receptor protein?, *Nucleic Acids Res.,* 10, 473, 1982.

Chapter 4

STRUCTURE OF THE 300-Å CHROMATIN FILAMENT

J. Widom

TABLE OF CONTENTS

I. INTRODUCTION

Each chromosome of a eukaryotic cell contains a single molecule of double-stranded DNA which has a contour length very much greater than the diameter of the cell nucleus. For example, an average human chromosome contains a DNA molecule with a contour length of centimeters! The chromosomal DNA is folded in a hierarchical series of steps by a family of highly conserved proteins that eventually produce a ~10,000-fold compaction of the DNA at metaphase.

In the first step of folding, two copies each of four different proteins (the "core histones" H2A, H2B, H3, and H4) aggregate together to form a compact globular core that has the physicochemical property of causing ~145 bp of DNA to wrap in 1 3/4 turns on its outer surface.[1,2] The particle that results is called a nucleosome core particle; it has the shape of a flat disk, with a diameter of ~110 Å and a thickness of ~57 Å. The structure of the nucleosome core particle is currently known at 7-Å resolution from crystallographic studies;[3] it can be expected that this will presently be extended to higher resolution. The location within the core particle of each of the proteins has been deduced.[3,4]

One molecule of a fifth histone protein (H1 or H5)[5] combines with the octamer of core histones to package an extra 10 bp at each end of the core particle DNA. This particle therefore contains 165 bp of DNA, presumably in two turns, and is called a chromatosome.[6] This packing motif is repeated at intervals along the full length of a chromosomal DNA molecule, with consecutive chromatosomes separated by a short distance of nonnucleosomal DNA called "linker DNA". (A chromatosome plus its linker DNA is called a nucleosome.) Chromatin can therefore be thought of as a long string of nucleosomes.

Each cell type has a characteristic average value for the length of linker DNA;[7] this can range from ~0 bp (in yeast[8] and neuronal chromatin[9]) to ~80 bp (in sea urchin sperm chromatin[10]). When looked at in detail, it is found that despite having a well-characterized average value, the actual linker length varies significantly from nucleosome to nucleosome within any given cell type.[11]

Throughout most of the cell cycle, eukaryotic chromosomes are maintained in an intermediate level of folding in which the strings of nucleosomes are folded in the form of filaments with a diameter of ~300 Å.[12-14] It is believed that the 300-Å filament is locally unfolded to allow the passage of RNA and DNA polymerases, and that it is further folded to produce the compact forms required for mitosis and meiosis.

Electron micrographs show that chromatin *in vitro* can be interconverted between several, apparently distinct folded states by adjustment of ionic conditions. These states include (1) an "unfolded" state, in which chromatin appears in electron micrographs as a continuous 100 Å nucleosome filament[15] or as a zig-zag chain of closely opposed nucleosomes,[16] (2) a "folded" state, in which strings of nucleosomes are folded intramolecularly into ~300 Å-wide filaments that resemble the chromatin filaments in interphase nuclei,[15,16] and (3) aggregates of 300-Å filaments that have certain features in common with chromatin packed in metaphase chromosomes, meiotic synaptonemal complexes, and in histone-containing sperm.[17]

It is of interest to determine the structures of each of these folded states of chromatin. From physical studies of chromatin *in vitro*, one can hope to gain insights into possible mechanisms by which chromosome structure might be regulated *in vivo*. Ultimately, one wishes to understand how chromosome structure and its regulation satisfy the functional requirements of gene expression, replication, recombination, and cell division.

These topics are the subject of a number of recent reviews,[18-24] to which the reader is referred for a general account. The present review will focus primarily on two points: (1) under what conditions does chromatin form native-like 300-Å filaments *in vitro*? — along with a discussion of the underlying physical chemistry — and (2) what is the structure of chromatin in the 300-Å filament state?

II. CHROMATIN FOLDING *IN VITRO*

A. ISOLATION OF NATIVE-LIKE 300-Å FILAMENTS

Chromatin suitable for physical and structural studies is typically isolated from purified cell nuclei using a procedure having two key steps.[25] First, the extremely long chromosomal DNA molecules are randomly digested *in situ* to the desired average size by a nuclease which preferentially attacks the linker DNA. Then, the nuclei are lysed and nucleosome oligomers are released into solution and separated from residual nuclear debris.

It is found that the yield of soluble chromatin is much greater if the nuclei are lysed in buffers at low ionic strength and in the absence of higher-valent cations. In these conditions, chromatin is in the unfolded ("100-Å filament") state[15,16] and so it must be refolded for studies of 300-Å filaments. This may be accomplished by the addition of appropriate concentrations of any of a variety of cations to the purified unfolded chromatin.

It is an important question whether such refolded 300-Å filaments are similar in structure (and not just in appearance) to the 300-Å filaments of chromatin *in vivo*; two lines of evidence (each of which will be discussed in more detail in a later section) suggest that they are. First, low-angle X-ray diffraction patterns from 300-Å filaments *in vitro*[17,26-31] show the same set of bands characteristic of chromatin in nuclei and in intact living cells.[32-35] This suggests that not only are 300-Å filaments *in vitro* similar in appearance to those *in vivo*, but that they also have a similar internal structure. The second line of evidence comes from studies of chromatin that has been prepared under conditions in which the 300-Å filaments should remain folded (as determined in studies of refolded chromatin). Electron microscopic[16] and hydrodynamic studies[36,37] have led to the conclusion that there are no significant differences between chromatin extracted in "folding" conditions and chromatin extracted in the unfolded state and subsequently refolded. Taken together, these data provide good evidence that *in vitro* refolding correctly restores native-like 300-Å filament structure to unfolded chromatin; it remains possible that subtle differences do exist, however, so the issue should not be forgotten.

B. CATION TITRATIONS

Before physical or structural studies of refolded 300-Å filaments are carried out, one wishes to identify conditions in which the refolding is complete. Refolding is induced by the addition of cations; therefore, one carries out titrations in which increasing concentrations of some cation are added to dilute chromatin solutions, and the resulting structure is probed by some structure-sensitive technique. Many techniques have been applied, including transmission and scanning-transmission electron microscopy, small-angle scattering of X-rays and neutrons, classical light scattering, hydrodynamic measurements of sedimentation coefficients and translational and rotational diffusion coefficients, and optical measurements of dichroism or birefringence of oriented molecules. Ideally, each of these techniques would identify similar conditions in which the further addition of cations produces no further change in chromatin folding; such conditions would then be appropriate for studies of 300-Å filaments.

Surprisingly, different physical techniques have given very different pictures of the refolding transition. It now appears that most conditions which stabilize 300-Å filaments also stabilize the aggregation of 300-Å filaments, and that samples undergo bulk precipitation before a true titration endpoint is reached. Nevertheless, one can identify ranges of conditions in which, by many criteria, the chromatin is in the form of 300-Å filaments whose structure is very similar to that of chromatin *in vivo*, and where aggregation is minimal. Chromatin in such conditions has been termed "optimally compacted".[38,39]

As one example of this behavior, let us consider the refolding of rat liver and chicken erythrocyte chromatin induced by Na[+]; it should be noted that similar results are obtained with a wide range of cations. Electron microscopy,[16,40] hydrodynamics,[36,37,40,41] and optical (electric dichroism)[37] studies suggest that the structure of the two chromatin types is very similar throughout the range of Na$^+$ concentrations, and so for the purposes of this discussion they may be considered to be equivalent. (A quantitative difference between the two chromatin types has been reported: folded chicken erythrocyte chromatin is found to be slightly more resistant to hydrodynamic shear at intermediate Na$^+$ concentrations.)[40]

Electron microscopy shows that the chromatin folds through a continuum of increasingly compact structures as the Na$^+$ concentration is increased, until the Na$^+$ concentration reaches ~60 mM. At this point, the chromatin appears as native-like ~300 Å-wide filaments, and further increases in the concentration of Na$^+$ appear not to induce any further changes in structure.[16] Similarly, small-angle neutron-scattering measurements show that the mass per unit length of chromatin (a quantitative measure of compaction) increases with increasing Na$^+$ concentration until reaching a limiting value at ~80 mM Na$^+$.[42] A number of optical experiments show limiting values of dichroism or birefringence, again indicating titration endpoints.[43,44]

A very different picture of the same transition comes from hydrodynamic and X-ray diffraction studies. Measurements of sedimentation velocity show that S increases without any sign of a plateau up to the point of bulk precipitation.[36,40] An example of such data for a Mg^{2+} titration is given in Figure 1. The increase in sedimentation coefficient correlates with the progressive folding of chromatin observed by other techniques, but for each of these experiments, the next point in the titration caused the chromatin to precipitate. A number of control experiments appear to have ruled out incipient aggregation as a cause of the ever-increasing S.[17,36] Similarly, measurements of translational diffusion coefficients show that, with increasing Na$^+$ concentration, D goes through a maximum (at ~75 mM Na$^+$) and then begins to drop again, indicating the onset of aggregation.[45] There is no plateau region and, thus, apparently no titration endpoint before the sample precipitates. Finally, X-ray diffraction patterns from a Na$^+$ titration show that X-ray bands due to (native-like) packing of nucleosomes within 300-Å filaments continue to increase in sharpness and prominence even when the concentration of Na$^+$ is increased above the point of bulk precipitation (Figure 2).[17]

Current evidence does not define the nature of the changes in chromatin structure that take place when titrations are continued past the 300-Å filament endpoint detected by electron microscopy or other techniques listed above. Even though X-ray diffraction patterns from such titrations do continue to sharpen as the Na$^+$ concentration is increased above the 60 to 80 mM Na$^+$ endpoint, the patterns nonetheless are all similar in that they show the same set of characteristic diffraction features. These data, together with the fact that the structure does not appear to change further, as judged by electron microscopy, suggest that further changes that do take place may be subtle. It appears that titration experiments identify a range of conditions in which chromatin is folded into native-like 300-Å filaments.

There are several other implications of these findings. From the perspective of physical chemistry, chromatin is a poorly behaved sample. It is difficult to carry out proper studies of a transition which does not go reversibly between two definite endpoints. For many studies of 300-Å filaments, in which no titration endpoint exists, it may be best to use conditions of "optimal compaction", but one must bear in mind that such material is on the verge of bulk precipitation and that small aggregates may be transiently forming. The lack of a titration endpoint also makes impossible certain solution experiments designed to distinguish between two-state vs. continuous models for chromatin refolding.

C. ROLE OF CATIONS

Electron microscopy has shown that millimolar Mg^{2+} causes chromatin to fold into native-like 300-Å filaments which are indistinguishable from those induced by 60 to 80 mM Na$^+$. This result has since been confirmed by hydrodynamic, optical, and diffraction tech-

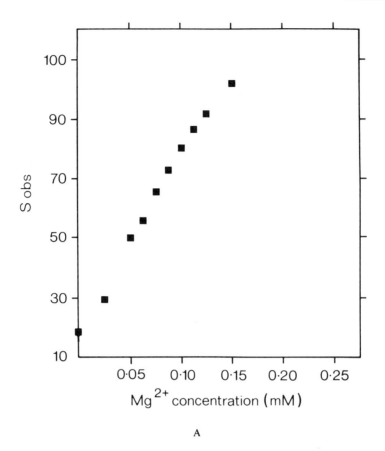

A

FIGURE 1.   Mg^{2+} titrations of chromatin monitored by sedimentation velocity, at three concentrations of Na$^+$. The chromatin is from chicken erythrocytes and is size-fractionated with an average of 75 nucleosomes. In (A) the concentration of Na$^+$ is 0.2 mM. In (B) the concentrations are 25 mM (■) and 75 mM (▲). For each titration, the next point (not shown) caused bulk precipitation; the arrow in (B) indicates a point which was possibly affected by aggregation. (From Widom, J., *J. Mol. Biol.*, 190, 411, 1986. With permission.)

niques and, more recently, it has been extended to include a large number of other multivalent cations: simple hydrated ions such as Ca^{2+}, Mn^{2+}, and Gd^{3+}; inert inorganic ions such as Co(NH$_3$)$_6$$^{3+}$; and polyamines such as spermidine (3+) and spermine (4+).[17,35,38,39] It is found that the valence of the cation is of primary importance in determining the concentration of that cation necessary to induce chromatin folding. As discussed above, monovalent cations are required at concentrations of 60 to 80 mM, divalent cations are typically effective at concentrations of 100 μM to 1 mM, trivalent cations at 10 to 100 μM, and tetravalent cations at 1 to 10 μM. The relatively wide concentration ranges given are due largely to experiments being done in the presence of a range of monovalent cation concentrations because of cation competition (see below). When the monovalent cation concentration is kept constant (and $<< 60$ mM), cations of the same valence are generally effective within a twofold concentration range. Exceptions to this rule may arise when a cation is particularly bulky or when the cation can bind nonionically to chromatin (e.g., Cu^{2+}).

These data are sufficient to define the role of cations in stabilization of the 300-Å filament.[17] *A priori,* one can list three possible mechanisms by which cations might act: (1) by binding to a particular site that has chemical selectivity, such as the metal-binding site in metalloenzymes, (2) by screening repulsions between small numbers of charges, such as

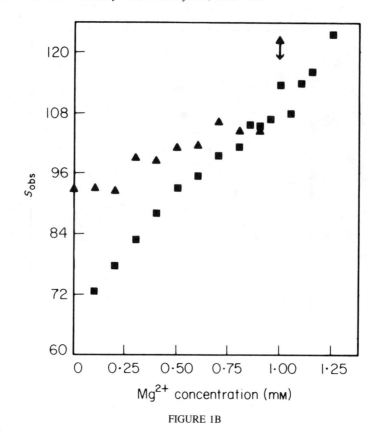

FIGURE 1B

amino acid sidechains, or (3) by acting as general DNA counterions, reducing the effective charge per phosphate and screening repulsions between adjacent DNA helices.

Model 1 is ruled out by the observed lack of sensitivity to the size or chemical nature of the cation. If model 2 were correct, chromatin folding would be governed by the ionic strength, in accord with the Debye-Hückel theory. However, it is found that the ionic strength necessary to refold chromatin decreases by three orders of magnitude as the valence of the cation increases from $+1$ to $+4$.

The "counterion condensation" theory governing model 3 has been worked out by Manning[46] and by Record and colleagues.[47] They find that the effectiveness of cations as DNA counterions depends primarily on the valence of the cation, with an effect much greater than for the Debye-Hückel theory. Such behavior is in qualitative accord with the results for chromatin refolding. Counterion-condensation theory further predicts that the concentration of a multivalent cation required to induce folding will depend on the monovalent cation concentration because of cation competition. As discussed below, this surprising prediction is verified experimentally. Thus, one concludes that cations induce chromatin refolding by reducing repulsions between DNA segments within 300-Å filaments.

D. PHASE DIAGRAM FOR CHROMATIN REFOLDING

In studies where multivalent cations are used to induce chromatin refolding, monovalent cations are often present too—sometimes in the form of a buffer (e.g., Tris $\rightleftharpoons$ Tris$^+$) or as the counterion of a buffer (e.g., phosphoric acid plus NaOH). When cations of two valences are present, titrations need to be carried out in two dimensions, varying the concentration of each cation systematically. Such experiments have been carried out most thoroughly for the Na$^+$/Mg^{2+} concentration plane.[17,38,43,45]

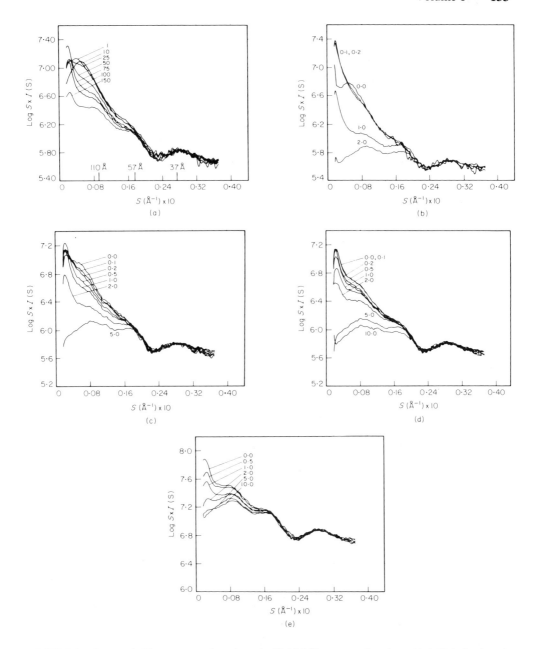

FIGURE 2.   Low angle X-ray patterns throughout the Na$^+$/Mg^{2+} concentration plane. (a) A Na$^+$ titration; the concentration of Na$^+$ for each curve is indicated. (b—e) Mg^{2+} titrations at fixed Na$^+$ concentrations: (b) 5 mM Na$^+$, (c) 25 mM Na$^+$, (d) 75 mM Na$^+$, (e) 150 mM Na$^+$. For (b—e) the Mg^{2+} concentration for each curve is indicated. I(s) is the measured intensity; S is related to the scattering angle (2θ) by the equation S = (2 sinθ)/λ, λ = 1.54 Å. (From Widom, J., *J. Mol. Biol.*, 190, 411, 1986. With permission.)

Data from electron microscopy, sedimentation velocity, and X-ray diffraction experiments are shown in Figures 1 to 3; these and other data are summarized in the form of a phase diagram in Figure 4. The electron micrographs and X-ray patterns fall naturally into three classes, designated I to III. In class I, the chromatin is unfolded or partly folded. The filament diameter (as judged from the micrographs) is quite irregular, but generally less than ~300 Å; the X-ray diffraction bands at 110 and 57 Å (which arise from the packing of nucleosomes within 300-Å filaments; see below) are weak or absent. In class II, nearly all

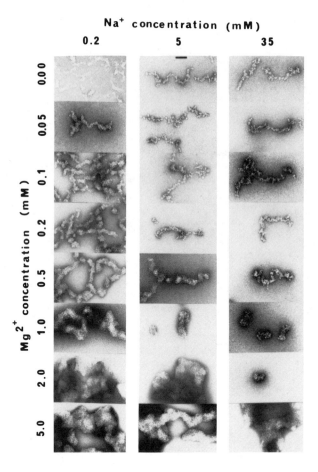

FIGURE 3.   Electron micrographs of chromatin throughout the Na^+/Mg^{2+} concentration plane. The chromatin is fixed and negatively stained. The bar represents 1000Å. (From Widom, J., *J. Mol. Biol.*, 190, 411, 1986. With permission.)

the chromatin is folded into ~300-Å-diameter filaments and the 110 and 57 Å X-ray diffraction bands are prominent. Hydrodynamic data suggest that most of the chromatin remains monomeric; aggregates visible in the corresponding electron micrographs may have arisen during adsorption or staining steps of sample preparation. Thus, class II corresponds to a region of "optimum compaction" for many physical studies. In class II_A, the chromatin appears similar in structure to optimally compacted chromatin, as judged by electron microscopy and X-ray diffraction; however, chromatin in these conditions is not soluble and precipitates even from very dilute solution. In class III, the chromatin is also insoluble. However, here the 300-Å filaments are very tightly aggregated so that electron density contrast between individual filaments is lost. As expected from such micrographs, the 300 to 500 Å X-ray diffraction band (due to the side-by-side packing of 300-Å filaments; see below) disappears and the very low angle scattering in general is greatly reduced (i.e., shifted to behind the beamstop).

Since chromatin folding may be a continuous process,[16,35,42] this is not a true phase diagram and the boundaries represent subjective divisions. Nevertheless, such a representation is useful for several reasons. It illustrates regions where physical studies of 300-Å filaments are best carried out; it reveals a striking correspondence between the behavior of 300-Å filaments *in vitro* and in metaphase chromosomes; and, as discussed below, it shows that the physical chemistry of cation competition dominates the ionic effects on chromatin

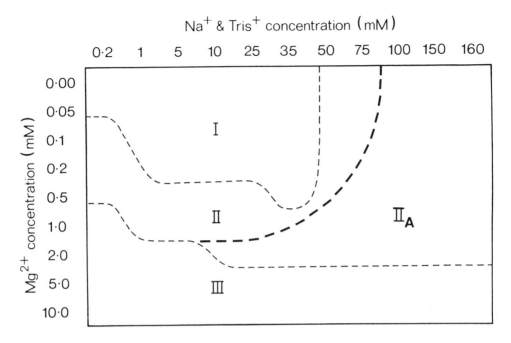

FIGURE 4.   Phase diagram for chromatin folding in the Na^+/Mg^{2+} concentration plane. Phase "boundaries" are approximate and are subjectively chosen. In region I, chromatin is unfolded or partly folded. In region II and II_A, chromatin is in the 300-Å filament state; in region II_A, these are insoluble. In region III, the 300-Å filaments are packed tightly together so that electron density contrast is lost. (Adapted from Widom, J., *J. Mol. Biol.*, 190, 411, 1986. With permission.)

folding. Finally, these data were of key importance in deciphering X-ray diffraction patterns from oriented chromatin samples.

E. CATION COMPETITION

The results of studies of chromatin folding throughout the Na^+/Mg^{2+} concentration plane, summarized in Figure 4, lead to the conclusion that Na^+ (and presumably any other monovalent cation) has dual effects: sufficiently high concentrtions of Na^+ ($\sim$ 60 to 80 m*M*) suffice to fold chromatin into native-like 300-Å filaments even in the absence of Mg^{2+}; however, at low concentrations of Na^+ (less than $\sim$45 m*M*), when Mg^{2+} (or some other multivalent cation) is present and stabilizing the 300-Å filament state, Na^+ has the opposite effect: it competes with the higher valence cation for binding to the chromatin and destabilizes the 300-Å filament state. Cation competition appears to underlie chromatin folding, conditions of optimal compaction, and the aggregation of 300-Å filaments.

The competitive effects of Na^+ and Mg^{2+} on the aggregation of 300-Å filaments are plainly evident in studies of chromatin solubility.[38,45] In the absence of any Mg^{2+}, an increase in the concentration of Na^+ above $\sim$100 m*M* causes the 300-Å filaments to precipitate even from very dilute chromatin solutions. In very low concentrations (e.g., 1 m*M*) of Na^+, increasing the concentration of Mg^{2+} above $\sim$0.6 to 0.8 m*M* causes 300-Å filaments to precipitate. If the effects of Na^+ and Mg^{2+} were additive, then increasing the concentration of Na^+ to, say, 40 m*M* would have the effect of decreasing the concentration of Mg^{2+} required to precipitate the chromatin. However, the opposite is found experimentally. In the presence of 40 m*M* Na^+ the concentration of Mg^{2+} required is increased, to 1.6 to 1.8 m*M*. Thus, when Na^+ is present at too low a concentration to precipitate 300-Å filaments on its own, it protects chromatin against Mg^{2+}-induced precipitation.

The same competitive effects are found on optimal compaction of chromatin.[38] When

Mg^{2+} is used to induce chromatin refolding, increasing the concentration of Na^+ increases the concentration of Mg^{2+} necessary for optimal compaction. One (more complicated) study reports the opposite conclusion,[39] but the primary data are not given and so it is difficult to assess the reasons for the discrepancy.

Low concentrations of Na^+ also appear to compete with Mg^{2+} for stabilization of individual 300-Å filaments (i.e., in the absence of any aggregation).[17] The sedimentation data in Figure 1 show that for chromatin in 100 μM Mg^{2+}, increasing the concentration of NaCl from 0.2 to 25 mM leads to a drop in S, indicating that Na^+ has destabilized the Mg^{2+}-induced 300-Å filaments. It is believed that chromatin in such conditions is folded into 300-Å filaments and not aggregated.[17,43]

It may seem paradoxical that Na^+ (or any other monovalent cation) can both stabilize and destabilize 300-Å chromatin filaments. However, several lines of reasoning make this seem less surprising. It is not evident that such behavior is in violation of any law or principle of thermodynamics. Furthermore, as discussed above, cation competition is expected theoretically whenever any cations interact with nucleic acid.[46,47] Finally, apparently identical behavior is observed for an analogous, but completely different system, the B-form $\rightleftharpoons$ Z-form transition of poly d(G-C).[48] When present individually, Na^+ at 2.5 M or $Co(NH_3)_6^{3+}$ at ~20 μM are each able to cause poly d(G-C) to isomerize from B-form to Z-form. However, when the Z-form is induced by $Co(NH_3)_6^{3+}$, Na^+ has the opposite effect: at concentrations of Na^+ too low to stabilize the Z-form on its own, Na^+ competes with $Co(NH_3)_6^{3+}$ and causes Z-form poly d(G-C) to convert back to B-form.

III. STRUCTURE OF THE 300-Å CHROMATIN FILAMENT

A. X-RAY STUDIES OF ORIENTED 300-Å FILAMENTS

Ever since the discovery of the nucleosome as the fundamental unit of chromosome structure, the question has arisen: how are nucleosomes packed in space within 300-Å chromatin filaments? Electron microscopic studies have led to the proposal of a number of distinct (i.e., mutually exclusive) models,[15,34,49-52] reviewed in References 18 to 24. The fundamental problem is that individual molecules, even on the same square of an EM grid, often appear different. Therefore, there is some flexibility in choosing which (if any) of the individual molecules best represents the 300-Å filaments of chromatin *in vivo*. For this reason, a number of other experiments have been done to distinguish between the various models.

One method that should distinguish definitively between various structural models is X-ray diffraction from oriented samples. Diffraction patterns from partially oriented samples have been obtained since 1967,[53-58] but it has not been possible to interpret these patterns because samples were prepared in ways now known to destroy the 300-Å filament or no evidence was presented that the chromatin was in that state. Recently, Widom and colleagues have obtained X-ray diffraction patterns from partially oriented samples of chromatin prepared without pulling fibers and in which the chromatin is shown to be in the 300-Å filament state.[28,29] These diffraction patterns allow one to distinguish between many of the previously proposed models.

It is possible to prepare long chromatin oligomers containing an average of ~150 to 200 nucleosomes; such chromatin folds into 300-Å filaments having axial ratios (length/width) of ~10:1. These filaments are highly negatively charged because only part of the DNA-phosphate negative charge is neutralized by histones or by condensed cations. One expects from physical chemistry that such molecules might spontaneously form nematic liquid crystals[59,60] which would be ideal for X-ray studies.

A wide range of ionic conditions have been investigated and it now appears that because of the flexibility of 300-Å filaments and their tendency to aggregate, nematic phases are not

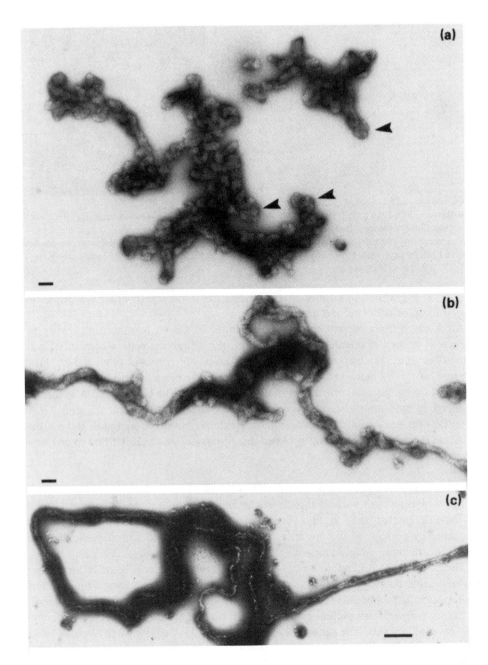

FIGURE 5.   Aggregates of 300-Å filaments, unfixed and negatively stained. The arrows in (a) indicate regions where 300-Å filaments appear to loop back and twist around themselves, similar to their packing in metaphase chromosomes; the bar is 1000Å. (b, c) Long rope-like aggregates which presumably produce the oriented specimens. In (b) the bar is 1000 Å; in (c) it is roughly 5000 Å. (From Widom, J., *J. Mol. Biol.*, 190, 411, 1986. With permission.)

formed. Fortunately, however, when macroscopic aggregates (which have formed from dilute solutions of long 300-Å filaments over a period of minutes to hours) are loaded into capillaries for X-ray examination, they frequently exhibit preferential orientation.[28] Electron micrographs of such samples are shown in Figure 5 (b, c). It appears that the long 300-Å filaments aggregate end-to-end and side-to-side, producing enormously long, rope-like structures which would be expected to align preferentially along the axis of a capillary. Because of the

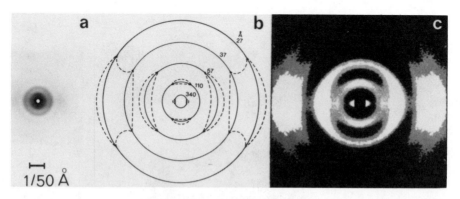

FIGURE 6.   Diffraction from oriented 300-Å filaments; the capillary direction is vertical. (a) Reproduction of original film (darkening is proportional to X-ray intensity). The light circle in the center is the shadow of the beamstop. (b) Schematic drawing illustrating regions of enhanced intensity. (c) Computer graphics display of same film, on same scale as (b); brightness is proportional to log [S·I(S)]. (From Widom, J. and Klug, A., *Cell*, 43, 207, 1985. With permission.)

presence of aggregates like those in Figure 5a, the X-ray specimens have properties in between those of solutions and oriented fibers.

An X-ray diffraction pattern of such a sample is shown in Figure 6. Owing to the presence of molecules oriented in all directions, the diffraction pattern consists of a series of rings—at spacings characteristic of 300-Å filaments *in vitro* and *in vivo*: 340, 110, 57, 37, and 27 Å.[17,26-35] However, the rings are significantly more intense in certain directions because of the preferential orientation of 300-Å filaments. The band at 110 Å is meridional (enhanced in the vertical, capillary direction); the bands at 340, 57, 37, and 27 Å are equatorial. The equatorially enhanced intensity extends inward from a peak at 57 Å toward 100 Å; on the equator, the region between the 37- and 27-Å bands is filled in and exhibits enhanced intensity.

B. INTERPRETATION OF DIFFRACTION PATTERNS

Diffraction patterns such as this do not uniquely define a three-dimensional structure, but, once deciphered, may suggest particular models and rule others out. The patterns are deciphered by correlating the presence of each diffraction band with a particular level of chromatin structure, using a phase diagram such as Figure 4 as a guide.

The 300- to 400-Å band in X-ray solution scattering patterns (Figure 2) correlates with the side-to-side packing of 300-Å filaments and is interpreted as arising from interference between laterally packed 300-Å filaments. Its equatorial location in oriented patterns confirms that 300-Å filaments are present in those samples and that they are oriented preferentially along the capillary axis. The bands at 110 Å and 57 to 100 Å appear when oligonucleosomes fold into the 300-Å filament state. The simplest explanation for the orientation of these bands, relative to the 340-Å equatorial band, is that nucleosomes are packed edge-to-edge vertically (approximately parallel to the 300-Å filament axis), giving meridional 110 Å diffraction, and radially around the axis, giving equatorial 57- to 100-Å diffraction. The bands at 37 and 27 Å are observed for all levels of folding and also in scattering from solutions of nucleosome core particles; this suggests that they are due to diffraction internal to nucleosome core particles.

This analysis of the oriented X-ray diffraction patterns leads to an arrangement of nucleosomes consistent with one of the models suggested by electron microscopy, the solenoid model of Finch and Klug,[15,16] depicted schematically in Figure 7. In this model, the string of nucleosomes is wrapped in a (one-start) contact helix (or "solenoid") having ~6 nucleosomes per turn. The nucleosomes are arranged radially, with their flat faces roughly

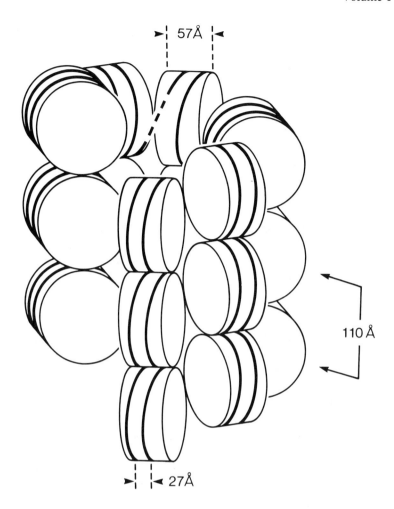

FIGURE 7.   The solenoid model. Important dimensions are indicated. For simplicity,
the model is drawn with six nucleosomes per turn and with nucleosomes untilted.
Variations in DNA linker length or nucleosome free regions may be accommodated
in an ordered, but not perfectly regular, structure. (From Widom, J. and Klug, A.,
Cell, 43, 207, 1985. With permission.)

parallel to the filament axis. The rotational setting of the nucleosomes about the nucleosomal-
disk axis is not known, but it is believed that the point at which DNA enters and leaves the
nucleosome (the proposed H1 binding site) is internal. The drawing is idealized; variations
in DNA linker length, or nucleosome free regions, may be accomodated in an ordered, but
not perfectly regular, structure. The number of nucleosomes per turn need not be integral
or even constant, and there may be some (small) tilt of the nucleosomes away from the
vertical orientation illustrated.

The preceding interpretation of the oriented patterns can be tested by computation. The
orientation of nucleosomes within 300-Å filaments is deduced from the orientation of the
internucleosomal diffraction bands at 110 and 57Å which arise from the packing of particles
having these external dimensions. This orientation should agree with that specified by the
intranucleosomal bands at 27 and 37 Å, if the structural origins of these were known.
Nucleosome core particles account for the majority of the mass of chromatin, and so diffraction from them dominates the diffraction from chromatin. Therefore, the intranucleosomal
contribution to oriented diffraction patterns can be calculated for different models, in which
nucleosomes have various orientations with respect to the 300-Å filament axis, using the 7-
Å electron density map of the core particle.[3]

It is found that there is good agreement between the orientation of 27- to 37-Å diffraction predicted for a solenoid-like orientation of nucleosome core particles, and the experimentally determined oriented diffraction.[28] A number of other orientations (characteristic of other models) lead to predicted diffraction which is inconsistent with observation.

C. EFFECTS OF VARYING LINKER LENGTH

Because the linker DNA represents only a minor fraction of the total mass of chromatin, it contributes very weakly to images or to diffraction patterns. Thus, currently available electron microscopic images and oriented X-ray patterns specify the packing of nucleosomes, but not their connectivity. Other models can be constructed that are consistent with these images or patterns, yet differ in their DNA linker paths from the solenoid in that laterally neighboring nucleosomes come from nonconsecutive locations along the DNA.[34,51] Different models of higher order structure imply different effects of a changed linker length. Therefore, one can take advantage of the 80-bp range in average linker lengths available from cells of different types to probe indirectly the path of the linker DNA.

An example of such an experiment is given in Figure 8, which shows densitometer traces along the equator and the meridian of diffraction patterns from partially oriented chicken erythrocyte and sea urchin sperm 300-Å filaments.[28,29] These two chromatin types have DNA repeat lengths of 212 and 240 bp, respectively,[10,61] and therefore the linker DNAs differ in length by ~28 bp or ~100 Å. Within experimental error, the position of the meridional "110-Å" diffraction band does not vary. This result places constraints on possible linker paths and is inconsistent with a number of models for 300-Å filament structure.

It is also important to determine the effects of a changed linker length on the diameter of the filaments; this is currently the subject of some controversy. An X-ray band (not seen in the oriented patterns)[28,29] that has been assigned by one group to diffraction from the diameter of rodlike chromatin filaments[34] has been assigned to the pitch of a loose superhelix by another.[62] One study reports that the apparent filament diameter measured from micrographs increases with linker length;[34] another reports that the diameter changes little or not at all.[29] The tendency of 300-Å filaments to aggregate must also be considered.[17] It is likely that this question will be cleared up in the next 1 to 2 years.

D. MODELS OF 300-Å FILAMENT STRUCTURE

1. Current Models

Three different models have received the greatest attention in the recent literature and are thought most likely to potentially be correct: the solenoid model,[15,26,28] the twisted-ribbon model,[51] and the crossed-linker model.[34] The solenoid model is described above and is illustrated in Figure 7. The twisted-ribbon model is based on a flat zigzag chain of nucleosomes with a dinucleosome repeating unit. This dinucleosome repeat ribbon is then given a tertiary coiling about the filament axis. The orientation of nucleosomes is solenoid-like, but the connectivity is vertical rather than lateral. In the crossed-linker model, a zigzag chain of nucleosomes is both twisted and compacted about the long axis of the chain, producing a two-start helix. The orientation of nucleosomes is again solenoid-like, but the connectivity is now transverse, between the front of the filament and the back.

The twisted-ribbon model as originally proposed appears to be ruled out by the oriented X-ray data. The model predicts a number of diffraction bands that are not observed and it predicts a DNA linker-length dependence to the meridional diffraction, which is also not observed. It is unclear how the model might be adapted to fit the X-ray data while remaining applicable to the full range of known linker lengths.

2. Status of the Solenoid Model

A large body of data appears to support the solenoid model (see References 18 to 24 for reviews); therefore, it may be worthwhile to examine three reports which appear to weigh

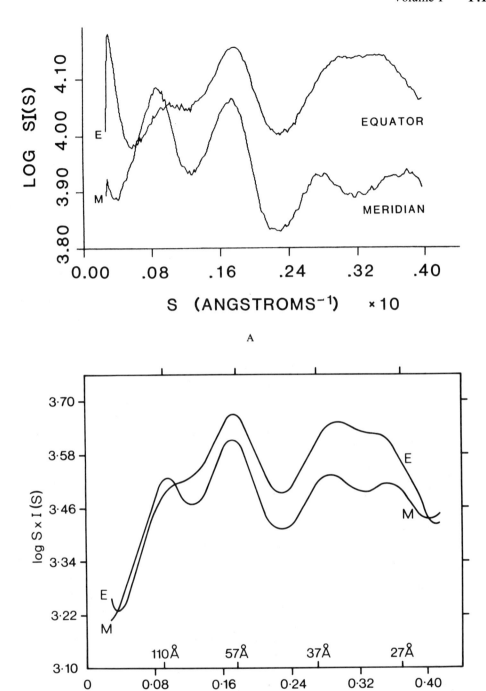

FIGURE 8.   Effects of changed DNA linker length. The figures show X-ray intensity data from the equator (E) and the meridian (M) of diffraction patterns from partially oriented 300-Å filaments. (A) From chicken erythrocyte, 212 bp repeat length; (B) from sea urchin sperm, 240 bp repeat. (Data from Widom, J. and Klug, A., *Cell*, 43, 207, 1985; Widom, J., Finch, J. T., and Thomas, J. O., *EMBO J.,* 4, 3189, 1985. With permission.)

most heavily against it. (The mechanism by which chromatin folds into a solenoidal state is a separate question which will be discussed elsewhere.)[24]

Certain nuclease digestion experiments show that oligonucleosomes containing even numbers of nucleosomes are produced preferentially during digestion of 300-Å filaments, suggesting that a dinucleosome might be an important structural unit.[63,64] This can be correlated with zigzag structures for unfolded chromatin observed in a number of electron microscopic studies.[16,51]

It is not clear what might be the origin within the solenoid model of a dinucleosome structural unit. Previously, one might have suggested that a head-to-head or tail-to-tail arrangement of H1 (or H5) molecules on consecutive nucleosomes would lead naturally to a dinucleosome repeating unit, but the arrangement of H5 has been analyzed experimentally and it is now thought to be polar — head-to-tail.[65] Therefore, proponents of the solenoid model must look elsewhere for a structural basis for dinucleosomes. Alternatively, one might argue that dinucleosomes could somehow result from a redistribution of proteins during nuclease digestion.[66]

A second key experiment which appears inconsistent with the solenoid model is a scanning transmission electron microscopy (STEM) study of the mass per unit length (m/l) of folded chromatin, which reported a value of 11.6 nucleosomes per 110 Å along the filament.[51] This is nearly twice the value ($\sim$6 to 7 nucleosomes per 110 Å) expected for the solenoid model.

The measurement of 11.6 nucleosomes per 110 Å was made with chicken erythrocyte chromatin in 150 mM Na$^+$. Other investigators find that chicken erythrocyte chromatin is not soluble in such high Na$^+$ concentrations (see Section II, above); if the STEM measurements included any aggregates, even just dimers, the measured mass per unit length would be erroneously high. A large number of different measurements of m/l made by other groups using a variety of techniques (including STEM) all give results of 6 $\pm$ 1 nucleosomes per 110 Å for chick erythrocyte chromatin in conditions where the chromatin is soluble.[23,42]

The evidence weighing most heavily against the solenoid model comes from the recent work of Langmore and colleagues[34] in which Fourier transforms of images of negatively stained 300-Å filaments have been analyzed. The transforms show evidence of helical symmetry; measured values of the pitch and the pitch angle, together with an assumption about the radius of stain penetration, lead to the conclusion that 300-Å filaments have two-start helical symmetry, not one-start as in the solenoid model.

These data are recent and have not yet been tested or confirmed by other studies. In the absence of other information, it appears that three points can be made in defense of the solenoid model. First, Pederson et al.[23] have pointed out that there is an alternative interpretation of the negative staining pattern. The apparent pitch angle may arise from the accumulation of stain in the groove between laterally adjacent nucleosomes, on successive turns of a single start (solenoidal) helix, rather than from staining of the groove between adjacent helical ramps of nucleosomes as had been assumed. Second, some of the images show evidence of stretching or curvature, which could possibly distort the measured pitch and pitch angle. The apparent pitch angle of 300-Å filaments from a different cell type is reported to be very sensitive to shear forces, which are encountered unavoidably during preparation of samples for electron microscopy.[15,16] Relatively small errors in the measured pitch angle would mean that single-start helices are not ruled out. A third point is that the conclusion that chromatin has a two-start helical structure requires a questionable assumption about the effective radius of stain contrast. Williams et al.[34] argue that the most reasonable value is the radius to the center of the nucleosomes, $\sim$95 Å. However, it is likely that stain can penetrate further in toward the center of the filament, down to the radius at which the nucleosomal disks meet edge to edge, $\sim$60 Å. If this were the case, the experimental measurements of pitch and pitch angle could also support the solenoid model.

3. Status of the Crossed Linker Model

The crossed-linker model is supported by measurements of the helical parameters discussed above and by measurements of filament diameter and m/l which are reported in the same study.[34] There is currently no evidence that directly contradicts this model; however, there is evidence which seems to require a small revision of the model and other evidence which leads one to question a key element of the model. Alternative interpretations of the helical transform data have been discussed above.

The linker-length dependence predicted by the crossed-linker model implies that chicken erythrocyte 300-Å filaments should have a mass per unit length of 8 to 9 nucleosomes per 110 Å. As discussed above, a large number of experiments show that the correct value is most probably 6 ± 1 nucleosomes per 110 Å. At least for this one case, it appears that the very high packing density characteristic of the crossed-linker model is not found *in vivo*. It may be that a less densely packed variant of the model could still give reasonable agreement with the helical diffraction data.

One of the key lines of evidence supporting the crossed-linker model is the very high mass per unit length (13.1 nucleosomes per 110 Å) found by STEM for 300-Å filaments of chromatin having a long DNA linker, from the echinoderm *Thyone* (sea cucumber) sperm.[34] These measurements were obtained in ionic conditions in which chromatin from sperm of another echinoderm (sea urchin) is known to aggregate.[67] As for the case of chicken erythrocyte chromatin, STEM determinations of mass per unit length are suspect if the chromatin is aggregated. Proponents of the crossed-linker model can point out that their micrographs of *Thyone* chromatin[34] appear to show monomeric filaments.

IV. HIGHER ORDER FOLDING

During meiosis and mitosis, chromosomes are folded into highly compact specialized structures. It is useful to characterize these structures by a "packing ratio", which is defined as the ratio of the contour length of a chromosomal DNA molecule to the length of that same DNA packaged in a meiotic or mitotic chromosome. For meiotic chromosomes in the form of synaptonemal complexes, the packing ratio is typically in the range of 300- to 1000-fold;[68] for mitotic chromosomes, the packing ratio is ~10,000-fold. For comparison, the solenoid model of 300-Å filaments predicts a packing ratio of ~40-fold for that level of structure. It is believed that both meiotic and mitotic chromosomes are constructed by further folding of chromatin in the 300-Å filament state.

For meiotic chromosomes, the evidence is limited and comes primarily from electron microscopic studies of the ultrastructure of synaptonemal complexes. These show 250- to 300-Å-diameter fibers, which are interpreted as chromatin in the 300-Å filament state, that loop out and rejoin the body of the synaptonemal complex.[13]

Two lines of evidence suggest that metaphase chromosomes are constructed by further folding of 300-Å filaments. Electron micrographs show stubby 500- to 700-Å-diameter projections emanating radially from the body of metaphase chromosomes. Thin sectioning studies[69-71] and image reconstruction[72] show that these projections are composed of a loop of 300-Å filament[73,74] that folds back and twists around itself. Very similar structures are formed spontaneously by long 300-Å filaments *in vitro*.[17,28,29] Figure 5a shows a number of such twisted 300-Å filament loops within a single aggregate and suggests that local metaphase-like packing may be a natural property of 300-Å filaments *in vitro*. Furthermore, there is a good correlation between the phase diagram shown in Figure 4 and the appearance of metaphase chromosomes;[71] if the concentration of Mg^{2+} or higher valence cations is too high, the 500- to 700-Å-diameter projections appear to merge together and can no longer be resolved. This correlates with a transition between region II (or IIA) and region III of the phase diagram.

X-ray solution scattering studies also indicate the presence of 300-Å filaments within metaphase chromosomes.[75,76] These show the same set of diffraction features characteristic of chromatin in interphase cells, which in turn is the same as that found for 300-Å filaments *in vitro*. This provides additional evidence that the 300-Å filaments which are seen in metaphase chromosomes are similar in internal structure to 300-Å filaments *in vitro*.

V. CONCLUSIONS AND PROSPECTS

The previous 5 years have brought significant advances in our understanding of chromosome structure. Perhaps the most spectacular advance has been the determination of the 7-Å structure of the nucleosome core particle. Significant advances have also been made at the next level of structure — the packing of nucleosomes within 300-Å filaments. Many previously proposed models have now been ruled out; two models currently seem most likely to be correct: the solenoid model and the crossed-linker model. These models are sufficiently different that it should be possible to distinguish between them in many ways. One clear-cut test will be a determination of the mass per unit length of chromatin having a long DNA linker, under conditions where the 300-Å filaments are not aggregated.

While progress has been made, much remains to be done. Since the rotational setting of nucleosomal disks in 300-Å filaments is not known with certainty, key information such as the locations of histone H1 and linker DNA is not yet available. One suspects that interactions other than those involving H1 may be important in stabilizing 300-Å filaments, but again there is little information available.

A wide range of biophysical experiments have been of importance and surely will continue to be so; four new areas also offer particularly exciting possibilities. Cryoelectron microscopy, in which specimens are examined in vitrified ice, without staining or shadowing,[77] may preserve intact the native structure of 300-Å filaments. Methods of image reconstruction could then give the three-dimensional structure directly. Another new development is the discovery of DNA sequences which position themselves uniquely on a histone octamer.[78-80] These can be cloned as tandem polymers and allow one to reconstitute nucleosome oligomers having an exactly conserved DNA linker length. Physical and structural studies of such well-defined molecules are likely to yield more information than equivalent studies of natural (random) chromatin. Recent advances in optical microscopy[81,82] and in electron microscopy[72,83] are providing new information about the most highly packaged forms of chromosomes; it will be of great importance to relate features observed in such images to the folded states of chromatin studied *in vitro*. Finally, the observation of condensed chromosomes in the yeasts *Saccharomyces cerevisiae*[82] and *S. pombe*[84] suggest the possibility of a genetic approach to eukaryotic chromosome structure.

ACKNOWLEDGMENT

The author is grateful to B. Widom for valuable discussions.

REFERENCES

1. **McGhee, J. D. and Felsenfeld, G.,** Nucleosome structure, *Annu. Rev. Biochem.,* 49, 1115, 1980.
2. **Wu, R. S., Panusz, H. T., Hatch, C. L., and Bonner, W. M.,** Histones and their modifications, *CRC Crit. Rev. Biochem.,* 20, 201, 1984.
3. **Richmond, T. J., Finch, J. T., Rushton, B., Rhodes, D., and Klug, A.,** Structure of the nucleosome core particle at 7Å resolution, *Nature, (London),* 311, 532, 1984.

4. **Klug, A., Rhodes, D., Smith, J., Finch, J. T., and Thomas, J. O.,** A low resolution structure for the histone core of the nucleosome, *Nature (London), 287,* 509, 1980.
5. **Bates, D. L. and Thomas, J. O.,** Histones H1 and H5: one or two molecules per nucleosome?, *Nucleic Acids Res.,* 9, 5883, 1981.
6. **Simpson, R. T.,** Structure of the chromatosome, a chromatin particle containing 160 base pairs of DNA and all the histones, *Biochemistry,* 17, 5524, 1978.
7. **Kornberg, R. D.,** Structure of chromatin, *Annu. Rev. Biochem.,* 46, 931, 1977.
8. **Thomas, J. O. and Furber, V.,** Yeast chromatin structure, *FEBS Lett.,* 66, 274, 1976.
9. **Pearson, E. C., Bates, D. L., Prospero, T. D., and Thomas, J. O.,** Neuronal nuclei and glial nuclei from mammalian cerebral cortex, *Eur. J. Biochem.,* 144, 353, 1984.
10. **Spadafora, C., Bellard, M., Compton, J. L., and Chambon, P.,** The DNA repeat lengths in chromatins from sea urchin sperm and gastrula cells are markedly different, *FEBS Lett.,* 69, 281, 1976.
11. **Prunell, A. and Kornberg, R. D.,** Variable center to center distance of nucleosomes in chromatin, *J. Mol. Biol.,* 154, 515, 1982.
12. **Davies, H. G. and Haynes, M. E.,** Electron microscope observations on cell nuclei in various tissues of a teleost fish: the nucleolus-associated monolayer of chromatin structural units, *J. Cell Sci.,* 21, 315, 1976.
13. **Ris, H. and Kubai, D. F.,** Chromosome structure, *Annu. Rev. Genet.,* 4, 263, 1970.
14. **Labhart, P. and Koller, T.,** Electron microscope specimen preparation of rat liver chromatin by a modified Miller spreading technique, *Eur. J. Cell Biol.,* 24, 309, 1981.
15. **Finch, J. T. and Klug, A.,** Solenoidal model for superstructure in chromatin, *Proc. Natl. Acad. Sci. U.S.A.,* 73, 1897, 1976.
16. **Thoma, F., Koller, T., and Klug, A.,** Involvement of histone H1 in the organization of the nucleosome and of the salt-dependent superstructures of chromatin, *J. Cell Biol.,* 83, 403, 1979.
17. **Widom, J.,** Physicochemical studies of the folding of the 100Å nucleosome filament into the 300 Å filament: cation dependence, *J. Mol. Biol.,* 190, 411, 1986.
18. **Butler, P. J. G.,** The folding of chromatin, *CRC Crit. Rev. Biochem.,* 15, 57, 1983.
19. **Staynov, D. Z.,** Possible nucleosome arrangments in the higher order structure of chromatin, *Int. J. Biol. Macromol.,* 5, 3, 1983.
20. **Thomas, J. O.,** The higher order structure of chromatin and histone H1, *J. Cell Sci. Suppl.,* 1, 1, 1984.
21. **Eissenberg, J. C., Cartwright, I. L., Thomas, G. H., and Elgin, S. C. R.,** Selected topics in chromatin structure, *Annu. Rev. Genet.,* 19, 485, 1985.
22. **Felsenfeld, G. and McGhee, J. D.,** Structure of the 30 nm chromatin fiber, *Cell,* 44, 375, 1986.
23. **Pederson, D. S., Thoma, F., and Simpson, R. T.,** Core particle, fiber, and transcriptionally active chromatin structure, *Annu. Rev. Cell Biol.,* 2, 117, 1986.
24. **Widom, J.,** Chromatin folding, *Annu. Rev. Biophys.,* 18, 365, 1989.
25. **Noll, M., Thomas, J. O., and Kornberg, R. D.,** Preparation of native chromatin and damage caused by shearing, *Science,* 187, 1203, 1974.
26. **Sperling, L. and Klug, A.,** X-ray studies on native chromatin, *J. Mol. Biol.,* 112, 253, 1977.
27. **Perez-Grau, L., Bordas, J., and Koch, M. H. J.,** Chromatin superstructure: synchrotron radiation x-ray scattering study on solutions and gels, *Nucleic Acids Res.,* 12, 2987, 1984.
28. **Widom, J. and Klug, A.,** Structure of the 300Å chromatin filament: x-ray diffraction from oriented specimens, *Cell,* 43, 207, 1985.
29. **Widom, J., Finch, J. T., and Thomas, J. O.,** Higher order structure of long repeat chromatin, *EMBO J.,* 4, 3189, 1985.
30. **Greulich, K. O., Wachtel, E., Ausio, J., Seger, D., and Eisenberg, H.,** Transition of chromatin from the "10 nm" lower order structure to the "30 nm" higher order structure, as followed by small angle x-ray scattering, *J. Mol. Biol.,* 193, 709, 1987.
31. **Bordas, J., Perez-Grau, L., Koch, M. H. J., Vega, M. C., and Nave, C.,** The superstructure of chromatin and its condensation mechanism I, *Eur. Biophys. J.,* 13, 157, 1986.
32. **Langmore, J. P. and Schutt, C.,** The higher order structure of chicken erythrocyte chromosomes, *in vivo, Nature (London),* 288, 620, 1980.
33. **Langmore, J. P. and Paulson, J. R.,** Low angle x-ray diffraction studies of chromatin structure *in vivo* and in isolated nuclei and metaphase chromosomes, *J. Cell Biol.,* 96, 1120, 1983.
34. **Williams, S. P. Athey, B. D., Muglia, L. J., Schappe, R. S., Gough, A. H., and Langmore, J. P.,** Chromatin fibers are left-handed double helices with diameter and mass per unit length that depend on linker length, *Biophys. J.,* 49, 233, 1986.
35. **Koch, M. H. J., Vega, M. C., Sayers, Z., and Michon, A. M.,** The superstructure of chromatin and its condensation mechanism III, *Eur. Biophys. J.,* 14, 307, 1987.
36. **Butler, P. J. G. and Thomas, J. O.,** Changes in chromatin folding in solution, *J. Mol. Biol.,* 140, 505, 1980.
37. **McGhee, J. D., Nichol, J. M., Felsenfeld, G., and Rau, D. C.,** Higher order structure of chromatin: orientation of nucleosomes within the 30 nm chromatin solenoid is independent of species and spacer length, *Cell,* 33, 831, 1983.

38. **Borochov, N., Ausio, J., and Eisenberg, H.,** Interaction and conformational changes of chromatin with divalent ions, *Nucleic Acids Res.,* 12, 3089, 1984.

39. **Sen, D. and Crothers, D. M.,** Condensation of chromatin: role of multivalent cations, *Biochemistry,* 25, 1495, 1986.

40. **Bates, D. L., Butler, P. J. G., Pearson, E. C., and Thomas, J. O.,** Stability of the higher order structure of chicken erythrocyte chromatin in solution, *Eur. J. Biochem.,* 119, 469, 1981.

41. **Butler, P. J. G.,** A defined structure of the 30 nm chromatin fibre which accommodates different nucleosomal repeat lengths, *EMBO J.,* 3, 2599, 1984.

42. **Gerchman, S. E. and Ramakrishnan, V.,** Chromatin higher-order structure studied by neutron scattering and scanning transmission electron microscopy, *Proc. Natl. Acad. Sci. U.S.A.,* 84, 7802, 1987.

43. **McGhee, J. D., Rau, D. C., Charney, E., and Felsenfeld, G.,** Orientation of the nucleosome within the higher order structure of chromatin, *Cell,* 22, 87, 1980.

44. **Harrington, R. E.,** Optical model studies of the salt-induced 10-30 nm fiber transition in chromatin, *Biochemistry,* 24, 2011, 1985.

45. **Ausio, J., Borochov, N., Seger, D., and Eisenberg, H.,** Interaction of chromatin with NaCl and MgCl$_2$, *J. Mol. Biol.,* 177, 373, 1984.

46. **Manning, G.,** The molecular theory of polyelectrolyte solutions with applications to the electrostatic properties of polynucleotides, *Q. Rev. Biophys.,* 11, 179, 1978.

47. **Record, M. T., Jr., Anderson, C. F., and Lohman, T. M.,** Thermodynamic analysis of ion effects on the binding and conformational equilibria of proteins and nucleic acids, *Q. Rev. Biophys.,* 11, 103, 1978.

48. **Behe, M. and Felsenfeld, G.,** Effects of methylation on synthetic polynucleotide: the B-Z transition in poly(dG-m⁵dC)·poly(dG-m⁵dC), *Proc. Natl. Acad. Sci. U.S.A.,* 78, 1619, 1981.

49. **Worcel, A., Strogatz, S., and Riley, D.,** Structure of chromatin and the linking number of DNA, *Proc. Natl. Acad. Sci. U.S.A.,* 78, 1461, 1981.

50. **Zentgraf, H. and Franke, W. W.,** Differences of supranucleosomal organization in different kinds of chromatin: cell type-specific globular units containing different numbers of nucleosomes, *J. Cell Biol.,* 99, 272, 1984.

51. **Woodcock, C. L. F., Frado, L.-L. Y., and Rattner, J. B.,** The higher-order structure of chromatin: evidence for a helical ribbon arrangement, *J. Cell Biol.,* 99, 42, 1984.

52. **McDowall, A. W., Smith, J. M., and Dubochet, J.,** Cryo-electron microscopy of vitrified chromosomes, *in situ, EMBO J.,* 5, 1395, 1986.

53. **Pardon, J. F., Wilkins, M. H. F., and Richards, B. M.,** Superhelical model for nucleohistone, *Nature (London),* 215, 508, 1967.

54. **Pardon, J. F., Richards, B. M., and Cotter, R. I.,** X-ray diffraction studies on oriented nucleohistone gels, *Cold Spring Harbor Symp. Quant. Biol.,* 38, 75, 1974.

55. **Pooley, A. S., Pardon, J. F., and Richards, B. M.,** The relation between the unit thread of chromosomes and isolated nucleohistone, *J. Mol. Biol.,* 85, 533, 1974.

56. **Carpenter, B. G., Baldwin, J. P., Bradbury, E. M., and Ibel, K.,** Organization of subunits in chromatin, *Nucleic Acid Res.,* 3, 1739, 1976.

57. **Baldwin, J. P., Carpenter, B. G., Crespi, H., Hancock, R., Stephens, R. M., Simpson, J. K., Bradbury, E. M., and Ibel, K.,** Neutron scattering from chromatin in relation to higher-order structure, *J. Appl. Cryst.,* 11, 484, 1978.

58. **Azorin, F., Martinez, A. B., and Subirana, J. A.,** Organization of nucleosomes and spacer DNA in chromatin fibers, *Int. J. Biol. Macromol.,* 2, 81, 1980.

59. **Bernal, J. D. and Fankuchen, I.,** X-ray and crystallographic studies of plant virus preparations, *J. Gen. Physiol.,* 25, 111, 1941.

60. **Onsager, L.,** *Ann. N.Y. Acad. Sci.,* 51, 627, 1949.

61. **Morris, N. R.,** A comparison of the structure of chicken erythrocyte and chicken liver chromatin, *Cell,* 9, 627, 1976.

62. **Bordas, J., Perez-Grau, L., Koch, M. H. J., Vega, M. C., and Nave, C.,** The superstructure of chromatin and its condensation mechanism. II. Theoretical analysis of the x-ray scattering patterns and model calculations, *Eur. Biophys. J.,* 13, 175, 1986.

63. **Burgoyne, L. A. and Skinner, J. D.,** Avian-erythrocyte chromatin degradation: the progressive exposure of the dinucleosomal repeat by bovine-pancreatic-DNAase-I-armed probes and free DNAase-I, *Nucleic Acids Res.,* 10, 665, 1982.

64. **Drinkwater, R. D., Wilson, P. J., Skinner, J. D., and Burgoyne, L. A.,** Chromatin structures: dissecting their mixed patterns in nuclease digests, *Nucleic Acids Res.,* 15, 8087, 1987.

65. **Lennard, A. C. and Thomas, J. O.,** The arrangement of H5 molecules in extended and condensed chicken erythrocyte chromatin, *EMBO J.,* 4, 3455, 1985.

66. **Thomas, J. O., Rees, C., and Pearson, E. C.,** Histone H5 promotes the association of condensed chromatin fragments to give pseudo-high-order structures, *Eur. J. Biochem.,* 147, 143, 1985.

67. **Thomas, J. O., Rees, C., and Butler, P. J. G.,** Salt-induced folding of sea urchin sperm chromatin, *Eur. J. Biochem.,* 154, 343, 1986.
68. **Lewin, B.,** *Gene Expression 2,* 2nd ed., John Wiley & Sons, New York, 1980.
69. **Marsden, M. P. F. and Laemmli, U. K.,** Metaphase chromosome structure: evidence for a radial loop model, *Cell,* 17, 849, 1979.
70. **Adolph, K. W.,** A serial sectioning study of the structure of human mitotic chromosomes, *Eur. J. Cell Biol.,* 24, 146, 1981.
71. **Adolph, K. W., Kreisman, L. R., and Kuehn, R. L.,** Assembly of chromatin fibers into metaphase chromosomes analyzed by transmission electron microscopy and scanning electron microscopy, *Biophys. J.,* 49, 221, 1986.
72. **Harauz, G., Borland, L., Bahr, G. F., Zeitler, E., and van Heel, M.,** Three-dimensional reconstruction of a human metaphase chromosome from electron micrographs, *Chromosoma,* 95, 366, 1987.
73. **Rattner, J. B. and Hamkalo, B. A.,** Higher order structure in metaphase chromosomes, *Chromosoma,* 69, 363, 1978.
74. **Rattner, J. B. and Lin, C. C.,** Radial loops and helical coils coexist in metaphase chromosomes, *Cell,* 42, 291, 1985.
75. **Pardon, J. F., Richards, B. M., Skinner, L. G., and Ockey, C.,** X-ray diffraction of isolated metaphase chromosomes, *J. Mol. Biol.,* 76, 267, 1973.
76. **Paulson, J. R. and Langmore, J. P.,** Low angle x-ray diffraction studies of HeLa metaphase chromosomes: effects of histone phosphorylation and chromosome isolation procedure, *J. Cell Biol.,* 96, 1132, 1983.
77. **Adrian, M., Dubochet, J., Lepault, J., and McDowall, A. M.,** Cryo-electron microscopy of viruses, *Nature (London),* 308, 32, 1984.
78. **Simpson, R. T. and Stafford, D. W.,** Structural features of a phased nucleosome core particle, *Proc. Natl. Acad. Sci. U.S.A.,* 80, 51, 1983.
79. **Ramsay, N., Felsenfeld, G., Rushton, B., and McGhee, J. D.,** A 145-base pair DNA sequence that positions itself precisely and asymmetrically on the nucleosome core, *EMBO J.,* 3, 2605, 1984.
80. **Simpson, R. T., Thoma, F., and Brubaker, J. M.,** Chromatin reconstituted from tandemly repeated cloned DNA fragments and core histones: a model system for study of higher order structure, *Cell,* 42, 799, 1985.
81. **Brakenhoff, G. J., van der Voort, H. T. M., van Sprousen, E. A., Linnemans, W. A. M., and Nanninga, N.,** Three-dimensional chromatin distribution in neuroblastoma nuclei shown by confocal scanning laser microscopy, *Nature (London),* 317, 748, 1985.
82. **Hiraoka, Y., Sedat, J. W., and Agard, D. A.,** The use of a charge-coupled device for quantitative optical microscopy of biological structures, *Science,* 238, 36, 1987.
83. **Belmont, A. S., Sedat, J. W., and Agard, D. A.,** A three-dimensional approach to mitotic chromosome structure: evidence for a complex hierarchical organization, *J. Cell Biol.,* 105, 77, 1987.
84. **Umesono, K., Hiraoka, Y., Toda, T., and Yanagida, M.,** Visualization of chromosomes in mitotically arrested cells of the fission yeast *Schizosaccharomyces pombe, Curr. Genet.,* 7, 123, 1983.

Section II. Eukaryotic Chromosomes: Mitosis; X and Y Chromosomes

INTRODUCTION

Chromosomes are more than just extremely long, linear arrays of genes. They are highly organized complexes of proteins and DNA with characteristic three-dimensional structures. And they are not static structures. Major chromosomal rearrangements occur as the cell cycle progresses from interphase through mitosis and again to interphase. The production of eggs and sperm during meiosis is accompanied by a halving of the diploid number of chromosomes. Three levels of chromosome organization can be recognized. Chromatin fibers, discussed in Section I, represent the shortest range of structure. In creating the fibers, histones interact with DNA to form nucleosomes, and the beads-on-a-string filament of nucleosomes then coils into the "30 nm" chromatin fibers. At a higher level of organization, the fibers are folded into mitotic chromosomes, as well as into interphase and meiotic chromosomes. Radially arranged loops have been shown by electron microscopy to be the primary mode of fiber packaging in all of these chromosome types. The final level of chromosome organization concerns the three-dimensional arrangement of chromosomes. In the mitotic cell, chromosomes are not randomly arranged but occupy particular locations on the metaphase plate and during anaphase and telophase.

Understanding the 3-D arrangement of chromosomes during mitosis is of great importance. This is because gene function is related to the spatial location of chromosomes. For example, exchange of genetic material between chromosomes may be significant in carcinogenesis and other disease states. In addition, changes in the configuration of chromosomes through mitosis are central to cell division. Comparing chromosome arrangement in malignant and normal cells should define the features of cell division unique to cancer cells. The 3-D computer reconstructions of chromosomes in mitotic HeLa cells presented in this section are, therefore, particularly valuable.

Mitotic chromosome formation is part of an intricate process that ensures the efficient distribution of replicated chromosomes to daughter cells. The chromosomes which congregate to form the metaphase plate must separate and move to opposite poles of the cell. Movement of chromosomes requires the involvement of tubulin and associated proteins of the mitotic spindle. Kinetochore microtubules are attached to chromosomes at the centromeres, and shortening of these microtubules draws the chromosomes toward the spindle poles.

Chromosomes display different sizes and shapes, but more profound differences reside in their genetic contents. For example, the genes present on the X and Y chromosomes fundamentally influence the physiology of these chromosomes. This can be seen in the phenomena of X-chromosome inactivation and Y-chromosome expression. X-chromosome inactivation refers to the inactivation, early in the development of the female embryo, of one of the two X chromosomes. Gene dosage compensation is the reason for this effect; female cells have two X chromosomes while male cells (XY) have only one. Inactivation results in essentially equal gene dosages for female and male cells, and this is important for normal development. Describing the effect is one thing, understanding the molecular basis for its establishment and maintenance is another. Current research, outlined in this section, is examining the involvement of chromosomal DNA and proteins, and particularly the role of DNA methylation. Expression of the Y chromosome has a significant role in spermatogenesis, as studies of *Drosophila*, described in this section, point out. The ability to combine investigations of genetics, molecular biology, ultrastructure, and cytology has made the fruit fly *Drosophila* a favored system for modern biology research. The knowledge that emerges by integrating different experimental approaches is more than a simply descriptive picture.

Chapter 5

THREE-DIMENSIONAL COMPUTER RECONSTRUCTIONS OF CHROMOSOMES IN HUMAN MITOTIC CELLS

Kenneth W. Adolph and Charles K. Knox

TABLE OF CONTENTS

I. INTRODUCTION

Chromosomes in mitotic cells have defined spatial locations and are not randomly distributed. The greatest degree of organization occurs during metaphase, when chromosomes are highly condensed and arranged for distribution to daughter cells. The spatial order of chromosomes is determined by their sizes and shapes. These conclusions emerge from three-dimensional (3-D) computer reconstructions, reported in this article, of chromosomes in mitotic HeLa cells. The results should be significant for understanding chromosome function, as are studies of nucleosomal fibers and chromosome substructure.

Spatial ordering of chromosomes is a particularly interesting problem because of the variability in the numbers and sizes of chromosomes for different organisms. HeLa cells are a human tissue-culture cell line with about 68 chromosomes. Normal diploid human cells have 46 chromosomes. Monkeys have 42, mice 40, and Chinese hamsters have only 22. Indian muntjac cells get by with 7. These chromosomes are all large and complex, but chromosome size can vary. *Drosophila melanogaster* mitotic chromosomes, for example, are much smaller.

Determining the 3-D arrangement of chromosomes is important because chromosome function is related to higher-order organization. Exchanges of genetic material between different chromosomes that are spatially close may be a common phenomenon, just as sister chromatid exchanges are frequent occurrences. Such exchanges between nonhomologous chromosomes may have a role in activating oncogenes in carcinogenesis. Furthermore, the spatial ordering of mitotic chromosomes is undoubtedly reflected in the higher-order structure of chromosomes in the interphase nucleus. And the organization of interphase chromosomes into domains or compartments is recognized as a major determinant in regulating DNA replication and gene transcription.

Three stages were involved in producing the 3-D reconstructions. Electron micrographs were first taken of consecutive thin sections through the volume occupied by the chromosomes. Cells were obtained by "mitotic shake-off" of monolayer cultures, and samples were prepared for EM after fixing intact, living cells. The second stage involved digitizing the electron micrographs, contouring the chromosome boundaries, and storing the data in a 3-D database suitable for rendering of the final images. These procedures were carried out nonmanually using computer-assisted image processing. Finally, the reconstructions were completed by applying 3-D computer graphics software to create surface-rendered views of the chromosomes.

The approach was used to determine changes in the distribution of HeLa chromosomes during cell division. Serial sections were obtained for cells at the metaphase, anaphase, and telophase stages of the cell cycle. The reconstructions revealed that the position and orientation of chromosomes depend on their size. The larger chromosomes (chromosomes 1, 2, 3,...) are arranged around the perimeter of the metaphase plate and are directed radially outward. The smaller chromosomes (22, 21, 20,...) are located close to the center of the metaphase plate and are more randomly disposed, with many perpendicular to the plate. This mode of chromosome packing is likely to be a general feature of mitotic cells.

II. EXPERIMENTAL PROCEDURES

A. PREPARATION OF MITOTIC CELLS FOR ELECTRON MICROSCOPY

Mitotic HeLa cells were obtained by mechanical detachment (mitotic shake-off) of cells growing as monolayers on plastic tissue culture plates. Cells were maintained in minimum essential medium containing fetal bovine serum (5%). The use of mitotic shake-off to obtain samples enriched in mitotic cells was essential. Cells treated with colchicine and nocodazole to arrest cells in mitosis were found, by serial sectioning and electron microscopy, to contain abnormal distributions of chromosomes in spherical structures. Detached cells were fixed by adding glutaraldehyde (0.8%) to the cells following resuspension in warm, serum-free growth medium.[1-3] After 1.0 h, samples were placed at 5°C for a further 12 h. Cells were washed with a buffer of 0.1 M sodium cacodylate, pH 7.0, and treated with osmium tetroxide (1%) for 1 h at 5°C. Following a washing with buffer, cells were stained with 1% uranyl acetate for 2 h and again washed.

Quetol 651 was employed as a dehydrating agent and embedding medium.[13] This water soluble, low viscosity medium is useful in providing ease of serial sectioning with high electron image contrast. Dehydration was accomplished through steps of 50% and 75% Quetol monomer in water, 100% Quetol monomer, and 100% Quetol polymerization mixture. The medium was cured for 16 h at 74°C as a thin layer; small squares of the material were then mounted for sectioning.

B. SERIAL SECTIONING AND ELECTRON MICROSCOPY

The Quetol-embedded mitotic cells were serially sectioned using a Reichert Ultracut E microtome with a diamond knife.[1-3] Ribbons of consecutive sections were collected on single-slot specimen support grids. The grid holes were covered with formvar films that were lightly coated with carbon. Sections were about 100 nm thick. The grids were post-stained with uranyl acetate (1%) in water and then with lead citrate, or with uranyl acetate alone. For image processing, it was desirable that the chromosomes were highly stained relative to other cellular material.

C. IMAGE PROCESSING

To determine the 3-D arrangement of chromosomes, the electron micrographs were digitized to gray-level images of 512 × 512 × 8 bits per pixel using a light box and a Dage MTI68 newvicon video camera coupled to a Model 75 image processor (International Imaging Systems, Milpitas, CA) and a Masscomp MC535 minicomputer. As each section was displayed on a video monitor, it was visually registered with the immediately preceding, already digitized section using the registration capabilities of the image processor. Each digitized image was subjected to a radiometric transform to enhance contrast and then thresholded to isolate the chromosomes. A rectangular 3 × 3 median filter operation was then applied to the thresholded image.[4] Median filtering removes single pixel noise and slightly smoothes object boundaries without dilating or eroding the object. Contours of the chromosomes were extracted by a boundary tracking or "contouring" program and stored in a 3-D database in a compact chain-coded format which included the x, y, z coordinates of each contour.[5]

D. THREE-DIMENSIONAL COMPUTER GRAPHICS

The 3-D database of contours was input to Wavefront 3-D modeling and rendering software (Wavefront Technologies, Santa Barbara, CA) running on an IRIS 2400T graphics workstation (Silicon Graphics, Mountain View, CA). On the IRIS, the entire set of contours could be viewed interactively from any chosen angle and at different scales. This allowed for slight corrections in section alignment and also for editing of contours which appeared to be artifacts in the original images. Wavefront software was used to connect the contours in each section with a polygonal mesh for subsequent representation and viewing as solid surfaces. Chromosome surfaces were rendered with artificially chosen colors and reflective properties simply for viewing purposes. Two methods of 3-D representation were employed. In the first, a quick solid-surface model was generated by "extruding" each section as a thick slice that abutted the next section. In the second, a more complete connection of vertices of neighboring contours was carried out to provide more realistic models. The latter method was used to reconstruct individual chromosomes and nuclei, whereas the thick-slice method was used for the entire metaphase plate and for the anaphase configuration. Images of the 3-D reconstructions were captured with a Dunn Instruments Multicolor 35-mm film recorder.

III. RESULTS

A. ELECTRON MICROSCOPY OF SERIAL SECTIONS OF MITOTIC HeLa CELLS

Characteristic electron micrographs of thin sections through metaphase HeLa cells are shown in Figures 1 and 2. The micrographs in Figure 1 are from the same consecutive series in which the sections cut into the face of the metaphase plate. Panel A shows a central section with chromosomes on all sides of the cell, while the section in panel B is away from the center and the chromosomes are confined to the lower right half of the cell. (This is due to the microtome knife cutting the metaphase plate not directly face-on, but at an angle.) The most evident feature of chromosome organization in metaphase cells is that the longer chromosome arms are located around the perimeter of the metaphase plate. The large chromosomes appear lined-up around the periphery as a result of their dense packing (panel B). Chromosome arms near the center of the sections are generally shorter. This could be because smaller chromosomes are located near the center of the metaphase plate, or because chromosomes in this region are oriented perpendicular to the face of the plate. Serial sections cut into the side of the plate, discussed below, show the former possibility to be true, but many chromosomes tend to be normal to the plate. Chromosomes are not extended in these sections, but are V-shaped, which suggests that the arms are folded back at the centromeres and directed away from the center of the cell. Examination of other, similarly sectioned metaphase plates reveals similar distributions of chromosomes, with the radial orientation of the longer chromosome arms being the dominant feature.

Figure 2 includes electron micrographs of three consecutive sections that intersect the side of the metaphase plate. It is clearly seen that the larger chromosomes are at the outside of the plate, while smaller chromosomes are toward the center. Such micrographs support the major conclusions derived from sections cut about 90° to these (Figure 1). However, there may be some preferential orientation of the shorter chromosomes perpendicular to the plane of the metaphase plate. Also, the radially oriented larger chromosomes may not be directed straight outward, but can be at an angle, as Figure 2 shows. Further evidence for this comes from examination of additional sections. Three-dimensional reconstructions were undertaken to clarify these aspects of chromosome arrangement during metaphase.

An advantage of the use of electron micrographs of thin sections as the basis of the reconstructions is that chromosome order in living, intact cells is preserved. Fixation captures the chromosomes in their *in vivo* mitotic configuration and serial sectioning allows the spatial distribution of chromosomes to be viewed without distortion. Sections of 100 nm are thin enough so that the smallest dimension of chromosomes (the arm width) is resolved in 5 to 7 sections. Individual chromosomes can therefore be adequately resolved in the reconstructions.

Besides resolving the 3-D arrangement of metaphase chromosomes, another major aim was to determine chromosome configuration changes during cell division. Electron micrographs of serial sections were therefore obtained for anaphase and telophase cells. Typical views of chromosomes during anaphase are seen in Figure 3. The microtome knife has entered the side of the anaphase configuration, and the chromosomes are moving toward the spindle pole at the top of the figure. The micrographs in panels A and B are separated by five sections, while (B) and (C) have seven sections between them. Chromatids have separated at this stage so that single chromosome arms are observed, and the chromosomes are close together and frequently in contact. The chromosome boundaries appear sharply defined since the nuclear envelope is beginning to form again around the chromosomes.

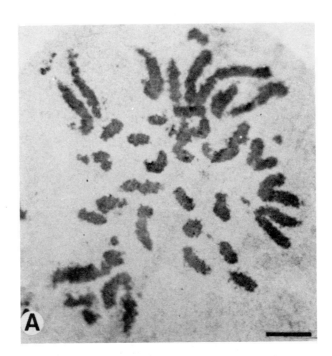

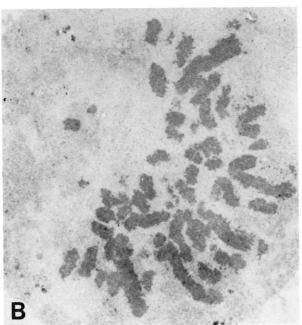

FIGURE 1.   Electron micrographs of thin sections through a metaphase
HeLa cell. The micrographs were selected from a consecutive series which
intersects the face of the metaphase plate. A central section is shown in
panel A, and a section cutting through the edge of the plate is shown in
panel B. The longer chromosome arms are seen to be situated around the
periphery of the metaphase plate and are radially oriented. Cells were
obtained by mitotic shake-off, fixed, and embedded in Quetol medium.
After cutting the series of 100-nm thick sections, grids were stained with
uranyl acetate and lead citrate. The magnification bar represents 2.0 μm.

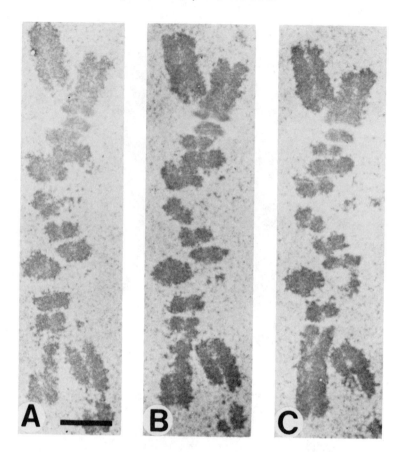

FIGURE 2.   Serial sections cutting into the side of a HeLa metaphase plate. The electron micrographs are about 90° to those in Figure 1. The three consecutive sections show the characteristic changes in electron density of chromosomes from section to section. Chromosomes at the ends of the plate are largely in the plane of the section and directed outward, while the preferred orientation of the smaller chromosomes in the middle is generally normal to the plate. Samples were prepared as in Figure 1. The bar is 2.0 μm.

The disk of chromosomes is rounded in the direction of movement and chromosome arms trail behind, giving the anaphase configuration a characteristic appearance. The sections in Figure 3 are just beginning to encounter the sister complement of chromosomes in the other half of the cell (not shown). Further along the series, as the sister configuration is more deeply penetrated, the chromosomes appear similar to Figure 3. Because of the close association of chromosomes in anaphase, it is not possible to resolve many individual chromosomes in 3-D reconstructions. But anaphase chromosomes, arranged as a structure intermediate to the metaphase plate and telophase nucleus, have unique and important features.

Individual chromosomes are no longer discernible in the telophase nucleus, and the nuclear envelope has coalesced into a continuous structure surrounding the decondensed chromosomes. Nevertheless, the nucleus is not spherical at this stage of cell division, but has a shape determined by the disk of chromosomes. The characteristic morphology of the telophase nucleus can be observed in the electron micrographs of Figure 4. The shape of the telophase nucleus is evidently determined by the structure of the metaphase plate. The disk-like structure, about 4 μm thick and 14 μm in diameter, is continued through separation of chromosomes, movement to opposite poles during anaphase, and formation of the nuclear membrane in telophase. The telophase nucleus is not a perfect disk, but has undulating faces

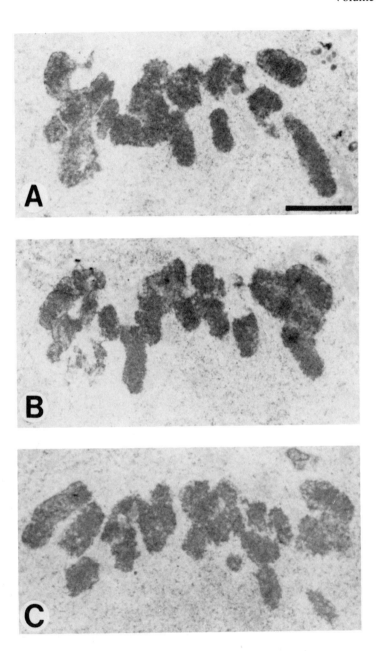

FIGURE 3.   Electron micrographs of serially sectioned anaphase chromosomes. The
anaphase configuration is intersected from the side. The sister complement of chromosomes is below each panel so that the direction of chromosome movement would
be to the top. Many chromosomes are touching and their boundaries are sharply
defined as a result of nuclear membrane formation around the chromosomes. Because
of chromosome movement to the spindle poles, the disk of chromosomes is rounded
and some chromosome arms extend behind. The sections are from the same consecutive series: (A) and (B) are separated by five sections, while seven sections are
between (B) and (C). Bar, 2.0 μm.

which typically are concave. Clumped chromatin is detected within the nuclei and may
represent the remnants of the decondensed chromosomes. The nuclei shown in Figure 4A
were used for 3-D reconstructions since this stage of the cycle can be precisely identified

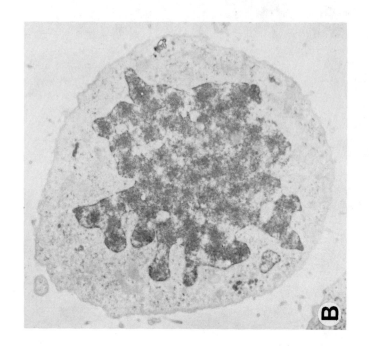

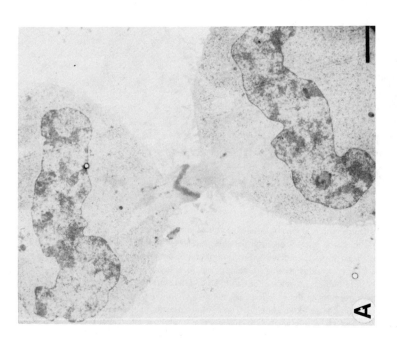

FIGURE 4. Thin sections of telophase nuclei. In panel A, daughter nuclei are shown; this stage of cell division is precisely defined by the midbody structure and deep furrowing between cells. Panel B shows a view about 90° to that of (A) and is close to the center of the nucleus. Telophase nuclei have an unusual structure established by the metaphase chromosomes serving as a template for nuclear envelope formation. The perimeter of the telophase nucleus is uneven and the ends are generally concave. Condensed clumps of chromatin are observed, but little additional substructure is evident. Bar, 2.0 μm.

as the point just preceding cell division. The cell is deeply furrowed and the midbody defines the two daughter cells. Figure 4B, which is from a different cell sectioned about 90° to the cells in Figure 4A, demonstrates that the perimeter of the developing nucleus is not circular, but has an irregular, ruffled margin. As with the faces of the telophase nucleus, the undulating nuclear perimeter is likely to be established by the positions of the decondensing chromosomes.

B. IMAGE PROCESSING: CONTOURING OF CHROMOSOMES AND ALIGNMENT OF SECTIONS

Figure 5 presents examples of stacked sections for metaphase and telophase cells. Contours in different sections were well-registered, indicating that distortion of the sections was not a significant problem. Resolving individual metaphase chromosomes was the greatest technical challenge since the chromosomes are densely packed in the metaphase plate and, not uncommonly, are in actual contact. The contours of particular chromosomes or nuclei progress reasonably smoothly from one section to another. The electron micrographs are projections through the 100-nm thick sections, which could contribute to unevenness between sections. The stacked sections point out that a major benefit of utilizing EM is that the entire volume occupied by individual chromosomes can be reconstructed (volume structuring). The results are therefore not confined to simply indicating the position and orientation of the central chromosome axis.

Contour representations such as the sections in Figure 5, are often considered the final result of a 3-D reconstruction since the human visual system can easily interpret such data. At this stage of the analysis, HeLa metaphase chromosomes are revealed to have a characteristic configuration and spatial distribution (Figures 5A and 5B) and telophase nuclei are seen as complex structures that are far from spherical (Figure 5C). Viewed from the top, the metaphase plate has a striking appearance. The visually dominant region of the metaphase plate is the periphery, where the long chromosome arms are observed to lie close to the plane of the plate and to be outwardly oriented. Chromosomes in the middle of the plate are shorter and, if anything, are perpendicular to the plane of the plate. For telophase nuclei, the irregular perimeter and flat, undulating faces are prominent features evident in the aligned sections.

Although images such as those in Figure 5 give a 3-D effect, the stacked sections represent the unprocessed 3-D data which is the basis for the full reconstructions.

C. THREE-DIMENSIONAL SURFACE RECONSTRUCTIONS

Solid models were constructed for the metaphase, anaphase, and telophase configurations. The 3-D reconstructions were undertaken to obtain an accurate and readily perceivable representation of chromosome changes during mitosis. The computer reconstructions would also be valuable for detailed measurements of chromosome dimensions, volumes, shapes, and positions. And they would be the foundation of further experiments concerning the location of particular chromosome regions, proteins, and genes.

The reconstruction procedure was carried out for contoured sections of metaphase, anaphase, and telophase HeLa cells. The aim with metaphase cells was to resolve individual chromosomes so that both the spatial disposition of chromosomes and their individual configurations could be observed. Chromosomes at this stage of the cell division cycle can be investigated and are of interest because they are greatly condensed and have characteristic and varied morphologies. Distinguishing individual chromosomes in anaphase is difficult because sister chromatids have separated and chromosomes are in contact, dispersing, and being coated by the nuclear membrane. However, a special aspect of anaphase chromosomes is that they are moving, and the chromosomes are closely associated with microtubules and other components of the mitotic spindle. The information derived from reconstruction of

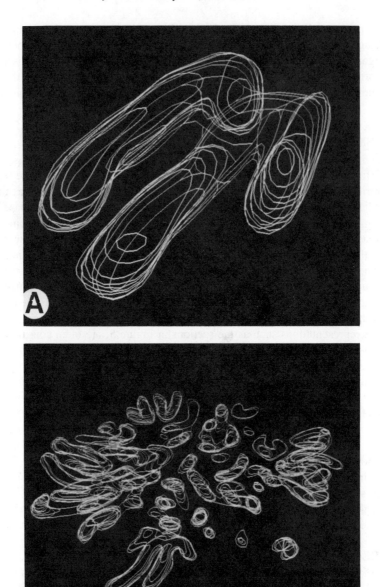

FIGURE 5.   Ordered computer images of digitized and contoured sections. (A) shows
the aligned sections of the metaphase chromosome in Plate 1C; (B) contains stacked,
consecutive sections corresponding to a metaphase plate similar to that in Figure 1;
(C) shows aligned sections for the lower telophase nucleus in Figure 4A. The contoured
sections in (B) are only part of the consecutive set; in (C), the sections extend through
the nucleus with every third section shown. Clearly revealed are the peripheral location
of the longer chromosome arms (panel B) and the unusual structure of the telophase
nucleus (panel C). (Copyright 1989 Kenneth W. Adolph.)

telophase cells is different since individual chromosomes can no longer be resolved and the
nuclear membrane surrounds the entire disk of chromosomes. The proto-nucleus of telophase
therefore has a structure which reflects the anaphase arrangement of chromosomes.

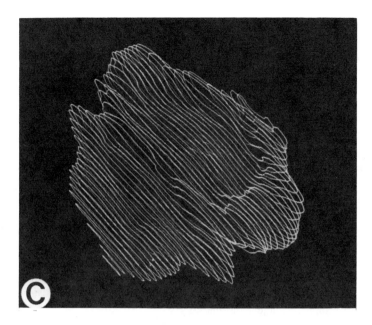

FIGURE 5C.

Representative views of 3-D reconstructions of the metaphase, anaphase, and telophase configurations are presented in Plates 1*, 2*, and 3*. As described technically in Experimental Procedures, the reconstructions were carried out by computationally connecting the contoured boundaries of adjacent sections and enclosing the volumes to create the solid structures. Spatial relationships of the distributions of chromosomes, which are otherwise difficult to discern, are apparent in the 3-D reconstructions.

The unique morphologies of chromosomes when assembled to form the metaphase cell are revealed in Plate 1. The 3-D reconstructions vividly show that the telomeres of the short and long arms of the peripheral chromosomes extend in the same radial direction (Plate 1 A to C). The long arms orient each chromosome with its telomere distal and its centromere proximal to the center of the metaphase plate. However, the telomere of the short arm is not proximal to the centromere, but is folded outward, similar to the long arm. These features are observed for each of the three chromosomes in panels A, B, and C. The three were chosen since they show different relative lengths for the two arms, B being the most metacentric. This unexpected observation is due to the flexibility of chromosome structure and the dense packing of chromosomes in assembling the plate. The reconstructions also show that the sister chromatids in panels B and C are just beginning to separate with the approach of anaphase. In addition, the contraction and lengthening of sister chromatids through metaphase are not entirely synchronous since the homologous chromosome arms are seen to have somewhat different lengths.

For the entire metaphase plate, chromosome position and orientation are determined by the lengths of the chromosome arms (Plate 1D). The 3-D view in panel D demonstrates the peripheral location and radial orientation of the larger chromosomes, reconstructions of which are in A to C. In panel D, the reconstructed chromosomes of A to C are shown at their *in vivo* locations within the entire metaphase plate. (Chromosome C lies on top of B, partially obscuring it from view.)

The anaphase configuration of chromosomes is a more amorphous structure than the metaphase plate. Chromosomes are touching and sister chromatids have separated, making recognition of individual chromosomes difficult. The 3-D reconstruction included in Plate

2 illustrates the intricate arrangement of anaphase chromosomes. With rounded edges in the direction of movement and trailing chromosome arms, the distribution of chromosomes is influenced by movement toward the spindle pole. Although the metaphase arrangement of chromosomes has been deformed, the relative positions of chromosomes should be unchanged so that the long chromosome arms should be around the rim of the anaphase structure. Such a feature is difficult to observe because of chromosome contact and nuclear membrane formation. The anaphase chromosome configuration is an intermediate structure that passes along aspects of the metaphase arrangement to the telophase cell.

As shown by the telophase nucleus reconstruction in Plate 3, the nucleus has an abundance of structural detail. Although separate chromosomes are no longer visible, the external morphology of the nucleus resembles the outer boundary of the anaphase and metaphase configurations. The overall dimensions are similar and the furrows and bulges on the disk-like structure correspond to the positions of chromosomes in the prior phases of cell division. Chromosome arrangement is responsible for nuclear morphology in telophase, but not the reverse. Rounding-up of nuclei in their later development may result from the further dispersion of chromatin. Certain basic features of the reconstructions can be described which are common to most telophase nuclei. These include an uneven nuclear perimeter, concave faces, and bulging regions on the faces. But variability, even between sister nuclei, is also characteristic of telophase nuclear morphology.

IV. DISCUSSION

The highest order of chromosome structure concerns the spatial arrangement of chromosomes, particularly during mitosis. The condensed chromosomes of the metaphase cell are highly organized in forming the metaphase plate. Understanding the mode of packing of metaphase chromosomes and the structural changes through anaphase and telophase were the aims of this investigation. The 3-D reconstructions revealed that, to achieve the maximum packing density of chromosomes, the larger chromosomes are positioned at the perimeter of the metaphase plate and are radially oriented. The investigation of this level of chromosome structure has been neglected for mammalian cells, but the spatial arrangement of chromosomes is of equal significance to the other levels of organization. Chromosome substructure, that is, the mode of folding the 30-nm fibers into the characteristic morphology of chromosomes, is the intermediate level of organization. The shortest range of structure deals with the interactions of histones with DNA to create nucleosomes, nucleosome filaments, and 30-nm fibers.

The use of thin sections as the basis for the reconstructions has a number of important advantages. One benefit is that the *in vivo* arrangement of chromosomes is preserved since intact, living cells are fixed for electron microscopy. Thin sectioning ensures that this arrangement is accurately recorded when electron micrographs are taken; chromosomes are not flattened, sheared, or otherwise distorted as in whole-mount electron microscopy. For similar reasons, this procedure is superior to scanning electron microscopy, which requires that the chromosome configurations be isolated, using relatively harsh conditions, from surrounding cellular material. Another advantage of combining thin sectioning with transmission EM is the high resolution that is provided. Future experiments will be concerned with determining the 3-D locations of centromeres, binding sites of monoclonal antibodies, and *in situ* hybridization of nucleic acid probes. This will be a major goal of the project and only EM can provide the high resolution that is necessary. A further benefit of thin sectioning is that the technique permits volume structuring. The 3-D results are not limited to simply giving the positions and lengths of the chromosome arms: the entire volume occupied by each chromosome is reconstructed.

Contouring of the chromosome boundaries in each section was accomplished without manual tracing of the boundaries. Starting with the digitized images, contours were determined with the image processing system by applying a thresholding function to highlight the highly stained chromosomes. A program then computed the contours. The procedure avoided having to impose qualitative decisions about the location of boundaries during the contouring process. And it permitted the best alignment of sections since allowance could be made for stretching, shrinkage, and other distortions to sections.

The reconstructions did not end with stacking the contours but were completed by the construction of solid boundaries representing the chromosomes. This will allow detailed characterization of the volume, configuration, substructure, and identity of the chromosomes. Human cells, even normal diploid cells, contain a large number of chromosomes, so that identification of chromosomes on the basis of size and shape alone will be difficult. But the detailed structural analysis of all chromosomes and the identification of certain chromosomes should be possible, particularly with information from further experiments mentioned above.

The completed reconstructions showed that the spatial location of metaphase chromosomes depends upon the length of the chromosome arms. What is the reason for the peripheral location and outward orientation of the larger chromosomes? This pattern of chromosome distribution appears to be a consequence of assembling chromosomes to form the metaphase plate. Efficient packing of chromosomes would result if the smaller chromosomes (those not much longer than they are wide) were arranged near the center of the plate. But the arms of the larger chromosomes are unwieldy. The dense packing of chromosomes would be disrupted less if the longer chromosome arms were distributed around the perimeter of the metaphase disk so as to protrude into the most available volume. The maximum density of chromosome packing would therefore result.

The same mode of chromosome organization will probably be found for other types of cells with different numbers, sizes, and shapes of chromosomes. The number of chromosomes would have to be sufficiently large so that general packing considerations, and not the particular sizes and shapes, determine the arrangement of chromosomes. The configurations of metaphase plates of cells with many fewer chromosomes are likely to be greatly influenced by the precise morphologies of the chromosomes. Examples of these cells would be Indian muntjac with seven chromosomes and *Drosophila melanogaster* with eight. However, for humans, monkeys, rats, and similar higher eukaryotes, the peripheral location and radial orientation of the longer chromosomes appear to be general features of the spatial distribution of metaphase chromosomes.

Packing considerations and chromosome morphology may not, however, totally determine the arrangement of chromosomes. Another factor that may be influential is the possible existence of specific connections between chromosomes. If DNA strands are continuous between chromosomes, this would clearly be a major determining factor in establishing the 3-D distribution. In addition, mitotic chromosomes are not isolated in the cell interior, but are associated with the spindle microtubules. The requirement for proper attachment of kinetochore microtubules to ensure the efficient separation of chromatids to daughter cells places major constraints on the spatial position and orientation of chromosomes.

The distribution of chromosomes established in forming the metaphase plate continues to be reflected in the anaphase and telophase configurations of chromosomes. The reconstructions of anaphase cells demonstrate that the distribution of chromosomes is deformed by movement toward the spindle poles, but that a plate-like structure is still present. Even in telophase, when the nuclear envelope reforms around the template of decondensing chromosomes, the chromosomes remain ordered as a disk.

No comparable results concerning the spatial distribution of mitotic chromosomes have been reported for other mammalian cells. But observations with various different systems have indicated the importance of the 3-D organization of chromosomes. Mitotic chromosomes of grasses such as barley were reconstructed from electron micrographs of serial sections by Bennett, Heslop-Harrison, and colleagues.[6,7] Chromosomes were identified by their volume and morphology, and the spatial positions of centromeres were determined. The 3-D coiling of the giant chromosome arms of *Drosophila melanogaster* polytene chromosomes was followed by optical sectioning microscopy (Agard, Sedat, and colleagues[8,9]). These unusual interphase chromosomes occupy mutually exclusive spatial domains and particular loci are in contact with the nuclear envelope. Insects have been the favored systems for other 3-D studies: distinct maternal and paternal compartments were found for meiotic chromosomes of the gnat *Sciara coprophila*[10] and an end-to-end order was seen for haploid chromosomes of *Ornithogalum virens*.[11]

Although no reconstructions of human mitotic chromosomes in intact cells have been published, informative results concerning chromosome rearrangements during mitosis have been obtained with scanning electron microscopy.[12] Observations of isolated chromosome clusters from HeLa cells showed characteristic structural changes from a metaphase disk to an anaphase cylinder and then to a telophase half-sphere. SEM reveals only external features and the chromosome structures must be extracted from cells, but the results are generally compatible with the 3-D reconstructions from serial sections.

The reconstructions demonstrate that the distributions of chromosomes in human mitotic cells are complex, yet ordered, and undergo a continuum of structural rearrangements through mitosis. This information should have important implications for an understanding of chromosome function, as further studies should prove.

V. SUMMARY

The distribution of chromosomes in mitotic HeLa cells was determined by serial sectioning and 3-D computer reconstruction techniques. Chromosome configurations were reconstructed for metaphase, anaphase, and telophase cells. Intact, living cells were fixed before preparation for electron microscopy. Thin sectioning was applied since, with this procedure, the spatial arrangements of chromosomes were preserved without distortion. Electron micrographs of consecutive sections revealed that metaphase chromosomes are not randomly organized, but their position and orientation depend on their sizes and shapes. The larger chromosomes (chromosomes 1, 2, 3,...) are distributed around the perimeter of the metaphase plate with the chromosome arms directed radially outward. The smaller chromosomes (22, 21, 20,...) are located closer to the center of the metaphase plate. These observations were strengthened and extended by computer-assisted image processing: electron micrographs were digitized, chromosome contours were aligned, and three-dimensional reconstructions were computed. The results demonstrated the peripheral location and radial orientation of the longer chromosomes. The short arms of these chromosomes did not extend toward the center of the plate, but were bent back and oriented radially outward. Three-dimensional reconstructions of chromosome configurations in anaphase and telophase cells were also carried out. Chromosomes were seen to be closely associated in anaphase, and movement of the chromosome plate to the spindle poles produced a rounded structure with some chromosomes extending behind. The morphology and dimensions of the telophase nucleus were apparently determined by the configuration of chromosomes so that the reconstructed telophase nucleus had an uneven and disk-like morphology.

ACKNOWLEDGMENTS

This investigation benefited from the expertise of Andre Abad in the art of serial sectioning. Thanks are also extended to Rod Kuehn for advice concerning electron microscopy. The facilities of the Biomedical Image Processing Laboratory of the Department of Cell Biology and Neuroanatomy were made available by a grant from the National Institutes of Health. This study was supported by a grant from the Graduate School, University of Minnesota.

REFERENCES

1. **Adolph, K. W.**, A serial sectioning study of the structure of human mitotic chromosomes, *Eur. J. Cell Biol.*, 24, 146, 1981.
2. **Adolph, K. W., Kreisman, L. R., and Kuehn, R. L.**, Assembly of chromatin fibers into metaphase chromosomes analyzed by transmission electron microscopy and scanning electron microscopy, *Biophys. J.*, 49, 221, 1986.
3. **Adolph, K. W.**, Arrangement of chromatin fibers in metaphase chromosomes, in *Chromosomes and Chromatin, Vol. 2*, Adolph, K. W., Ed., CRC Press, Boca Raton, FL, 1988, 3.
4. **Rosenfeld, A. and Avinash, C. K.**, *Digital Picture Processing*, Academic Press, 1982, 261.
5. **Castleman, K. R.**, *Digital Image Processing*, Prentice-Hall, Englewood Cliffs, NJ, 1979, 316.
6. **Heslop-Harrison, J. S. and Bennett, M. D.**, Prediction and analysis of spatial order in haploid chromosome complements, *Proc. R. Soc. London Ser. B*, 218, 211, 1983.
7. **Heslop-Harrison, J. S., Smith, J. B., and Bennett, M. D.**, The absence of the somatic association of centromeres of homologous chromosomes in grass mitotic metaphases, *Chromosoma*, 96, 119, 1988.
8. **Agard, D. A. and Sedat, J. W.**, Three-dimensional architecture of a polytene nucleus, *Nature (London)*, 302, 676, 1983.
9. **Hochstrasser, M., Mathog, D., Gruenbaum, Y., Saumweber, H., and Sedat, J. W.**, Spatial organization of chromosomes in the salivary gland nuclei of Drosophila melanogaster, *J. Cell Biol.*, 102, 112, 1986.
10. **Kubai, D. F.**, Nonrandom chromosome arrangements in the germ line nuclei of Sciara coprophila males: the basis for nonrandom chromosome segregation on the meiosis I spindle, *J. Cell Biol.*, 105, 2433, 1987.
11. **Ashley, T. and Pocock, N.**, A proposed model of chromosomal organization in nuclei at fertilization, *Genetica*, 55, 281, 1981.
12. **Welter, D. A., Black, D. A., and Hodge, L. D.**, Nuclear reformation following metaphase in HeLa S3 cells: three dimensional visualization of chromatid rearrangement, *Chromosoma*, 93, 57, 1985.
13. **Abad, A. and Kuehn, R.**, personal communication.

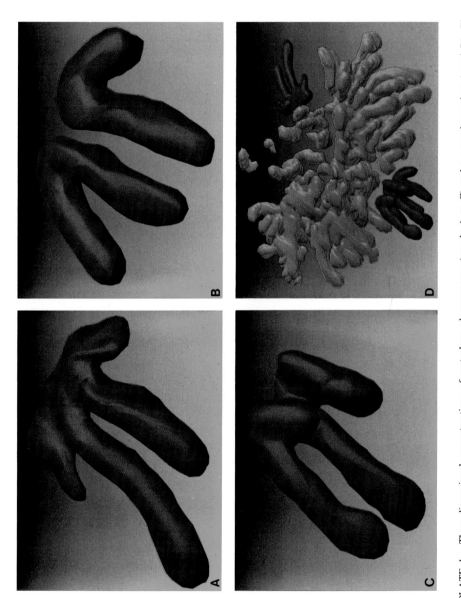

PLATE 1. Three-dimensional reconstructions of metaphase chromosomes (panels A to C) and a metaphase plate (panel D). The chromosomes in D with a deep-red color are the same as those in A to C. (Chromosome B is partially hidden by C.) (Copyright 1988 Kenneth W. Adolph.)

PLATE 2. The anaphase configuration of chromosomes. The disk-shaped structure is rotated 55° counterclockwise from an edge-on view about a vertical axis in A, while in B the rotation is 305°. (Copyright 1988 Kenneth W. Adolph.)

PLATE 3. Three-dimensional reconstruction of a telophase nucleus. The two sides of the nucleus are shown, with an angle of rotation from edge-on of 120° counterclockwise in A and 300° in B. (Copyright 1988 Kenneth W. Adolph.)

Chapter 6

THE KINETOCHORE AND ITS ROLES DURING CELL DIVISION

Ronald D. Balczon and B. R. Brinkley

TABLE OF CONTENTS

I. INTRODUCTION

In the past, the kinetochore had been viewed as simply being a structurally differentiated site within the primary constriction that serves to attach chromosomes to spindle microtubules during mitosis and meiosis. That idea has changed, however, and the kinetochore is now thought of as being an active participant in many of the events of cell division. In addition to its attachment function, recent evidence suggests that the kinetochore may also play a role in both spindle formation and in the chromosome movements that occur during mitosis.[1-3] Together, these properties of the kinetochore help to insure the precise chromosomal distributions that occur during the process of cell division.

Until recently, the kinetochore was described strictly in ultrastructural terms. The applications of new technologies, such as immunocytochemistry and molecular biology, have finally allowed investigators to begin to biochemically dissect the kinetochore. For instance, a family of kinetochore proteins from mammalian chromosomes has been identified using immunoblotting techniques.[4-10] With time, it should be possible to define the individual roles that these proteins play during normal kinetochore functioning. In addition to beginning to characterize the proteins that make up the kinetochore, DNA sequences have been identified in yeast that confer mitotic and meiotic stability to replicating plasmids that contain these sequences.[11-22] These centromeric DNAs, and the methodologies used in identifying these sequences, may be useful in isolating similar DNA regions in mammalian chromosomes.

In this chapter we will summarize the recent advances that have occurred in our knowledge of kinetochore biochemistry as well as describe the experiments that have led to the shifting of ideas concerning kinetochore function during cell division.

A. CENTROMERE VS. KINETOCHORE

The terms centromere and kinetochore have been used synonomously by light microscopists, geneticists, and cell biologists for years.[23-25] In light of our current understanding of chromosome structure and our knowledge of chromosome interaction with spindle microtubules, the indiscriminate use of these two terms appears to be incorrect. The term centromere refers to a broad region of the chromosome that includes the microtubule binding site as well as the primary constriction and its associated heterochromatin. The centromeric heterochromatin has several unique properties that distinguish it from the remainder of the heterochromatin. It is highly condensed, resistant to both nuclease digestion and hypotonic swelling, and genetically silent.[26-28] In addition, the centromeric heterochromatin in many, but not all, species is composed of highly repetitive satellite DNA sequences.[29] The function of the centromeric heterochromatin remains completely unknown, although it has been postulated that it may be involved in either the synapsis and recombination events of meiosis[30] or in maintaining the sister chromatids in paired form until the onset of anaphase.[31,32]

The term kinetochore, on the other hand, refers to the specific structure on chromosomes that spindle fibers attach to and is the region through which mitotic and meiotic forces act to move the chromosomes to their respective poles. Ultrastructurally, the kinetochore has been identified as a trilaminar disc located at the primary constriction of most eukaryotic chromosomes (Figures 1 and 2). Although closely apposed to the centromeric heterochromatin, distinct biochemical differences between centromeres and kinetochores have been reported. Specifically, antibodies from the sera of patients with the CREST (CREST = calcinosis, Raynaud's phenomemon, esophageal dismotility, sclerodactyly, and telangiectasia) variation of the autoimmune disease scleroderma have been shown to specifically label the kinetochore, rather than the entire centromere, by immunoelectron microscopy.[33] Work is presently underway to characterize these kinetochore-specific CREST antigens.

It should be pointed out that some species (e.g., yeast) do not contain a morphologically identifiable kinetochore. In these cases, the spindle microtubules appear to terminate directly

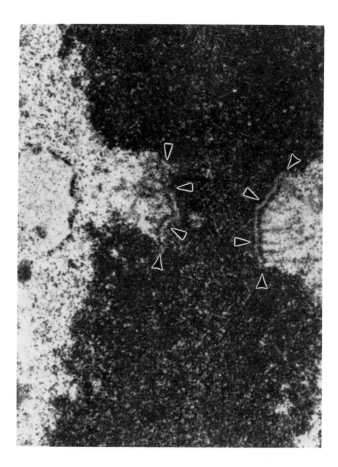

FIGURE 1.   Ultrathin section of a metaphase chromosome of an Indian muntjac cell. The trilaminar plate arrangement of the kinetochore (arrows) can clearly be seen, with spindle microtubules terminating on both of the kinetochores.

on the chromatin fibers. In these instances, the distinction just made between centromeres and kinetochores can be misleading. In general, however, the term kinetochore can be used to describe the structural components that are assembled onto the centromeric DNA that are involved in attaching chromosomes to spindle microtubules, while the term centromere refers to the DNA itself.[21]

B. ULTRASTRUCTURE

The kinetochore has been studied extensively at the ultrastructural level. In thin sections of mitotic chromosomes, the kinetochore is observed to be a trilaminar structure within the clefts of the primary constriction composed of an outer layer approximately 40 to 60 nm thick, a 25- to 30-nm-thick middle layer, and an inner layer that is also 40 to 60 nm thick. The inner layer of the kinetochore plate sits on the surface of the centromeric heterochromatin and is firmly attached to it. The spindle microtubules terminate in the outer plate (Figure 1), which also contains a fuzzy, fibrillar surface known as the corona.[34,35] The biochemical composition of the corona is unknown, but the fine fibrils do have the ability to bind tubulin,[36] the building-block protein subunit of spindle microtubules. The middle layer is composed of loosely-organized chromatin fibrils that appear to link the outer and inner plates.

The three-dimensional organization of the kinetochore varies, depending upon the cell

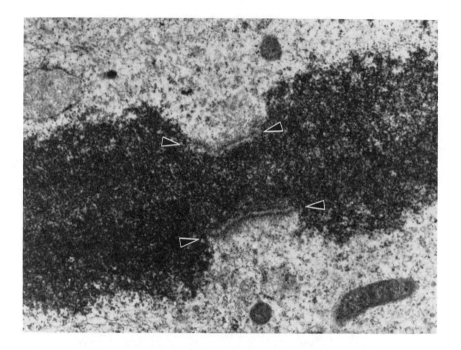

FIGURE 2. The kinetochores from a metaphase-arrested Indian muntjac cell. This cell had been treated with 0.06 μg/ml colcemid for 6 h. The kinetochores (arrows) appear slightly recessed into the centromeric heterochromatin.

cycle stage.[24] During interphase, the kinetochores can be visualized in the nucleus of mammalian cells using CREST serum (Figure 3). The interphase prekinetochores are organized as spherical structures 0.22 μm in diameter, and the number of these CREST-positive interphase structures correlates with the known chromosome number for a particular cell.[33] The kinetochores are duplicated during late S or early G_2, and as the cell enters prophase the kinetochores can be seen to occupy positions on opposite sides of the primary constriction. By metaphase, the kinetochores are organized into the trilaminar structure described in the previous paragraph, and this structure persists through mitosis. In addition to varying in organization during the cell cycle, kinetochores also differ in size, depending on the particular chromosome that they are part of.

The ultrastructure of the kinetochore is influenced by several mitotic poisons that are known to disrupt microtubules (e.g., colchicine). Following colchicine treatment, the kinetochore appears as a single layered plate on the surface of the centromeric heterochromatin (Figure 2). In fact, it has been demonstrated by immunoperoxidase staining using CREST serum that the trilaminar kinetochore is present after colchicine treatment, but that the inner layer is recessed into the centromeric heterochromatin, giving the impression of only a single plate.[33] In addition, colcemid-treated kinetochores appear to be much larger than untreated metaphase kinetochores.[37] The significance of alterations in kinetochore structure caused by colchicine and other mitoclastic drugs is unknown.

Readers who are interested in kinetochore morphology should refer to the excellent review article by Rieder.[24]

C. KINETOCHORE TYPES

Kinetochores show a considerable amount of variability, depending upon the species studied.[24] Some lower eukaryotes, such as yeast, do not appear to contain a morphologically distinct kinetochore on their chromosomes. In general, however, chromosomes from higher

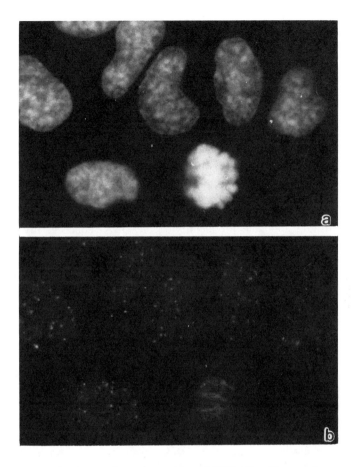

FIGURE 3.   Immunofluorescent staining of CHO cells with serum from a patient with the autoimmune disease scleroderma CREST. (a) Staining of nuclei and chromosomes with Hoechst dye 33258; (b) the corresponding nuclei and chromosomes stained with CREST serum. Both the interphase prekinetochores and the mitotic kinetochores are stained by the CREST serum.

eukaryotic organisms do contain a morphologically distinct kinetochore and these kinetochores can be classified as being either diffuse or localized.

Diffuse kinetochores are found in many higher plant groups and some insects. In these species, the chromosomes lack a primary constriction and the spindle microtubules attach along the length of the chromosome. As a result, when the chromosomes are moving toward the spindle poles during anaphase, they remain perpendicular to the spindle, as opposed to the more conventional V-shape of chromosomes that are associated with microtubules at a single localized site. Electron microscopy has shown that diffuse kinetochores can, in fact, be recognized as distinct, lightly staining plates located along the length of these chromosomes with spindle microtubules attached to each individual plate.[24,25] Little is known about the biochemical organization of diffuse kinetochores.

The kinetochore in most organisms is located at a specific site along the chromosome. These localized kinetochores can be easily identified by the prominent primary constriction on metaphase chromosomes. In mammals and some lower eukaryotes, the kinetochores appear as the trilaminar disk structure that was described in Section I.B. In most plants, the kinetochore appears ultrastructurally as a diffuse ball of poorly staining material contained within a cup of chromatin at the primary constriction (ball and cup kinetochores).[24] Unless specifically instructed, the reader should assume that we are referring to the typical trilaminar kinetochore whenever the term kinetochore is used.

II. KINETOCHORE EVOLUTION

The partitioning of genetic material to daughter cells during cell division does not occur by chance, and specific machinery has evolved to guarantee that this process occurs with unerring accuracy. This segregation machinery can range in complexity from the relatively simple DNA-cortical membrane arrangement in prokaryotes to the highly complex and elaborate processes of the eukaryotic mitotic apparatus. Regardless, a consistent feature of each system is a specific DNA region whose function is the attachment of the genetic material to the segregation apparatus. In this section, these centromeric DNA regions will be described and an hypothesis will be presented that may explain how the complex kinetochores of higher plants and animals evolved from the simpler kinetochores of lower eukaryotes.

A. BACTERIA AND YEAST

The most fruitful approach for identifying the DNA sequences involved in assuring proper segregation has been the use of the plasmid systems of bacteria and yeast. A specific locus, referred to as *par* (*par* = partition), has been identified in *Escherichia coli* that appears to be functionally identical to the centromere/kinetochore of higher eukaryotes.[38-40] The par sequences are *cis*-acting elements that are required for equipartition of plasmids in exponentially growing cell populations, and plasmids that lack the *par* locus are distributed randomly between daughter cells.[38]

The most useful organism for studies of centromere structure has probably been the yeast *Saccharomyces cerevisiae*. The development of genetic and molecular techniques using *S. cerevisiae* has allowed for in-depth analysis of the structure and function of the yeast centromere. In particular, the identification of centromere sequences (CEN), autonomously replicating-sequence plasmids (ARS), and telomere sequences has enabled investigators to produce minichromosomes *in vitro* and to test their viability *in vivo*. These experiments have provided much information on the mechanisms involved in proper centromere behavior.

Several functional yeast centromeric DNA sequences have been isolated based on the ability of the CEN sequences to stabilize ARS plasmids during mitosis and meiosis.[11-22] The ARS plasmids are generally unstable and are lost at a rate of approximately 30% per generation.[13] The insertion of a CEN sequence into an ARS plasmid improves the loss frequency of the plasmids to 10^{-2} to 10^{-3} per cell division.[11,15] Although this is a significant increase in stability, it is still about 100 times as high as the loss frequency of yeast chromosomes. The reason for this disparity is still not completely understood, but may be due to the small size of these plasmids relative to chromosomes.[15]

The structure of the yeast centromere has been studied extensively, and several characteristics of the CEN sequences have been identified. The most prominent feature of CEN sequences is the presence of three conserved regions called elements I, II, and III. Element II is the largest domain (78 to 86 bp) and is greater than 90% A + T. This sequence is flanked on one side by sequence element I, a conserved 8-bp region, and on the other side by element III, a highly conserved 25-bp sequence that appears to be essential for proper CEN function.[13-19] Another property of the CEN sequences is that they are interchangeable. The centromere of chromosome III (CEN 3) can be replaced by CEN 4, CEN 5, or CEN 11 without affecting the meiotic and mitotic stability of chromosome III.[22]

Micrococcal nuclease and DNase I have been used to map the chromatin structure of the yeast centromere regions.[13] These experiments identified a 220- to 250-bp nuclease-resistant centromere core that contained elements I to III. The structure of this centromere core was examined by chromatin mapping studies and found to be folded into a unique chromatin particle 15 to 20 nm in diameter. This is in contrast to conventional chromatin which consists of 160 bp of DNA wrapped around histone proteins to give a cylindrical nucleosome particle with a diameter of 11 nm. Since the inside diameter of a microtubule is about 20 nm, Bloom and Carbon[12,17] proposed that the nuclease-resistant core of the yeast centromere may serve as a microtubule attachment site and therefore act as a structurally primitive kinetochore.

Microtubules do not bind to centromeric DNA alone, and a major interest has been to identify and purify the proteins that interact with centromeric DNA. Bloom et al.[13,16,41] have identified a protein fraction that binds with high specificity to CEN DNA from yeast. However, the roles that these proteins play in kinetochore/centromere function have not been determined. Characterization of these CEN binding proteins will provide considerable information on the organization of the yeast kinetochore as well as information on how the kinetochore attaches to spindle fibers during mitosis.

The relationship between the complex kinetochore of mammalian chromosomes and the simple kinetochore of yeast is not yet understood. If one visualizes the kinetochore of yeast (which binds one spindle microtubule) as being the most basic "unit" kinetochore, then it is not too difficult to visualize the mammalian kinetochore (which binds several microtubules) as being a direct amplification of the basic yeast structure.[22] Regardless of the obvious differences between the simple kinetochores of yeast chromosomes and those of higher eukaryotes, key components should be similar because all kinetochores are involved in interacting with spindle fibers. In this respect, an understanding of the yeast kinetochore and the yeast kinetochore-spindle fiber interface should be invaluable to our understanding of the mammalian kinetochore.

B. MUNTJAC KINETOCHORES

If the large kinetochores of higher eukaryotes did, in fact, evolve from simpler kinetochore units, then recent studies involving a family of Asiatic deer may help to explain how this might have occurred. The Indian muntjac (*Muntiacus muntjac vaginalis*) and the Chinese muntjac (*Muntiacus muntjac reevesi*) are virtually identical phenotypically and are capable of producing healthy F_1 hybrids.[42] However, the two species are karyotypically quite distinct (Figure 4). The Chinese muntjac has a chromosome number of 2N = 46, and all of the chromosomes of the complement are small, telocentric elements. The Indian muntjac, on the other hand, has a diploid number of either 2N = 6 (♀) or 2N = 7 (♂).[42,43] It is believed that the Chinese muntjac is the ancestral species, and that through multiple centric and tandem fusions the karyotype of the Indian muntjac has arisen. On the basis of various cytogenetic, taxonomic, and biochemical evidence, it has been concluded that the fusions involved centromeric DNA and that the structural genes have been maintained.[42-45] This has resulted in the morphology of the two deer being unchanged.

Brinkley and co-workers[45] have investigated the kinetochores in these muntjac species. Using scleroderma CREST serum, these workers were able to demonstrate what they termed "compound kinetochores" on the chromosomes of the Indian muntjac. As Figures 4A, B show, the kinetochores of the Indian muntjac range in size from approximately 1 to 6 μm in length. In contrast, the kinetochores of the Chinese muntjac are much smaller (Figure 4C, D). When the total volume occupied by the complete complement of kinetochores of the two species was measured by electron microscopy, no significant difference in the amount of kinetochore material could be found between the two types of muntjac, indicating that kinetochore components had not been lost during evolution even though a considerable change in the chromosome number occurred.

In addition to the similarity in kinetochore volume that was noted in the two species, Brinkley et al.[45] also reported an interesting "kinetochore cycle" in muntjac cells. After staining with CREST serum, six or seven discrete fluorescent spots could be seen in the nucleus of Indian muntjac cells as the cells enter into G_1 (Figure 5A). As the cell progresses toward S-phase, the interphase kinetochores begin to unravel into multiple subunits that are arranged like beads on a string. The number of fluorescent foci that can be identified are

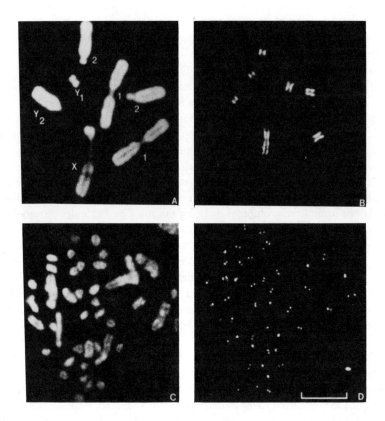

FIGURE 4.   Chromosomes of the male Indian muntjac (A) and female Chinese muntjac (C) stained with Hoechst dye 33258. The corresponding chromosomes stained with CREST serum — Indian muntjac (B) and Chinese muntjac (D). (From Brinkley, B. R., Valdivia, M. M., Tousson, A., and Brenner, S. L., *Chromosoma*, 91, 1, 1984. With permission.)

more than would be expected from the chromosome number. As the cells progress toward mitosis, the fluorescent foci recondense back into the six or seven discrete, doubled fluorescent kinetochores (Figure 5B to E).

In contrast, when the interphase nuclei of Chinese muntjac cells in G_1 were examined following staining with CREST antibodies, 46 discrete fluorescent foci could be identified corresponding to the chromosome number (Figure 5F). As these cells progress through the cell cycle, the fluorescent foci begin to aggregate into clusters resembling the bead-like aggregates observed in the nuclei of Indian muntjac cells (Figure 5G to I). As the cells approach mitosis, the bead-like clusters disaggregate into 46 pairs of randomly arranged fluorescent spots (Figure 5J).

From the above data, Brinkley et al.[45] concluded that the Indian muntjac evolved from the Chinese muntjac with no loss of kinetochore material taking place during the evolution process. These investigators also suggested several hypotheses to explain how the evolution of these compound kinetochores could have occurred. Basically, these workers proposed that the clustering of kinetochores that is seen in interphase in muntjac cells may have, in some way, facilitated the breakage and Robertsonian fusion of centromeres with a concomitant loss of centromeric heterochromatin. The deletion of portions of heterochromatin during fusion is consistent with DNA sequence data from Schmidtke et al.,[44] who found that there is a $10\times$ difference in the amount of intermediate repetitive DNA between the two subspecies, but virtually no difference in single-copy sequences. Another tenet of this hypothesis

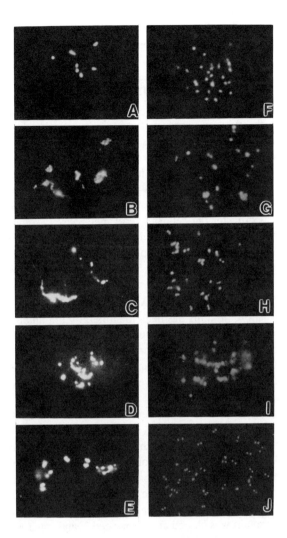

FIGURE 5.   The interphase prekinetochore cycle in nuclei of Indian muntjac (A to E) and Chinese muntjac (F to J). In early interphase, seven prominent spots can be seen in the nucleus of the Indian muntjac (A). As interphase proceeds, each of the seven spots becomes enlarged (B) and finally bead-like (C and D). As the cell prepares for mitosis, the prekinetochores are duplicated and condensed (E). A similar interphase cycle is observed in the nuclei of Chinese muntjac. In early interphase, the individual kinetochores are distinct (F). As interphase progresses, the prekinetochores become clustered (G and H) and finally bead-like (I). The prekinetochores are doubled and dispersed as the cell prepares for mitosis (J). Notice the similarity in arrangements between D and I. Clustering of prekinetochores could have contributed to the evolution of compound kinetochores, as proposed by Brinkley et al.[45] (From Brinkley, B. R., Valdavia, M. M., Tousson, A., and Brenner, S. L., *Chromosoma*, 91, 1, 1984. With permission.)

is that the chromosome arms and associated genes would be retained, although the gene linkages would be altered. This is consistent with the DNA sequence data of Schmidtke et al.[44] as well as with the G-banding pattern data for muntjac chromosomes reported by Shi et al.[42]

In summary, the muntjac kinetochores may help to explain how larger kinetochores may have evolved from simpler "unit" kinetochores via fusion and deletion of centromere material. It should be pointed out that both kinetochore clustering during interphase and compound kinetochores are characteristics that are not unique to the muntjacs. Brinkley et al.[46] reported that the kinetochores of Sertoli cells and spermatocytes in mice cluster during interphase. In addition, Bajer[47] has reported that *Haemanthus* kinetochores may also be composed of multiple subunits. Thus, it is possible that compound kinetochores occur often in nature.

III. KINETOCHORE PROTEINS

As mentioned in the previous section, the majority of what has been learned about centromere/kinetochore DNA sequences has been elucidated using the yeast system. In contrast, much more is known about the proteins from mammalian kinetochores than is known about kinetochore proteins in yeast and other lower eukaryotes. This is because of a discovery that occurred in a rheumatology clinic in the 1980s. Tan and co-workers[48,49] reported that sera from patients with the autoimmune disease scleroderma CREST contained antibodies that were specific for the centromere of metaphase chromosomes. Soon after, it was demonstrated that these autoantibodies do not bind to the entire centromere, but that they specifically recognize the kinetochore plates.[33] Since that time, numerous labs have been using CREST sera as probes for the protein components of the mammalian kinetochore.

A. CHARACTERIZATION OF CREST ANTIGENS

A family of kinetochore proteins has been identified by immunoblotting with sera from CREST patients. Proteins ranging in molecular weights from 14 to 140 kDa have been identified on Western blots of mammalian chromosomes using CREST sera.[4-10] In this lab, the most consistently identified antigens on immunoblots of either HeLa or CHO chromosomes are proteins of M_rs of 18 and 80 kDa, respectively, although the sera from one patient also recognizes antigens at 21 and 52 kDa, respectively (Figure 6). The reason for the differences in circulating antibodies in the sera of the CREST patients is unknown at this time.

Of the kinetochore proteins, the 18- and the 80-kDa proteins have been studied the most extensively. The 18-kDa protein shows many of the characteristics of a histone-like molecule. Among these properties are that the 18-kDa protein can be extracted from chromosomes by $2 M$ NaCl[8] and it sediments with nucleosome-sized particles.[9] The 80-kDa protein, on the other hand, resists salt extraction and appears to be part of the chromosome scaffold.[5] Recently, the 80-kDa protein was cloned and sequenced,[51] and was found to be a very acidic protein with two long domains near the carboxy terminus composed almost entirely of aspartic and glutamic acid residues.

The function of the CREST antigens is not known, although two reports have suggested that some of these proteins may be involved in microtubule attachment. In one report,[4] affinity-purified CREST antibodies were able to inhibit microtubule nucleation in a lysed cell system. In these experiments, cells were arrested in mitosis and then lysed with nonionic detergents. The chromosomes were then either incubated with pure tubulin or preincubated with CREST antibodies prior to incubation with tubulin. Microtubules were nucleated from the kinetochores in the untreated controls, while no microtubule growth was observed from kinetochores that had been pretreated with several of the affinity-purified CREST antibodies.

More recently, Balczon and Brinkley[52] were able to isolate a tubulin-binding fraction from kinetochores of CHO cells. These investigators incubated purified CHO chromosomes with phosphocellulose-purified 6S tubulin under conditions that allowed for the tubulin to bind specifically to the kinetochores. The tubulin was then crosslinked to the kinetochores using a nearest-neighbor crosslinker, the chromosomes were solubilized, and the tubulin-kinetochore protein complexes were isolated by antitubulin affinity chromatography. These protein complexes were then analyzed by both SDS-PAGE and immunoblotting. In addition to tubulin, proteins of M_rs of 24, 52, and 80 kDa were contained within the protein complex (Figure 7). In some preparations, a protein of molecular weight of 110 kDa was also observed.

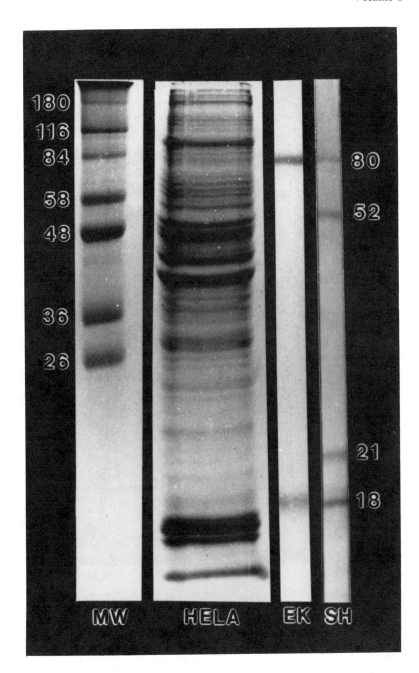

FIGURE 6.   Characterization of CREST sera by immunoblotting. When nitrocellulose blots of proteins from HeLa chromosomes are probed with CREST sera, some variability can be observed. All of the sera that we have in this lab recognize proteins at M_rs of 18 and 80 kDa (as demonstrated using serum from patient E. K.). One serum (from patient S. H.) recognizes proteins of M_rs of 21 and 52 kDa, respectively, in addition to the 18- and 80-kDa antigens. The reasons for this variability between CREST sera are not known at this time.

On immunoblots with CREST serum, the 80-kDa protein was immunoreactive. These results suggest that the 80-kDa CREST antigen is contiguous to kinetochore-bound tubulin and may, in fact, be the tubulin-binding protein within the mammalian kinetochore. Balczon and Brinkley[51] also demonstrated that chromosome scaffolds were able to bind tubulin. This

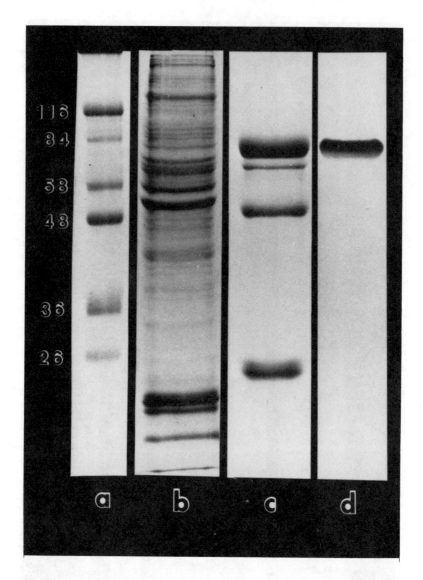

FIGURE 7.   Identification of a tubulin-binding complex from HeLa chromosomes. Tubulin was cross-linked to HeLa chromosomes and the tubulin-kinetochore protein complex purified using the methods outlined by Balczon and Brinkley.[52] Lane b shows a Coomassie-stained gel of whole HeLa chromosomes, and lane c is a Coomassie-stained gel of the tubulin-binding complex isolated from these chromosomes. Lane d is an immunoblot of lane c using CREST serum from patient E. K. (see Figure 6). The 80-kDa band from lane c is immunoreactive.

result suggests that the 18-kDa protein, which is extracted during scaffold preparation, is not necessary for tubulin binding to the mammalian kinetochore and, as a result, probably plays no role in the attachment of kinetochores to spindle microtubules.

B. KINETOCHORE PURIFICATION

In addition to purifying and characterizing the individual kinetochore proteins, a major aim in studying and understanding the kinetochore is to isolate intact kinetochores. Recently, Valdivia and Brinkley[8] succeeded in fractionating this region of the metaphase chromosome. These investigators began with isolated HeLa chromosomes and then digested the chro-

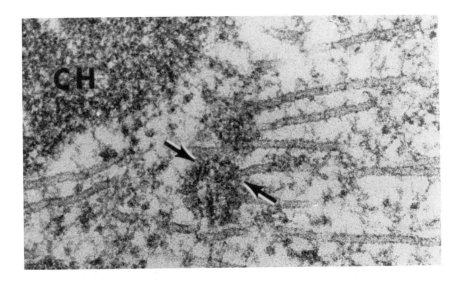

FIGURE 8.   Following certain drug treatments, kinetochores can be induced to specifically detach from chromosomes. Following detachment, the kinetochores retain the ability to interact with spindle microtubules. The mitotic chromosome (CH) is in the upper left. (From Brinkley, B. R. and Shaw, M. W., in *Genetic Causes of Neoplasia,* Williams & Wilkins, Baltimore, 1970, 313.)

mosomes with enzymes. They followed this with various extractions to selectively remove chromosomal proteins and residual nucleic acids and then centrifuged on a metrizamide density gradient. Fractions from the gradient that were immunoreactive when assayed with CREST serum were collected and characterized. When analyzed by immunoperoxidase electron microscopy, the purified structures appeared relatively similar in morphology to the kinetochores seen originally on intact chromosomes. This was in spite of the fact that >99% of the chromosomal DNA and >95% of the chromosomal protein had been removed by this procedure. Unfortunately, the kinetochores isolated in this manner were unable to bind tubulin or to nucleate microtubules *in vitro*. This was probably due to the harsh extraction conditions employed.

More recently, Brinkley and co-workers[53] reported a method that may allow for the isolation of intact and functional kinetochores from CHO cells. This procedure is built upon an earlier observation that kinetochores can be selectively detached from chromosomes *in vivo* by treatment with certain drugs (Figure 8),[54] and incorporates the cell cycle manipulations reported by Schlegel and Pardee.[55] Basically, when caffeine was added to CHO cells that had been arrested at the G_1/S boundary of the cell cycle, the cells entered into mitosis prematurely without completing DNA synthesis. When these mitotic cells were observed by either immunofluorescence or electron microscopy, the chromosomes were observed to be displaced from the spindle with the exception of the entire complement of kinetochores, which were attached to the spindle microtubules. At anaphase, the kinetochores apparently segregated to opposite spindle poles in a fashion characteristic of normal mitosis. Efforts are presently underway to purify these detached kinetochores.

In summary, sera from patients with scleroderma CREST have provided a tool to begin the biochemical dissection of the mammalian kinetochore. The CREST sera have been used to identify a family of kinetochore proteins and to screen expression libraries for the genes that code for these proteins. The CREST sera should also be useful in purifying functional kinetochores.

IV. MITOSIS

A good deal of work on mitosis today is focused on the kinetochore and the kinetochore-microtubule interface. For the past century, the traction-fiber theory has been the predominant idea in the field of mitosis. According to this theory, chromosomes are pulled to the spindle poles during anaphase by forces that act on the kinetochore microtubules, with the kinetochores playing nothing more than a passive attachment role. This does not appear to be the case, however. Recent evidence suggests that, in addition to functioning as a microtubule attachment site, the kinetochore is also an active participant in spindle formation and the events of chromosome segregation.

A. MICROTUBULE BEHAVIOR

Before we begin a discussion of the role of the kinetochore during mitosis, it would be useful to first very briefly review microtubule behavior and the dynamic instability model of microtubule assembly.[56] Microtubules are 24-nm-wide cylindrical polymers composed of tubulin. Microtubules are also polar structures with a fast growing end (the plus end) and a slow growing end (the minus end). When microtubules are nucleated from centrosomes, either *in vitro* or *in vivo*, the negative end is proximal to the centrosome, while the plus end is distal to it.[57,58]

With the exception of the stable microtubules found in cilia and flagella, microtubules are labile structures capable of rapid assembly and disassembly. This dynamic behavior of microtubules is necessary for the rapid cytoskeletal reorganizations that occur during the cell cycle and in response to certain extracellular stimuli. The dynamic instability model of Mitchison and Kirschner[56,59] explains how the lability of microtubules could function in mitosis and morphogenesis.

The dynamic instability model is based on the *in vitro* observations made by Mitchison and Kirschner.[56,58] These investigators incubated pure tubulin with axonemes and isolated centrosomes under microtubule assembly conditions. What they observed was that centrosomes had the ability to nucleate microtubules at concentrations of tubulin below the critical concentration necessary for spontaneous assembly, and that all of the assembled microtubules had their plus ends distal to the centrosomes. In addition, they observed that when microtubule populations at steady state were diluted, some of the microtubules in the population elongated, while others depolymerized catastrophically. Their interpretation of these observations was that microtubules coexist in shrinking and growing populations and that this dynamic instability is a general property of microtubules. This dynamic instability behavior of microtubules has been directly visualized *in vitro* using dark-field microscopy.[60]

Mitchison and Kirschner also studied the kinetics of assembly of microtubules from kinetochores *in vitro*.[36] With kinetochores, they observed a complex kinetics of microtubule assembly, with some microtubules being nucleated with their plus ends distal to the kinetochore, while others were nucleated with their minus ends distal to the kinetochore. This mixed polarity of microtubules nucleated by kinetochores is inconsistent with polarity determinations of the microtubules in a half spindle.[57] They also observed that kinetochores could not nucleate microtubules at tubulin concentrations below the steady-state concentration. This also appears to be inconsistent with what would be necessary for the kinetochore to assemble microtubules during spindle formation *in vivo*.

An interesting observation made by Mitchison and Kirschner is that kinetochores can rapidly and efficiently capture preformed microtubules. Furthermore, these captured microtubules, which are bound on one end by a kinetochore and the other by a centrosome, are stable to the dilution-dependent disassembly that was described earlier. These results suggest an interesting model for spindle morphogenesis that will be described in more detail later.

B. MICROTUBULE ATTACHMENT TO KINETOCHORES

The mechanical attachment of kinetochores to spindle microtubules is well documented.[61] Besides being clearly visualized by electron microscopy (Figure 1), the attachment between the kinetochore and spindle microtubules can be demonstrated by micromanipulation. When chromosomes are pulled with a microneedle, the kinetochore remains attached to the spindle fiber, while the remainder of the chromosome is stretched and deformed.[61] In addition, chromosomes or chromosome fragments that lack a kinetochore are unable to attach to the mitotic spindle. In one study,[62] cells were grown in the presence of fluorodeoxyuridine to induce chromosome breakage. When the cells entered mitosis, only centromeres and chromosome fragments containing a centromere were able to attach to the spindle; chromosome arms and fragments lacking a centromere did not associate with spindle microtubules. In a more recent study by Brinkley et al.,[53] cells were treated with hydroxyurea and caffeine to induce detachment of kinetochores from chromosomes. At mitosis, only the detached kinetochores were associated with spindle microtubules, while the remainder of the chromatin was scattered around the periphery of the cell. Thus, proper attachment between spindle microtubules and the kinetochore insures the correct distribution of the chromosome complement between daughter cells during cell division.

C. THE ROLE OF THE KINETOCHORE IN SPINDLE FORMATION

As more is learned about the formation of the mitotic spindle, it is becoming increasingly apparent that the kinetochore and chromosomes play a very important part in spindle morphogenesis. As mentioned previously, kinetochores have the ability to capture and stabilize microtubules that have been nucleated from centrosomes and, under certain circumstances, have the ability to nucleate their own microtubules. In addition, kinetochores also appear to play an important role in the prometaphase chromosome migrations. Recent studies have begun to shed light on how the kinetochore and chromosomes are able to achieve all of this.

The origin of kinetochore microtubules is uncertain, and there are two schools of thought on this. According to one view, the kinetochore initiates the assembly of the kinetochore microtubules. This idea is based on the fact that kinetochores can assemble microtubules *in vitro* when chromosomes are mixed with exogenous tubulin under conditions which favor microtubule assembly.[36,63-66] Kinetochores can also nucleate microtubules *in vivo* during recovery from drug treatment. When cells are blocked with microtubule poisons such as colchicine, and then allowed to recover, microtubules can be observed emanating from the kinetochores.[24,67]

According to the second hypothesis on the origin of kinetochore fibers, the kinetochore microtubules are nucleated by the centrosomes and captured by the kinetochores. The ability of kinetochores to efficiently capture microtubules has recently been demonstrated *in vitro* by Mitchison and Kirschner.[1] These investigators allowed purified tubulin to assemble from centrosomes *in vitro* and then added isolated chromosomes. The chromosomes were able to become attached to the tubules, and these captured microtubules were stabilized against dilution-dependent disassembly. The notion that centrosomes nucleate the kinetochore microtubules is further supported by the work of Rieder and Borisy.[68] These investigators treated cells with cold temperature to disassemble microtubules, and during the recovery from this treatment microtubules were observed growing from centrosomes well before microtubules were ever observed in association with kinetochores. Studies of spindle microtubule polarity also support the idea that centrosomes nucleate the kinetochore microtubules. When the polarity of spindle microtubules was observed using specialized electron microscopy techniques, it was demonstrated that virtually all of the microtubules in a half spindle are of the same polarity.[57] This is in contrast to the polarity of microtubules nucleated from kinetochores *in vitro*. As mentioned in Section IV.A, when microtubules are nucleated from kinetochores *in vitro*, some of the microtubules have their plus ends distal to the kinetochore, while others have their negative ends distal to the kinetochore.

Studies of microtubule assembly using sea urchin embryos also support the notion that kinetochores do not efficiently nucleate microtubules *in vivo*. In one set of experiments, monopolar spindles were induced in sea urchin blastomeres with the chromosomes oriented such that one kinetochore from each chromosome faced the centrosome, while the other kinetochore faced away from the pole toward the cell cortex.[69] When the kinetochores were examined by electron microscopy, only the kinetochores facing the spindle pole had microtubules associated with them, while the kinetochores facing away from the spindle pole were devoid of microtubules. A possible conclusion from this was that the kinetochores that were devoid of microtubules were unable to nucleate fibers, while the kinetochores that were associated with microtubules had captured fibers that had been nucleated by the single spindle pole. In another type of experiment, sea urchin eggs were fertilized in the presence of microtubule poisons and the inhibitors were removed just prior to the time when mitosis would begin.[70,71] Pronuclear fusion was blocked under these conditions and at the onset of mitosis the two pronuclei were separated by great distances. As the spindle assembled, it formed in association with the chromosomes from the male pronucleus. (Note: at fertilization in the sea urchin, the centrioles enter the egg with the male pronucleus.) In contrast, the chromosomes from the female pronucleus, which were not associated with a centrosome, were devoid of microtubules. Therefore, the kinetochores from sea urchin chromosomes were not sufficient to nucleate microtubules or to organize a spindle.

Taken together, the evidence suggests that the majority, if not all, of the kinetochore microtubules nucleated *in vivo* during a normal mitosis are assembled from the centrosome and captured by kinetochores. This conclusion is based primarily on microtubule polarity evidence as well as the fact that kinetochores can and do capture microtubules well *in vitro*. This is not to say that kinetochores never nucleate microtubules *in vivo*, and it appears fairly certain that at times, such as during drug recovery, kinetochores do nucleate microtubules efficiently *in vivo*. It should be pointed out that in insect spermatocytes, evidence has been presented for the existence of fully functional spindles that lack centrosomes at either one or both poles.[72-74] However, more work is needed before one can conclude that these spindle microtubules are being nucleated by kinetochores. Interested readers are referred to a review on this work by Nicklas.[74]

With the assumption that the centrosome nucleates microtubules and the kinetochore captures them, one can begin to hypothesize as to how spindle formation occurs (see Kirschner and Mitchison, Reference 59). When the nuclear envelope breaks down at the onset of mitosis, the chromosomes are not in any particular orientation. The nucleation capacity of the spindle poles apparently increases at this time and numerous microtubules are formed. Because of dynamic instability, one would assume that most of these microtubules rapidly disassemble and more microtubules are rapidly formed. By chance, some of the microtubules could contact kinetochores and be stabilized, while the remainder of the microtubules continue to cycle through assembly and disassembly. With time, both of the kinetochores on a chromosome would have captured several microtubules and the two kinetochores would be facing opposite spindle poles.

The above scheme certainly helps to visualize how the original attachment of microtubules to kinetochores occurs, but it does not begin to address the question of how the metaphase plate arrangement of chromosomes arises. As mentioned in the previous paragraph, the chromosomes are randomly distributed throughout the cytoplasm at nuclear envelope breakdown. These chromosomes then make an attachment to microtubules and eventually make their way to the metaphase plate. During these prometaphase movements, the chromosomes are undergoing an oscillatory type of motion as they appear to be pushed and pulled between the two spindle poles.[75] Both the nature of this movement and how it results in a metaphase array of chromosomes are unknown.

Two recent findings may help to explain the prometaphase congression of chromosomes. Mitchison and Kirschner[1] have reported evidence for a kinetochore ATPase that might be involved in these movements, and Salmon and co-workers[76-78] have suggested that ejection forces from astral microtubules as well as pulling forces generated at the kinetochore-microtubule interface may be responsible for this oscillatory motion. Whether the putative ATPase described by Mitchison and Kirschner is one of the forces described by Salmon et al. has not yet been determined.

In the experiments of Mitchison and Kirschner,[1] kinetochores were first allowed to capture preformed microtubules, and then the microtubules were elongated at the kinetochore end by adding exogenous tubulin. When ATP was added, the kinetochores were able to translocate along the newly formed microtubule segments. Surprisingly, the direction of motion was away from the centrosome and toward the plus end of the microtubule. This direction of motion is exactly opposite of what would occur during anaphase, and Mitchison and Kirschner proposed that this ATP-driven movement may be involved in establishing the metaphase plate. Much more work is needed, however, before the nature of this ATP-dependent translocation can be completely understood.

The hypothesis of Salmon and co-workers is based on their microscopic observations in newt lung cells. These investigators studied the formation of mitotic spindles in these cells, and occasionally observed chromosomes that were only associated with one spindle pole. These mono-oriented chromosomes underwent normal oscillatory prometaphase movements, and when they were examined by EM it was observed that only the kinetochore that faced the spindle pole was attached to microtubules.[77] Therefore, the saltatory prometaphase movements of these chromosomes were not due to microtubules arising from the other spindle pole. These investigators then severed the arms from mono-oriented chromosomes using laser microbeam irradiation. The cut arms immediately shot outward from the spindle poles toward the cell periphery, while the remainder of the chromosome remained attached to the kinetochore microtubules. From these results, it was concluded that polar ejection forces exist, possibly tied to the assembly properties of astral microtubules, that contribute to the prometaphase congressions. According to the hypothesis presented by these workers, the eventual formation of the metaphase plate is the sum result of the polar ejection forces and forces acting between the kinetochore and spindle fibers at the kinetochore-microtubule interface.[76-78]

In summary, chromosomes and kinetochores appear to play a very active role in the formation of the mitotic spindle and one can now begin to hypothesize as to how spindle assembly occurs. At the onset of mitosis, microtubules are nucleated from the centrosomes. Some of these microtubules become associated with kinetochores and are selectively stabilized, while the remainder of the microtubules may continue through cycles of assembly and catastrophic disassembly. As mentioned in Section 3.A, a complex of kinetochore proteins that bind to tubulin has been isolated from mammalian chromosomes and it will be interesting to see which of the proteins in that complex is involved in the microtubule capture and stabilization phenomenon. Kinetochores and chromosomes also appear to play a prominent role in the oscillatory movements that occur during prometaphase, and experiments are underway that will attempt to explain the forces involved.

D. ANAPHASE MOVEMENT

As previously mentioned, the traction fiber theory viewed the kinetochore as playing little more than a passive attachment role during chromosome movement. That idea is changing, however, and recent experimental evidence suggests that the kinetochore also participates in microtubule disassembly during anaphase and may either contain the mitotic motor or somehow activate or regulate a motor on the microtubules during anaphase.

Mitchison et al.[3] provided the first direct evidence that microtubule disassembly occurs at the kinetochore during anaphase. These investigators microinjected labeled tubulin into cells during metaphase and then fixed the cells. When these cells were observed by EM, labeled microtubule segments were detected extending poleward from the kinetochore with the labeled segments of the same microtubules being distal to the unlabeled segments. If cells were treated in the same way, but allowed to enter anaphase before fixation, the labeled microtubule segments were shorter or absent. These experiments identified the kinetochore as being the site of microtubule assembly at metaphase, and it was further concluded that the kinetochore is the site of disassembly during anaphase.

Gorbsky et al.[2] have reported that chromosome movement during anaphase occurs along stationary microtubules. These investigators injected fluorescently tagged tubulin into cells and allowed the labeled tubulin to be incorporated into the spindle microtubules. They then irradiated the cells with a band of laser light to bleach a small region of the fluorescent kinetochore microtubules and subsequently observed the anaphase movement of the chromosomes relative to the bleached band and spindle pole by immunofluorescence microscopy. According to the traction fiber theory, the chromosomes and the bleached microtubule region should both move toward the pole at the same velocity. However, when mitosis in these cells was observed, the bleached region remained stationary, while the chromosomes moved toward or through the bleached zone as they moved toward the pole in anaphase. Their interpretation of this was that kinetochore microtubules are not pulling the chromosomes poleward, but rather that the chromosomes move along stationary spindle microtubules and that depolymerization of these microtubules occurs at the kinetochore. They further concluded that the kinetochore is an active participant in generating the mitotic force that propels the chromosomes to the pole.

One of the puzzling questions in the field of mitosis is that of how the forces necessary for chromosome segregation are generated. The two most prevalent ideas on the nature of these motive forces are that they are either produced during microtubule disassembly or that they are due to a mechanochemical ATPase molecule. As the first hypothesis states, chromosome movement would be driven directly by the depolymerization of the kinetochore microtubules.[79] According to this hypothesis, the assembled microtubule contains stored energy within its structure. This stored energy is presumably a result of the GTP hydrolysis that occurs during microtubule assembly. During microtubule depolymerization, this energy would be released and the kinetochore would be able to transduce this microtubule disassembly into the force for chromosome movement by some as yet unknown mechanism. According to the second proposed mechanism, the movement of the chromosomes during anaphase would be due to an ATPase molecule, like kinesin or dynein. Evidence for an ATPase molecule in the kinetochore has been provided by Mitchison and Kirschner.[1] As mentioned previously, these investigators described an ATP-dependent chromosome movement using an *in vitro* system. The location of this putative ATPase is not known, but it may either be part of the kinetochore or be associated with the microtubules and activated by the kinetochore during anaphase.

Any model of kinetochore organization certainly has to consider how the kinetochore is able to fulfill the roles of attachment site, regulator of spindle fiber assembly and disassembly, and mitotic motor. An attractive possibility as an explanation for kinetochore organization, and one that is getting much attention, is a sliding collar structure in which the kinetochore is viewed as being made up of several sleeves or collars, with the spindle fibers being inserted into the lumen of the sleeves.[80] According to this hypothesis, each sleeve would make multiple contacts along the surface of the microtubule while leaving the microtubule end free for assembly and disassembly. As the microtubule disassembled, the kinetochore would move poleward along the wall of the microtubule and, thus, insert more of the microtubule into the sleeve. The nature of the attachments between the microtubule and collar is not known, but two mechanisms have been proposed (see Figure 9). According to the first hypothesis, the interactions between the collars and microtubules would be due

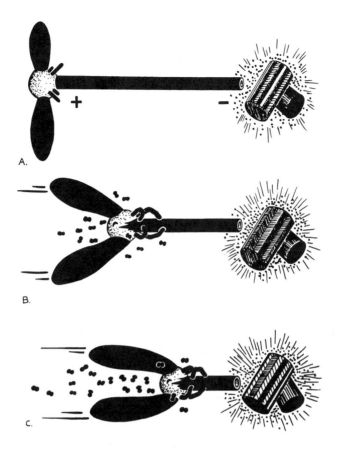

FIGURE 9.   A schematic representation of kinetochore-microtubule interaction. The centrosome is to the right and the chromosome is to the left in
each panel. One of the sleeves of the kinetochore is drawn as finger-like
projections located at the centromere region. (A) The '' + '' end of a microtubule is captured by the kinetochore. (B and C) The chromosome moves
toward the spindle pole during anaphase. As the chromosome moves, tubulin
subunits fall off of the '' + '' end of the microtubule and the kinetochore
"walks" along the surface lattice of the microtubule. As the kinetochore moves
toward the spindle pole, the microtubule is moved deeper into the sleeve. The
nature of both the kinetochore-microtubule attachment and the "walking"
motion is unknown.

to a weak attractive force.[81] Depolymerization of the spindle microtubules would result in
a diffusion of the chromosome along the microtubule toward the spindle pole. The movement
of the chromosome would be driven by the lateral interactions between the microtubule and
the wall of the sleeve. According to the second proposal, the attachment of the chromosome
to the spindle microtubule would be due to a mechanochemical ATPase molecule. As already
mentioned, Mitchison and Kirschner[1] have provided evidence for an ATPase activity in the
kinetochore. The attachments between the wall of the microtubules and the collars as well
as chromosome movement would be due to the ATPase. As the microtubule disassembles,
the ATPase molecules would walk along the microtubule toward the spindle pole.[82]

V. SUMMARY

The kinetochore is only one component of a complex mitotic apparatus composed of
kinetochore and interpolar microtubules, spindle poles, membranes, and a host of regulatory

ions and molecules that are only beginning to be characterized. Much of the recent evidence suggests, however, that the kinetochore is one of the more important pieces of the mitotic puzzle.

For years, it was believed that the kinetochore played little more than an attachment role during cell division. This idea is changing, and it appears that the kinetochore performs several different functions during spindle formation and chromosome segregation. The kinetochore has the ability to efficiently capture microtubules, appears to be important in the prometaphase congression of chromosomes, regulates microtubule assembly and disassembly, and may either contain or somehow regulate the motor that drives the anaphase chromosome movements. How the kinetochore is able to fulfill these various roles is still unknown, although an understanding of the biochemistry of the kinetochore-microtubule interface should contribute much to our understanding of these phenomena.

If one visualizes the kinetochore-microtubule interface along the simple schematic of microtubules → tubulin → kinetochore proteins → DNA binding proteins → centromeric DNA, it is easy to see how rapidly our knowledge of the kinetochore has increased over the past few years. Certainly much was known about microtubules and tubulin, but only recently have we been able to begin to understand the biochemistry and molecular biology of the centromere/kinetochore region. With yeast, centromeric DNA sequences have been isolated that confer mitotic and meiotic stability to plasmids. Although very little is known about kinetochore DNA sequences in higher eukaryotes, it is possible that the larger kinetochores of mammals are just elaborations of the basic yeast kinetochore. In this respect, the CEN sequences from yeast should be invaluable to our understanding of kinetochore biology. The discovery of autoantibodies in the sera of patients with scleroderma CREST has allowed for much to be learned about the proteins of the mammalian kinetochore. Several proteins have been identified on immunoblots of whole chromosomes using CREST sera, and the gene that codes for one of these proteins has been cloned and sequenced. The CREST sera will also be useful for purifying the kinetochore proteins and, hopefully, functional kinetochores.

In conclusion, it is becoming increasingly apparent that the kinetochore plays a pivotal role in the events of cell division. With time, the kinetochore will be biochemically and molecularly dissected, and the roles that the individual biochemical constituents play during mitosis and meiosis will be understood.

ACKNOWLEDGMENTS

We would like to thank Cynthia Webster for carefully typing the manuscript and Albert Tousson for allowing us to use his outstanding illustration of kinetochore-microtubule interaction. This work was supported by grant numbers 5 R01 CA41424 from the HHS-NCI to B. R. B. R. D. B. is an NIH-sponsored postdoctoral fellow (fellowship number F32-GM10477).

REFERENCES

1. **Mitchison, T. J. and Kirschner, M. W.,** Properties of the kinetochore *in vitro*. II. Microtubule capture and ATP-dependent translocation, *J. Cell Biol.,* 101, 766, 1985.
2. **Gorbsky, G. J., Sammak, P. J., and Borisy, G. G.,** Chromosomes move poleward in anaphase along stationary microtubules that coordinately disassemble from the kinetochore, *J. Cell Biol.,* 104, 9, 1987.
3. **Mitchison, T., Evans, L., Schulze, E., and Kirschner, M. W.,** Sites of microtubule assembly and disassembly in the mitotic spindle, *Cell,* 45, 515, 1986.

4. **Cox, J. V., Schenk, E. A., and Olmsted, J. B.,** Human anticentromere antibodies — distribution, characterization of antigens, and effect on microtubule organization, *Cell,* 35, 331, 1983.
5. **Earnshaw, W. C., Halligan, N., Cooke, C., and Rothfield, N.,** The kinetochore is part of the metaphase chromosome scaffold, *J. Cell Biol.,* 98, 352, 1984.
6. **Guldner, H. H., Lakomek, H. J., and Bautz, F. A.,** Human anticentromere sera recongnizes a 19.5 kD non-histone chromosomal protein from HeLa cells, *Clin. Exp. Immunol.,* 5, 13, 1984.
7. **Earnshaw, W. C. and Rothfield, N.,** Identification of a family of human centromere proteins using autoimmune sera from patients with scleroderma, *Chromosoma,* 91, 313, 1985.
8. **Valdivia, M. M. and Brinkley, B. R.,** Fractionation and initial characterization of the kinetochore from mammalian metaphase chromosomes, *J. Cell Biol.,* 101, 1124, 1985.
9. **Palmer, D. K., O'Day, K., Wener, M. H., Andrews, B. S., and Margolis, R. L.,** A 17 kD centromere protein (CENP-A) copurifies with nucleosome core particles and with histones, *J. Cell Biol.,* 104, 805, 1987.
10. **Kingwell, B. and Rattner, J. B.,** Mammalian kinetochore/centromere composition: a 50 kDa antigen is present in the mammalian kinetochore/centromere, *Chromosoma,* 95, 403, 1987.
11. **Clarke, L. and Carbon, J.,** Isolation of a yeast centromere and construction of functional small circular chromosomes, *Nature (London),* 287, 504, 1980.
12. **Bloom, K. S. and Carbon, J.,** Yeast centromere DNA is in a unique and highly ordered structure in chromosomes and small circular minichromosomes, *Cell,* 29, 305, 1982.
13. **Bloom, K. S., Fitzgerald-Hayes, M., and Carbon, J.,** Structural analysis and sequence organization of yeast centromeres, *Cold Spring Harbor Symp. Quant. Biol.,* 47, 1175, 1982.
14. **Clarke, L. and Carbon, J.,** Genomic substitutions of centromeres in *Saccharomyces cerevisiae, Nature (London),* 305, 23, 1983.
15. **Murray, A. W. and Szostak, J. W.,** Construction of artificial chromosomes in yeast, *Nature (London),* 305, 189, 1983.
16. **Bloom, K. S., Amaya, E., Carbon, J., Clarke, L., Hill, A., and Yeh, E.,** Chromatin conformation of yeast centromeres, *J. Cell Biol.,* 99, 1559, 1984.
17. **Carbon, J.,** Yeast centromeres: structures and function, *Cell,* 37, 351, 1984.
18. **Blackburn, E. H. and Szostak, J. W.,** The molecular structure of centromeres and telomeres, *Annu. Rev. Biochem.,* 53, 163, 1984.
19. **Hieter, P. A., Mann, C., Snyder, M. P., and Davis, R. W.,** Mitotic stability of yeast of chromosomes: a colony color assay that measures nondisjunction and chromosome loss, *Cell,* 40, 381, 1984.
20. **Clarke, L. and Carbon, J.,** The structure and function of yeast centromeres, *Annu. Rev. Genet.,* 19, 29, 1985.
21. **Murray, A. W. and Szostak, J. W.,** Chromosome segregation in mitosis and meiosis, *Annu. Rev. Cell Biol.,* 1, 289, 1985.
22. **Ng, R., Cumberledge, S., and Carbon, J.,** Structure and function of centromeres, *Yeast Cell Biol. UCLA Symp. Mol. Cell. Biol.,* 33, 225, 1986.
23. **Schrader, F.,** *Mitosis,* 2nd ed., Columbia University Press, New York, 1953, 1.
24. **Rieder, C. L.,** The formation, structure, and composition of the mammalian kinetochore and kinetochore fiber, *Int. Rev. Cytol.,* 79, 1, 1982.
25. **Godward, M. B. E.,** The kinetochore, *Int. Rev. Cytol.,* 94, 77, 1985.
26. **Rattner, J. B., Krystal, G., and Hamkalo, B. A.,** Selective digestion of mouse metaphase chromosomes, *Chromosoma,* 66, 259, 1978.
27. **Brinkley, B. R., Cox, S. M., and Pepper, D. A.,** Structure of the mitotic apparatus and chromosomes after hypotonic treatment of mammalian cells *in vitro, Cytogenet. Cell Genet.,* 26, 165, 1980.
28. **Ris, H. and Witt, P. L.,** Structure of the mammalian kinetochore, *Chromosoma,* 82, 153, 1981.
29. **John, B. and Miklos, G. L.,** Functional aspects of satellite DNA and heterochromatin, *Int. Rev. Cytol.,* 58, 1, 1979.
30. **Miklos, G. L. and John, B.,** Heterochromatin and satellite DNA in man — properties and prospects, *Am. J. Hum. Genet.,* 31, 264, 1979.
31. **Stubblefield, E.,** The structure of mammalian chromosomes, *Int. Rev. Cytol.,* 35, 1, 1973.
32. **Vig, B. K.,** Sequence of centromere separation — an analysis of mitotic chromosomes from long-term cultures of *Potorus* cells, *Cytogenet. Cell Genet.,* 31, 129, 1981.
33. **Brenner, S. L., Pepper, D., Berns, M. W., Tan, E., and Brinkley, B. R.,** Kinetochore structure, duplication, and distribution in mammalian cells — analysis by human autoantibodies from scleroderma patients, *J. Cell Biol.,* 91, 95, 1981.
34. **Brinkley, B. R. and Stubblefield, E.,** The fine structure of the kinetochore of a mammalian cell *in vitro, Chromosoma,* 19, 28, 1986.
35. **Jokelainen, P. T.,** The ultrastructural and spatial organization of the metaphase kinetochore in mitotic rat cells, *J. Ultrastruct. Res.,* 19, 19, 1967.

36. **Mitchison, T. J. and Kirschner, M. W.,** Properties of the kinetochore *in vitro*. I. Microtubule nucleation and tubulin binding, *J. Cell Biol.,* 101, 755, 1985.

37. **Luykx, P.,** Cellular mechanisms of chromosome distribution, *Int. Rev. Cytol. Suppl.,* 2, 1, 1970.

38. **Meacock, P. A. and Cohen, S. N.,** Partitioning of bacterial plasmids during cell division: a *cis*-acting locus that accomplishes stable plasmid inheritance, *Cell,* 20, 529, 1980.

39. **Austin, S. and Abeles, A.,** Partition of unit copy miniplasmids to daughter cells. I. P1 miniplasmids and F miniplasmids contain discrete, interchangeable sequences sufficient to promote equipartition, *J. Mol. Biol.,* 169, 373, 1983.

40. **Austin, S. and Abeles, A.,** Partition of unit copy miniplasmids to daughter cells. II. The partition region of miniplasmid P1 encodes an essential protein and a centromere-like site at which it acts, *J. Mol. Biol.,* 169, 373, 1983.

41. **Amaya, E., Kenna, M., and Bloom, K.,** Isolation of the centromere chromatin complex from yeast, *J. Cell Biol.,* 103, 413a, 1986.

42. **Shi, L., Ye, Y., and Duan, X.,** Comparative cytogenetic studies on the red muntjac, Chinese muntjac, and their F1 hybrids, *Cytogenet. Cell Genet.,* 26, 22, 1980.

43. **Wurster, D. H. and Bernirschke, K.,** Indian muntjac *Muntiacus muntjac*: a deer with a low diploid chromosome number, *Science,* 168, 1364, 1970.

44. **Schmidtke, J., Brennecke, H., Schmid, M., Nietzel, H., and Sperling, K.,** Evolution of muntjac DNA, *Chromosoma,* 84, 187, 1981.

45. **Brinkley, B. R., Valdivia, M. M., Tousson, A., and Brenner, S. L.,** Compound kinetochores of the Indian muntjac: evolution by linear fusion of unit kinetochores, *Chromosoma,* 91, 1, 1984.

46. **Brinkley, B. R., Brenner, S. L., Hall, J. M., Tousson, A., Balczon, R. D., and Valdivia, M. M.,** Arrangements of kinetochores in mouse cells during meiosis and spermiogenesis, *Chromosoma,* 94, 309, 1986.

47. **Bajer, A. S.,** Substructure of the kinetochore and reorganization of kinetochore microtubules during early prometaphase in *Haemanthus* endosperm, *Eur. J. Cell Biol.,* 42, 23, 1987.

48. **Tan, E. M., Rodnan, G. P., Garcia, I., Moroi, Y., Fritzler, M. J., and Peebles, C.,** Diversity of antinuclear antibodies in progressive systemic sclerosis — anti-centromere antibody and its relationship to CREST syndrome, *Arthritis Rheum.,* 23, 617, 1980.

49. **Moroi, Y., Peebles, C., Fritzler, M. J., Steigerwald, J., and Tan, E. M.,** Autoantibody to centromere (kinetochore) in scleroderma sera, *Proc. Natl. Acad. Sci. U.S.A.,* 77, 1627, 1980.

50. **Valdivia, M. M., Maul, G. G., Jimenez, S. A., Kidd, V. J., and Brinkley, B. R.,** Putative kinetochore cDNA clone isolated by using human autoantibodies from scleroderma CREST patients, *J. Cell Biol.,* 103, 491a, 1986.

51. **Earnshaw, W. C., Sullivan, K. F., Machlin, P. S., Cooke, C. A., Kaiser, D. A., Pollard, T. D., Rothfield, N. F., and Cleveland, D. W.,** Molecular cloning of cDNA for CENP-B, the major human centromere autoantigen, *J. Cell Biol.,* 104, 817, 1987.

52. **Balczon, R. D. and Brinkley, B. R.,** Tubulin interaction with kinetochore proteins: analysis by *in vitro* assembly and chemical crosslinking, *J. Cell Biol.,* 105, 855, 1987.

53. **Brinkley, B. R., Zinkowski, R. P., Mollon, W. L., and Rao, P. N.,** *In situ* detachment of kinetochores from mammalian chromosomes: spindle assembly and kinetochore segregation in the absence of chromosomes, *J. Cell Biol.,* 105, 176a, 1987.

54. **Brinkley, B. R. and Shaw, M. W.,** The ultrastructural aspects of chromosome damage, in *Genetic Causes and Neoplasia, 23rd Annu. Symp. Fundamental Cancer Research,* M. D. Anderson Hospital, Williams & Wilkins, Baltimore, 1970, 313.

55. **Schlegel, R. and Pardee, A. B.,** Caffeine-induced uncoupling of mitosis from the completion of DNA replication in mammalian cells, *Science,* 232, 1264, 1986.

56. **Mitchison, T. and Kirschner, M. W.,** Dynamic instability of microtubule growth, *Nature (London),* 312, 237, 1984.

57. **Euteneuer, U. and McIntosh, J. R.,** Structural polarity of kinetochore microtubules in PtK1 cells, *J. Cell Biol.,* 89, 338, 1981.

58. **Mitchison, T. and Kirschner, M. W.,** Microtubule assembly nucleated by isolated centrosomes, *Nature (London),* 312, 232, 1984.

59. **Kirschner, M. W. and Mitchison, T.,** Beyond self assembly: from microtubules to morphogenesis, *Cell,* 45, 329, 1986.

60. **Horio, T. and Hotani, H.,** Visualization of the dynamic instability of individual microtubules by darkfield microscopy, *Nature (London),* 321, 605, 1986.

61. **Ellis, G. W. and Begg, D. A.,** Chromosome micromanipulation studies, in *Mitosis/Cytokinesis,* Zimmerman, A. M. and Forer, A., Eds., Academic Press, New York, 1981, 155.

62. **Taylor, J. H., Haut, W. F., and Tung, G.,** Effect of fluorodeoxyuridine on RNA replication, chromosome breakage, and reunion, *Proc. Natl. Acad. Sci. U.S.A.,* 48, 190, 1962.

63. **McGill, M. and Brinkley, B. R.,** Human chromosomes and centrioles as nucleating sites for *in vitro* assembly of microtubules from bovine brain tubulin, *J. Cell Biol.,* 67, 189, 1975.

64. **Telzer, B. R., Moses, M. J., and Rosenbaum, J. L.,** Assembly of microtubules onto kinetochores of isolated mitotic chromosomes of HeLa cells, *Proc. Natl. Acad. Sci. U.S.A.,* 72, 4023, 1975.

65. **Snyder, J. A. and McIntosh, J. R.,** Initiation and growth of microtubules from mitotic centers in lysed mammalian cells, *J. Cell Biol.,* 67, 744, 1975.

66. **Gould, R. R. and Borisy, G. G.,** Quantitative initiation of microtubule assembly by chromosomes from Chinese hamster ovary cells, *Exp. Cell Res.,* 113, 369, 1978.

67. **Brinkley, B. R., Stubblefield, E., and Hsu, T. C.,** The effects of colcemid inhibition and reversal on the fine structure of the mitotic apparatus of Chinese hamster cells *in vitro, J. Ultrastruct. Res.,* 19, 1, 1967.

68. **Rieder, C. L. and Borisy, G. G.,** The attachment of kinetochores to the prometaphase spindle in PtK1 cells — recovery from low temperature treatment, *Chromosoma,* 8a, 693, 1981.

69. **Mazia, D., Paweletz, N., Sluder, G., and Finze, E. M.,** Cooperation of kinetochore and poles in the establishment of monopolar mitotic apparatus, *Proc. Natl. Acad. Sci. U.S.A.,* 78, 377, 1981.

70. **Schatten, H., Schatten, G., Petzelt, C., and Mazia, D.,** Effects of griseofulvin on fertilization and early development in sea urchins — independence of DNA synthesis, chromosome condensation, and cytokinesis from microtubule-mediated events, *Eur. J. Cell Biol.,* 27, 74, 1982.

71. **Sluder, G. and Rieder, C. L.,** Experimental separation of pronuclei in fertilized sea urchin eggs — chromosomes do not organize a spindle in the absence of centrosomes, *J. Cell Biol.,* 100, 897, 1985.

72. **Church, K., Nicklas, R. B., and Lin, H. P. D.,** Micromanipulated bivalents can trigger mini-spindle formation in *Drosophila melanogaster* spermatocyte cytoplasm, *J. Cell Biol.,* 103, 2765, 1986.

73. **Steffen, W., Fuge, H., Dietz, R., Bastmeyer, M., and Muller, G.,** Aster-free spindle poles in insect spermatocytes — evidence for chromosome-induced spindle formation, *J. Cell Biol.,* 102, 1679, 1986.

74. **Nicklas, R. B.,** Chromosomes and kinetochores do more in mitosis than previously thought, in *Chromosome Structure and Function: The Impact of New Concepts,* Gustafson, J. P., Appels, R., and Kaufman, R. J., Eds., Plenum Press, New York, in press.

75. **Bajer, A.,** Functional autonomy of monopolar spindle and evidence for oscillatory movement in mitosis, *J. Cell Biol.,* 93, 33, 1982.

76. **Hays, T. S. and Salmon, E. D.,** Poleward traction force on a kinetochore at metaphase is a product of the number of kinetochore microtubules and fiber length, *J. Cell Biol.,* 101, 6a, 1985.

77. **Rieder, C. L., Davison, E. A., Jensen, L. C. W., Cassimeris, L., and Salmon, R. D.,** Oscillatory movements of mono-oriented chromosomes and their position relative to the spindle pole result from the ejection properties of the aster and half spindle, *J. Cell Biol.,* 103, 581, 1986.

78. **Cassimeris, L. U., Walker, R. A., Pryer, N. K., and Salmon, E. D.,** Dynamic instability of microtubules, *Bioessays,* 7, 149, 1987.

79. **Inoue, S. and Sato, H.,** Cell motility by labile association of macromolecules. The nature of spindle fibers and their role in chromosome movement, *J. Gen. Physiol.,* 50, 259, 1967.

80. **Margolis, R. L. and Wilson, K.,** Microtubule treadmills — possible molecular machinery, *Nature (London),* 293, 705, 1981.

81. **Hill, T. L.,** Theoretical problems related to the attachment of microtubules to kinetochores, *Proc. Natl. Acad. Sci. U.S.A.,* 82, 4404, 1985.

82. **Mitchison, T. J.,** The role of microtubule polarity in the movement of kinesin and kinetochores, *J. Cell. Sci. Suppl.,* 5, 121, 1986.

Chapter 7

X. INACTIVATION IN MAMMALS, AN UPDATE

Stanley M. Gartler and Elisabeth Keitges

TABLE OF CONTENTS

I. INTRODUCTION

Animals appear to be exquisitely sensitive to aneuploidy. In humans, a duplication no larger than 0.5% of the genome on chromosome 21 causes Down's syndrome.[1] Such aneuploid phenotypes appear to be due to the cumulative effects of a number of duplicated genes on a chromosome and not to the effects of single regional genes.[2] In fact, of the several thousand individual gene mutations studied in several higher animal species, not a single one results in an aneuploid phenotype. Exactly how segmental aneuploidy results in disturbed phentoypes remains a molecular puzzle, but one implication is clear: normal development in animals requires general chromosomal balance. The requirement for balance applies to the X as well as to the autosomes, with the added necessity that the expression level in females and males be equal. This latter requirement, called dosage compensation, stems from the evolutionary origin of forms with heteromorphic sex chromosomes from a general homomorphic ancestor.[3-5] The other sex chromosome, the Y, seems immune from aneuploid effects; probably due to its generally low genetic content. In order to bring about dosage compensation, genetic systems had to evolve to change the levels of X-linked gene expression in the sexes. In *Drosophila,* most, if not all, X-linked genes that are subject to dosage compensation have an adjacent sequence which responds to a signal that results in a twofold enhancement of transcription in males.[6]

In mammals, the dosage compensation system appears quite different; a single X-linked site responds in the female to a signal that results in the inactivation of one of the two X chromosomes. In fact, the mammalian system operates in such a manner that all X chromosomes over one, in an otherwise euploid organism, are inactivated, thus buffering the mammal from the effects of X chromosome aneuploidy.[4,7] The goal of this review is twofold: (1) to bring the reader up to date with a brief survey on the genetic biology of X chromosomal inactivation, and (2) to consider the attempts of the last 5 years to unravel the molecular biology of this system.

II. THE GENETIC BIOLOGY OF X CHROMOSOME INACTIVATION

The original hypothesis of mammalian X chromosome inactivation was based on observations of variegation in mice heterozygous for X-linked markers affecting coat characteristics.[8] It was assumed that early in development random inactivation turned off one or the other X chromosome in a female somatic cell, and that once inactivation occurred, it became a fixed part of the somatic heredity of the cell. Thus, the mammalian female became a mosaic with some cells expressing the paternal X and others the maternal X chromosome. For characters with a clonal pattern of development, like hair, variegation was regularly observed in animals heterozygous for X-linked genes.

Phenotypic expression of inactivation was accompanied by two major cytological features: (1) the formation of a distinctive heterochromatic structure (sex chromatin or Barr body) visible in the interphase cell and formed by the inactive X chromosome, and (2) asynchrony of replication of the active and inactive X chromosomes. The inactive X chromosome also may appear precociously condensed at prophase and often appears to exhibit a peculiar bend on the long arm at metaphase which has been hypothesized to correspond to the location of the inactivation center.[9,10] Furthermore, the original observations relied heavily on X autosome translocations in which some of the attached autosomal genes also exhibited variegation.[11] These results indicated that inactivation could spread into an adjoining autosomal region. This hypothesis turned out to apply with considerable accuracy to the embryo proper and its derivative adult cells and tissues. However, in earlier developmental stages some rather unexpected facts emerged which lead to a consideration of the ontogeny of X inactivation.

A. DEVELOPMENTAL BIOLOGY OF X CHROMOSOME INACTIVATION

In the zygote and for the first few cleavages, there is no dosage compensation. Apparently, very few embryonic genes are expressed at this stage, but for two X-linked genes, hypoxanthine phosphoribosyl transferase (*Hprt*) and alpha galactosidase (*alpha-gal*), female embryos express twice the level of enzyme activities as male embryos.[12-16] When dosage compensation and X inactivation first occurs, it takes place in the mouse trophectoderm in the early mouse blastocyst (about $3^1/_2$ days)[17] The remainder of the embryo shows no dosage compensation. Furthermore, in the trophectoderm, inactivation is nonrandom, with the paternal X chromosome being preferentially inactivated.[18] Dosage compensation next occurs in the primitive endoderm, which is also extraembryonic like the trophectoderm, and again inactivation is nonrandom with the paternal X chromosome being preferentially inactivated. At about $5^1/_2$ d in the mouse, the cells of the epiblast in the very late blastocyst are inactivated in a random manner.[7] This is the X inactivation event predicted in the original hypothesis and the one that leads to variegation of coat characteristics in heterozygous animals.

The onset of dosage compensation appears to be correlated with early differentiative events; in fact, it would seem that differentiation may be the signal for the onset of dosage compensation.[17] It certainly cannot be the other way around since in the male, differentiation proceeds without the development of dosage compensation. When random inactivation occurs in the epiblast, it apparently affects all cells, including the primitive germ cells. Since we know that the oocyte does not contain a differentiated or imprinted X regarding X inactivation, it follows that the inactive X in oogonia must be reactivated some time prior to conception. In fact, a considerable amount of evidence now indicates that the inactive X in female germ cells is reactivated at the entry to meiosis.[19-21] This fact is quite remarkable since in all other normal tissues the inactivation state is quite stable, even to inductive attempts at reactivation. We do not know the basis of the germ-line reactivation and limitations of material will make molecular approaches quite difficult.

Though the male with a single X chromosome does not undergo X inactivation in somatic cells, there is a male germ-line process which appears related to female X chromosome inactivation. At the entry to spermiogenesis, the single X in the male is precociously condensed and inactivated.[22] Conceivably, this process may act as a sort of imprinting mechanism to predetermine the preferential inactivation of the paternal X chromosome in extraembryonic tissues.

This pattern of X chromosome inactivation has been primarily documented in the mouse, though the broad outline certainly applies to humans. In fact, the female germ cell pattern of inactivation and reactivation was first documented in human material. However, there is controversy regarding the pattern of inactivation of extraembryonic tissues in humans; three groups have reported preferential paternal inactivation in human extraembryonic tissues.[23,23a,24] Another group reported no preferential paternal inactivation[25] and then later reported incomplete dosage compensation in human chorionic villi.[26] It seems possible that these differences may reflect different methods of tissue sampling and/or changes in extraembryonic inactivation with development.

B. RANDOM VS. PREFERENTIAL INACTIVATION

As we have just noted, preferential inactivation of the paternal X chromosome occurs in extraembryonic membranes. Presumably, this distinction involves some sort of parental imprinting in which either the paternal X is modified during spermiogenesis to insure inactivation or the maternal X is modified during oogenesis to prevent inactivation. The requirement for inactivation overrides the imprinting signal since in parthenogenetic embryos, inactivation takes place in extraembryonic tissues.[27,28] Since the paternal X is precociously repressed during spermiogenesis, it is tempting to think in terms of the paternal X being imprinted for preferential inactivation. However, when one considers the fact that the paternal

X is active during early cleavages before it is preferentially inactivated in the trophectoderm, it may be simpler to think in terms of imprinting of the maternal X to prevent inactivation.

In the case of X chromosome abnormalities, nonrandom inactivation can occur in somatic cells. The basic explanation of nonrandomness in these instances is cell selection. That is, X inactivation occurs initially in a random fashion, but selection eliminates those cells with a karyotypic imbalance. For example, in X chromosome deletions it is the deleted X which is always inactivated.[29] An important exception to this particular case is when the inactivation center on the long arm is deleted; then, the remaining chromosomal material cannot be inactivated. These patterns have been shown to be due to selection of the most balanced karyotype rather than to preferential inactivation.[30-32] In balanced X autosome translocations, inactivation of the translocated X leads to spreading of inactivation into the adjoining autosomal material, thus creating a partial autosomal monosomic condition. Therefore, the normal X is seen to be nonrandomly inactivated. The converse is true in unbalanced X autosome translocations where spreading of inactivation into the partially trisomic autosome confers a selective advantage, and the translocated X is nonrandomly inactivated.[33]

C. STABILITY OF THE X INACTIVATION STATE

Gross reactivation such as the programmed reactivation of the female germ line at oogenesis is easy to detect. Not only are a large number of cells involved, but the whole X chromosome is reactivated. On the other hand, low rates of reactivation involving only one gene are extremely difficult to detect without a clonal system and a selectable or easily screenable marker. Two such systems have been developed in the mouse and in both cases evidence has been presented for some degree of reactivation with age. Cattanach reported some years ago that an autosomal gene on an X autosome translocation, which affected coat pigmentation, exhibited evidence of reactivation with age.[34] Since regular X-linked genes affecting the coat phenotype did not seem to show such changes, these observations were interpreted as suggesting that the autosomal portions of the translocation were not inactivated in the same way as the X chromosome segment. However, recently it has been reported that X-linked ornithine transcarbamylase (*OTC*) showed a similar reactivation pattern with age, as determined histochemically.[35] These observations are most interesting, but they remain to be confirmed and supported by more definitive evidence. In humans, studies of clonal myomas of the uterus in heterozygotes for glucose-6-phosphate dehydrogenase electrophoretic variants indicate that inactivation is quite stable regardless of age.[36]

In contrast to *in vivo* studies, it is possible to screen for very low levels of reactivation in cell culture systems. Normal cell cultures are extremely stable, with only a single reactivant having been reported. It is likely that the spontaneous reactivation rate in normal cells is of the order of 1×10^{-8}.[37] The first strong evidence for consistent instability in any system was the rodent-human hybrid cell containing an inactive human X chromosome. Kahan and DeMars made the first such hybrid and showed that the *Hprt* gene, on the inactive X, spontaneously reactivated at a rate of 1×10^{-6} in these cells.[38,38a] Since the hybrids segregated human chromosomes, these workers were able to show that the inactivation state was not related to the presence or absence of a particular autosome. Later, Mohandas et al.[39] showed that 5 azacytidine (5AC) could enhance reactivation rates in such cells by 1000-fold. More recently, Ellis et al.[40] have reported 5AC-induced reactivation rates of the *Hprt* and glucose-6-phosphate dehydrogenase (*G6pd*) genes in a hybrid cell line of close to 50%. There have been a number of further reports showing instability of the inactive X in a variety of hybrid cells and transformed cells.[41] It seems clear from these studies that the inactive X in a hybrid or transformed cell background becomes potentially very unstable. The molecular basis for this instability is not immediately apparent, but different possibilities will be considered in our analysis of the initiation of X chromosome inactivation.

D. IS THE ENTIRE X CHROMOSOME INACTIVATED?

It was originally hypothesized that the entire X chromosome was inactivated. However, a variety of clinical observations led to the suggestion that the distal tip of the short arm of the human X might escape inactivation. It was argued that a dosage effect in this region might be necessary for normal female sexual differentiation.[42] Recent studies have demonstrated that in both humans[43-45] and mice[46,47] a true X-Y pairing region exists with homologous sequences and obligatory recombination. It would seem that genes within this region would not require or exhibit dosage compensation and, in fact, this is the case for the *Mic2* locus in humans[48] and the murine steroid sulfatase (*Sts*) gene.[49] The *Sts* gene in humans is proximal to and just outside the pairing region and partially escapes inactivation.[50-53] In humans, the pairing region has been estimated to be of the order of 5×10^6 bp, which is about 3% of the X chromosome.[43] A physical estimate of the size of the pairing region in the mouse has not been made, though genetically it may be larger.

There is no *a priori* reason to assume that genes in any other region of the X chromosome are not subject to inactivation. However, one must keep in mind that relatively few X-linked genes have been shown to exhibit a pattern of X chromosome inactivation. It should be pointed out that differences between active and inactive X-linked genes, such as chromatin structure, appear to be confined to expressed genes. Nonexpressed regions, as far as can be told at present, are not distinguishable between the active and inactive X chromosomes.[54,55]

E. IS INITIATION OF INACTIVATION UNI- OR MULTIFOCAL?

Since X inactivation is chromosomal, it has been widely assumed that the system is initiated from a single site. It was difficult to conceive of a multifocal initiation mechanism giving rise to one chromosome active and one inactive. Cattanach developed the concept of an inactivation center and mapped it in the mouse by selecting for variegation changes in X-linked markers affecting coat phenotype.[56] Cytological observations further supported the unifocal hypothesis in that only one part of an X chromosome translocation could be inactivated. In humans, the short arm of the X chromosome, when translocated to an autosome, cannot become inactivated, supporting cytological localization of an inactivation center proximal on the long arm.[7,57] Because reactivation was multifocal,[39,40] some workers made the extrapolation that initiation of inactivation must also be multifocal. However, the reactivation data do not contradict the inactivation center concept; rather, they show that inactivation is a multistep process, with the unifocal inactivation center representing the first step. The complicated, differential replication kinetics of the active and inactive chromosomes was also used to argue for a multifocal initiation hypothesis. How the inactivation center operates and its molecular identification are the key questions of X inactivation.

F. LOCALIZATION OF THE INACTIVATION CENTER

Cattanach first developed the concept of a single inactivation center which he termed *Xce* and carried out selection experiments in mice to try and localize it.[56] His rationale was that mutations in a *Xce* gene should affect the ability of the X carrying the mutant to be inactivated and so he selected for changes in variegation of X-linked coat characters. He achieved a limited degree of success, obtaining lines that would regularly inactivate at levels different from random and mapped the *Xce* locus in the vicinity of the phosphoglycerate kinase gene *Pgk*. This was an important concept and it would be most interesting if an X with a completely deficient *Xce* could be obtained. Inactivation would be completely nonrandom, which might have some effect on the female; and if the inactivation center also functions during male spermiogenesis, such a mutant should probably affect male fertility sex-chromatin bodies in duplicated Xs[58] and bends in the inactive X[9] have been interpreted as placing the *Xce* proximal on the long arm of the human X.

III. MOLECULAR BIOLOGY OF X CHROMOSOME INACTIVATION

In this section, we will consider what is known at the molecular level about differences between active and inactive X chromosomes. We will also consider problems that remain, which are many, and possible approaches to their solution. It is convenient to think of the inactivation process in terms of initiation, spreading, and maintenance. Whether this concept turns out to be realistic remains to be seen, but for didactic purposes it seems quite useful.

A. INITIATION OF X INACTIVATION

One of the striking features about mammalian X inactivation is that in cells with euploid autosomal complements, all but one X chromosome is inactivated, regardless of the number of X chromosomes. For example, in a 5X cell there are four inactive X chromosomes. How the cell arrives at this state is still a molecular puzzle, but it has to be assumed that the number of inactivating sites is determined in some way by the ratio of autosomes to X chromosomes.

The first experiment to reveal some insight into the molecular basis of X inactivation was the transformation experiment of Liskay and Evans, in which it was demonstrated that DNA from the inactive X chromosome did not function in transfection experiments.[59] This result was interpreted as indicating a DNA modification on the inactive X. This important observation was confirmed[60,61] and extended to show that 5AC-reactivated X chromosome DNA became functional in transformation,[62,63] supporting a role for methylation in X inactivation. Later, it was reported that DNA from the inactive X of extraembryonic tissue was able to function in transformation,[64] suggesting that the initial event of X inactivation does not involve DNA modification. Thus, inactivation in extraembryonic tissues differs in two ways from inactivation in the embryo proper: (1) nonrandom vs. random inactivation and (2) nonmodification vs. DNA modification. A common element in inactivation of extraembryonic and true embryonic tissues is sex chromatin formation; this leads to the idea that the initial event in X inactivation is sex chromatin formation without DNA modification, and it may be the only level of X chromosome repression in extraembryonic tissues. Later, when inactivation occurs in the true embryonic cells, sex chromatin formation is again the initial event, but it is followed by differential DNA modification in the form of methylation. Thus, X inactivation in somatic cells would have two levels of repression, with differential methylation acting as a type of fail-safe mechanism. A recent study by Lock et al.[65] has reported evidence for differential methylation following inactivation during murine development.

Our own work has focused on sex chromatin as the important initiating event in X inactivation. High voltage electron microscopy supported much earlier ideas of a unique nuclear envelope attachment site as part of the mechanism of sex chromatin formation.[66] This observation in turn led to the hypothesis that the basis for the instability of the inactive human X in rodent-human hybrids was related to problems of forming normal sex chromatin in a primarily rodent cell.[67] If attachment to a nuclear envelope site was an important step in sex chromatin formation, then a predominantly rodent nuclear envelope might not be a good substrate for human sex chromatin formation. Thus, in hybrid cells the first level of repression of the inactive X, heterochromatization, would not be functioning normally, and genes on the inactive X could be reactivated by simple alterations in methylation. It is clear from a number of studies that methylation patterns, both in cell culture and the body, may not be terribly stable and, of course, with 5AC treatment, very significant changes in methylation can be achieved. In the normal cell, on the other hand, even though demethylation may occur, sex chromatin structure is not necessarily affected and repression is still in effect. Studies from our lab with a human X chromosome centromere probe in normal and hybrid cells have produced results supportive of this hypothesis.[67a] Whereas in normal human cells the sex chromatin centromere signal is predominantly compact, peripheral in location, and always over a sex chromatin body, this is not the case in hybrid cells. In the majority of instances in hybrid cells, the human X centromeric signal is not associated with a recognizable sex chromatin structure, and is often disperse and not peripheral in location. These results support a two-step model of inactivation in which sex chromatin formation is followed by differential methylation and also serves to explain the instability of inactive X-linked genes in rodent-human hybrid cells.

A simple extension of the above results is to assume that the inactivation center is equivalent to the X chromosome attachment site which promotes sex chromatin formation. Several approaches are possible to try to identify and isolate such a site. Purely genetic studies in which selection is applied for alterations in variegation are one possibility. As pointed out earlier, Cattanach initiated such work and identified a locus, Xce, as a candidate for such a site.[56] Unfortunately, because of the limited selection pressure that can be applied to this system, the genetic identification of Xce is not of much value for molecular studies. A modified reverse genetic approach is possible in which large pieces of X chromosome DNA in the vicinity of the Pgk gene are transfected into recipient cells capable of undergoing X chromosome inactivation. The aim would be to integrate into an autosome and look for a sex chromatin-like structure and/or inactivation of genes adjoining the Xce. Several such recipient cells are available, including teratocarcinomas and transgenic mouse systems. The major assumption underlying the successful completion of such experiments is the autonomous behavior of Xce. It could be that special signals are required for spreading of the sex chromatin body information; if so, they would certainly be restricted to the X chromosome and such experiments would yield only negative results.

A third approach is largely biochemical in which one focuses on a search for unique chromosomal proteins or nuclear envelope proteins in cells with multiple sex chromatin bodies. Comparisons would, of course, be made with cells lacking sex chromatin and candidate proteins would be studied for binding affinities. Two groups have studied nuclear proteins from metaphase chromosomes in this way, but reported negative results.[68,69] Recently, a third group reported the possible identification of a nonhistone nuclear protein which may be associated with the inactive X.[69a] Another approach would be to look for DNA-binding proteins from nuclear envelope extracts with selected pieces of X chromosome DNA in the vicinity of Pgk, which is near the proposed Xce. All these approaches suffer from the very high possibility of negative results, but these paths may be the only way to isolate the initiating factor of X inactivation.

B. SPREADING OF INACTIVATION

Spreading of inactivation is a chromosomal event. With the exception of the active pairing region, the entire X is inactivated, beginning from the inactivation center and spreading through some 10^8 bp of DNA and never acting in *trans* on the other X chromosome. We would like to address two questions in this section: (1) how the inactivation signal spreads along the X chromosome and (2) how inactivation spreads into adjoining ectopic sites.

1. Spreading Along the X

We have postulated that inactivation of the X has at least two levels of control: the primary inactivation event, sex chromatin formation, and a secondary DNA modification system involving methylation. Unfortunately, little is known about the primary inactivation event, but several cytological and molecular features differentiate the active and inactive X chromosomes and their genes. The inactive X forms a heterochromatic sex chromatin body and replicates later than the active X. In true embryonic and adult cells, X-linked housekeeping genes are further differentiated by methylation and endonuclease sensitivity patterns. The GC-rich regions at the 5' end of these genes are hypermethylated and nuclease insensitive on the inactive X, while their alleles in the same cell are hypomethylated and nuclease hypersensitive.[70-76] How this cytological and molecular pattern differentiating the two X chromosomes in the same cell spreads along some 10^8 bp is the essence of the spreading problem.

As indicated in the previous section, we have hypothesized that sex chromatin formation is the initial and only basis of inactivation in extraembryonic tissues and the initial event in the embryo proper. Chromosome condensation is thought to be a highly cooperative process; therefore, it is possible that a change in the inactivation center by a protein or enzyme such as a topoisomerase could result in a domino effect along the entire X. However, it has also been argued that these types of interactions are weak. The question arises whether this initial change at the inactivation center is sufficient, or whether X-specific signals are recognized and possibly propagate the process. If the inactive X chromosome in the form of sex chromatin is compartmentalized on the nuclear envelope as some observations suggest,[77] then unique spreading signals might not be required for either complete sex chromatin formation or differential DNA modification. However, in the absence of some form of compartmental-ization, it would appear essential that there be propagation signals along the X chromosome to maintain the *cis*-active inactive X configuration in the same cell. This latter possibility is intuitively appealing, although repeated efforts to find an X-specific repetitive sequence with the required distribution have failed.[78-80] Arguments for a signal sequence include the fact that the process appears to be X specific. We never observe an entire autosome inac-tivated, although there would be strong selection against such an event. Secondly, spreading of inactivation into autosomal material attached to the X is not stable or complete.[81,82] Finally, within the confines of a single X, inactivation appears to be able to recognize X-linked material and inactivate it even if it must jump over the active pairing region, as in the recombinant X described by Mohandas et al.[83] or the Cattanach translocation with an insertion of autosomal material into the X.[56]

The 5' sequences of several X-linked genes have been compared and no conserved sequences which might act as an inactivation recognition sequence were found.[84-89] If the sites exist, they must be outside the sequenced regions and possibly control large areas instead of particular genes. Riggs et al.[78] developed the enhancer way-station model which utilizes the concept of regional control of spreading. They propose that the GC-rich islands found near some X-linked genes which are differentially methylated may act as enhancers and propagation signals for spreading. The two events act together in that a chromatin configuration change makes the enhancers available for methylation and, in turn, change the chromatin conformation in the next domain. They propose that the islands need to be regularly spaced along the X, not necessarily at the 5' end of each gene. The enhancer function of these sequences amplifies the signal as well as making sure inactivation spreads in a *cis* manner. Because GC-rich islands are not X specific, autosomal genes may or may not have similar sequences present which could account for the variable spreading seen in X autosome translocations. It may be helpful now to look more extensively at the evidence for spreading signals by asking whether the X can inactivate a non-X chromosomal piece of material attached to it. Does the inactivation process recognize the material as foreign and, if so, how?

2. Spreading into Ectopic Sites

The earliest observations from which the hypothesis of X chromosome inactivation were based include the behavior of a murine autosomal pigment marker translocated into the X chromosome (Cattanach's translocation).[90] In animals heterozygous for the translocation and pigment markers, variegation was hypothesized to result from inactivation of the dominant pigment marker when the translocated X was seen to be heteropycnotic. This could only occur if spreading of inactivation into the autosomal insert had taken place.

Observations in human X autosome translocations also suggested that inactivation may spread into the autosomal segment in unbalanced X autosome translocations. An example is a case described by Allderdice et al.[91] where, in an unbalanced X;14 translocation, inactivation had spread from the X into the adjoining translocated autosome and prevented the usual lethal effect of trisomy 14. However, in other translocations it became clear that the entire autosomal segment may not be inactivated as some patients were clearly phenotypically abnormal, such as the case of Palmer et al.,[92] where a much smaller region appeared to remain completely active. The abnormalities seen in these translocations ranged from reproductive failure to multiple congenital anomalies (MCA) and features of the partial trisomy involved. Overall, the data show that 78% of unbalanced X autosome translocations present with MCA or mental retardation.[82]

Therman et al.[58] put forth several possible alternatives to account for the abnormal phenotypes observed: (1) damage could occur before X inactivation at the blastocyst stage, (2) an effect may be exerted after activation of the X chromosome in the germ line, or (3) abnormal chromosome constitutions may be effective during the period of selection. Although these are possibilities, we feel that most of the abnormalities stem from incomplete spreading of inactivation which may be neither continuous nor complete. It should be pointed out that there are some difficulties inherent in the methods used to document molecular spreading of inactivation. Allocycly is a characteristic which differentiates the active and inactive X chromosomes: the inactive X initiates and terminates replication later than the rest of the complement. Bromodeoxyuridine (BrdU) incorporation into the late- or early-replicating regions of the chromosomes is the preferable way to demonstrate allocycly. Spreading of inactivation in X autosome translocations is documented as a change in the replication timing of the translocated autosome. Variation in normal replication timing of autosomes and bands within the autosomes makes documenting a change in replication kinetics difficult and necessitates strict controls. Ideally, the most precise way to demonstrate spreading of inactivation would be to establish its effect on autosomal loci. This has been done in two translocations, the Cattanach translocation in the mouse[93] and an X;21 translocation in man,[94] where a correlation between late replication and gene inactivity was shown.

3. Is Inactivation Continuous?

Russell originally described spreading of inactivation as a continuous gradient,[95] but several observations have suggested this may not be true. In Cattanach's translocation, it was found that genes on either end of the insert are more frequently inactivated than genes in the center of the insert. This could result if either inactivation could jump over the insert or inactivation is unstable in the insert and reactivation occurs. Mohandas et al.[96] documented an inactivated interstitial segment within an active autosomal region in an X;13 translocation, and Keitges et al.[82] reported what appeared to be skipping of inactivation in bands in the autosomal segment in X autosome translocations. Most convincingly, Mohandas et al. described a recombinant X chromosome with a duplication of the region Xq26.3-Xqter and deletion of Xp22.3-Xpter, where the duplicated region was positioned into the active pairing region. They showed using molecular probes that the *Sts* and *Mic2* genes were still present; however, more distal pseudoautosomal sequences were deleted. Dosage studies of the duplicated *G6pd* gene from mouse-human hybrids which retained the rec(X) demonstrated that both copies were inactivated and STS and MIC2 were expressed. Inactivation in this case seemed to skip over what was left of the active pairing region and inactivate the duplicated region of Xq. There is no evidence that the duplicated segment contains the inactivation center.

Related to this issue are two studies in which genes were introduced into the X via transgenic mice. The first was a minigene alpha-fetoprotein (AFP) construct which is normally autosomal. The authors found the gene was inactivated in fetal liver when present on the inactive X.[97] The second was the transferrin gene from the chicken, an organism which does not have X chromosome inactivation.[98] In this case, the gene did not appear to be inactivated. The authors suggest that the different results may be due to the nature, size, or position of the insert into the chromosome. These results are reminiscent of the variation of spreading of inactivation into the autosomal segments seen in X autosome translocations.

4. Is Inactivation Complete?

The replication studies in unbalanced translocations showed that the translocated X is most often inactivated, although in some cases a mixture of cells was found with either the normal or translocated X inactivated. The replication pattern of the autosome again suggested variation in the degree of spreading of inactivation from absent to complete, and tissue differences in the replication pattern have been reported.[99] There is evidence that, within the capabilities of spreading of inactivation, selection for the most balanced karyotype does occur.[100] Those translocations with failure of spreading of inactivation were most likely balanced translocations with a cell line with the translocated X inactivated. X autosome translocations with partial autosomal trisomy are more likely to have spreading of inactivation. It does not seem likely that selection can direct the extent of spreading, but rather that inactivation may spread to varying degrees in different cells and the most balanced cells survive. This variation in the pattern of spreading may be telling us something about the molecular differences between inactivation in the X and spreading of inactivation into autosomal material.

The question is, why does inactivation spread into some translocations and not others? Could the particular autosome involved influence spreading? A comparison of the autosome involved in those translocations with and without spreading of inactivation is shown in Figure 1.[81,82,91,92,94,103-128] There is no evidence in man that would indicate that the autosome is a factor, but there are very few translocations of each particular autosome and there is nothing to say that the breakpoint within an autosome, or specific DNA sequences of the autosome, might not be important. Another possibility is that the X chromosome breakpoint may affect the potential for spreading, i.e., the farther away from the inactivation center, the less likely for spreading to be complete. A comparison of the X chromosome breakpoint and spreading vs. no spreading is shown in Figure 2.[81,82,91,92,94,101-132] Again, there is no evidence that the X chromosome breakpoint is significantly different in those translocations with spreading and those without spreading of inactivation. A possible exception was pointed out by Palmer et al.,[92] in which autosomal segments translocated into the active pairing segment might be more likely to remain active. We know that this is the case with other genes translocated into the pairing region, but detailed molecular studies would be necessary to demonstrate that the breakpoint was in the pairing region.

5. Genes Inserted into the Active Pairing Region

Genes such as the sex reversed mutation (*Sxr*) and *Mov15*, when translocated into the region, usually remain active.[47,133] The mechanism which allows the tip of the X chromosome to remain active on the inactive X chromosome is not known. Burgoyne et al.[134] tested the hypothesis that pairing in male meiosis was sufficient to protect the pairing region from inactivation by producing mice homozygous for the *Sxr* gene and carrying the Searles translocation. They found *Sxr* could still be inactivated, even though it could now pair in meiosis in the preceding generation. So the mechanism which protects the active region is still unknown. However, it should be pointed out that genes in the region can occasionally become inactivated, as in the *Sxr* gene in the Searles translocation[135,136] and the murine *Sts* gene in tissue culture.[137] In addition, the X-linked genes *Sts* and *Xg*, which are normally

Effect of the translocated autosome
on spreading of inactivation.

| | no. of translocations | |
Autosome	spreading of inactivation	no spreading of inactivation
1	• •	
2	•	
3	•	• •
4	• •	
5		•
6	•	•
7	•	
8	•	•
9	• •	• • • •
10	• •	•
11		• •
12		
13	• • • •	
14	• •	
15	• • •	•
16		
17		• • •
18	•	
19		
20		
21	• • •	
22	• • •	• • •
Totals	29	19

FIGURE 1.   Comparison of the autosome involved in those translocations with and without spreading of inactivation into the autosome.[81,82,91,92,94,101-128] The autosome involved did not differ significantly in the two groups; Chi square$_{22}$ = 24.02, 0.5 > p > 0.1.

present in the pairing region, can occasionally be inactivated when present on a structurally abnormal X.[138,139] The most likely explanation is that this is due to aberrant methylation. These studies show that genes in the pairing region are capable of becoming inactivated, but under normal conditions something confers protection to inserted autosomal genes and the X-linked genes normally in the pairing region.

6. Molecular Interpretation of Spreading

A recent study of the murine steroid sulfatase gene is especially illuminating regarding molecular interpretations of spreading. The locus is in the pairing region with functional alleles on both the X and Y chromosomes.[46,140] As expected, the gene under normal conditions completely escapes inactivation.[49] However, most, if not all, established murine cell lines are STS deficient, even though they originate from STS$^+$ mice.[137] Through the development of a selective system enabling detection and selection of STS$^+$ revertants from STS$^-$ populations, it was possible to show that the established murine cell lines were STS$^-$ because of hypermethylation. Massive reactivation was induced by treatment of STS$^-$ lines with the demethylating agent 5AC. These results show that the locus has the signals for methylation and consequent inactivation, but somehow the locus is protected from cellular methylases. Our model supposes that it is the first step in inactivation, sex chromatin formation, that acts as a guide for proper differential methylation of active and inactive X-linked genes.

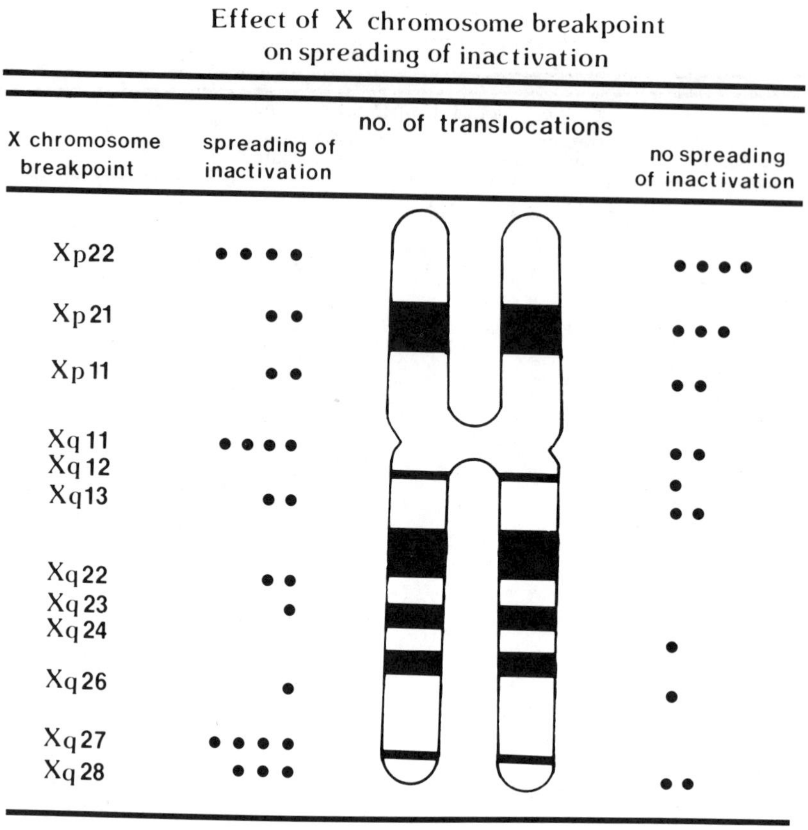

Effect of X chromosome breakpoint on spreading of inactivation

X chromosome breakpoint	no. of translocations		
	spreading of inactivation		no spreading of inactivation
Xp22	• • • •		• • • •
Xp21	• •		• • •
Xp11	• •		• •
Xq11	• • • •		• •
Xq12			•
Xq13	• •		• •
Xq22	• •		
Xq23	•		
Xq24			•
Xq26	•		•
Xq27	• • • •		
Xq28	• • •		• •

FIGURE 2. Comparison of the X chromosome breakpoint involved in those translocations with and without spreading of inactivation into the autosome.[81,82,91,92,94,101-132] The breakpoint involved did not differ significantly in the two groups; Chi square$_{13}$ = 8.33, $0.9 > p > 0.5$.

Under normal conditions, proper chromatin conformation protects the murine *Sts* genes from aberrant methylation. However, under abnormal, prolonged culture, we assume some breakdown in steric hindrance and a resulting hypermethylation. Several cases of hypermethylation and inactivation of autosomal genes maintained under prolonged culture conditions also have also been reported.[141-144] This observation of *Sts* inactivation leads to the idea that spreading requires two sets of signals, one for sex chromatin formation and the other for DNA methylation. The methylation signals are, of course, not X chromosome specific, but what gives them some specificity is the requirement for proper chromatin conformation in order for methylation to take place. Support for this spreading model comes from recent observations in transgenic mice in which the autosomal *AFP* gene has been inserted into the X chromosome. As mentioned earlier, this gene is inactivated in the liver when it is present on the inactive X chromosome. However, when the gene is in the inactive X in extraembryonic cells,[145] it escapes inactivation. Our model would explain this pattern by the fact that only sex chromatin formation is acting in extraembryonic cells and the autosomal *AFP* genes would not have the required specific signals. In the liver, the gene is presumably inactivated by ubiquitous methylation signals and resultant hypermethylation.

The picture that appears to emerge is of two levels of spreading; one is determined by the presence of heterochromatization signals unique to the X chromosome and a second is the presence of the more ubiquitous GC-rich islands which permit spreading of methylation. Therefore, when an X chromosomal segment that is normally subject to inactivation is translocated to, say, the pairing region, as in the rec(X) described by Mohandas et al.,[83] it is still inactivated because it presumably contains signals for heterochromatization. However, when an autosomal segment is inserted into the X chromosome, it may or may not become inactivated, depending on the presence or distribution of GC-rich islands in the region. On the other hand, if the autosomal segment is translocated into the active pairing region, then it is likely to remain active since this region is normally protected from methylation. The complexity of the system is apparent. Confirmation of these observations awaits further studies.

C. MAINTENANCE OF X CHROMOSOME INACTIVATION

The question we want to consider in this section is how a pair of homologous chromosomes in the same cell is maintained in an active and inactive configuration throughout somatic development. This is a most unusual, if not unique, situation in biology. It is a case of intracellular mosaicism at the chromosome level. As we have indicated, the inactivation process appears to involve at least two distinct steps: the first occurring in extraembryonic membranes and involving only sex chromatin formation without any apparent DNA modification, and the second taking place in true embryonic cells where differential DNA methylation follows sex chromatin formation.

With this two-step model, we must invoke two maintenance steps, one for sex chromatin formation and one for differential methylation. Through every cell division, the same X must form a sex chromatin structure and the genes on the X must remain hypermethylated on their 5′ end. Without DNA modification, it is not immediately apparent what type of memory can be involved in sex chromatin formation. If specific attachment to the nuclear envelope is involved, then some part of the nuclear envelope must remain with the inactive X chromosome to ensure inactivation in the following cycle. This is essentially the problem of initiation which we discussed earlier, and not any easier to solve in this particular form. Of some interest in this respect, a recent report has identified a possible inactivating protein unique to the inactive X chromosome.[69a]

In contrast to sex chromatin maintenance, the situation with respect to differential methylation is very attractive. In fact, it was the properties of maintenance methylase that led to three independent proposals that differential methylation was a possible mechanism underlying X chromosome inactivation.[146-148] DNA methylation can dramatically affect gene expression by altering DNA protein-binding properties. Then, once a methylation pattern is established, it serves as a marking device in conjunction with a maintenance methylase to maintain a particular pattern indefinitely. The essence of the idea is that maintenance methylase works very effectively on half-methylated sites and very ineffectively on nonmethylated sites. Therefore, once a pair of alleles or chromosomes are distinguished by their methylation pattern, that pattern should be stable through cell division.

As pointed out before, there is a considerable amount of evidence that methylation is involved in X inactivation, which we will summarize now. Transformation experiments with purified DNA from the inactive X have indicated that it is modified in such a way that it will not function in a DNA-mediated transformation system.[59,60] On the other hand, DNA from the active X[61] and DNA from the inactive X that has been reactivated by 5AC will function in transfection,[62,63] implying that the DNA modification on the inactive X involves hypermethylation. Even more impressive are the experiments with 5AC-induced reactivation of rodent-human hybrids containing the inactive human X chromosome, where reactivation rates of 10^{-3} or more have been reported.[39] Individual genes on the inactive X in normal human cells probably reactivate spontaneously at rates of 10^{-8} or less, while in hybrid cells the spontaneous reactivation rates may approach 10^{-5}.[37,38] In one case, 5AC treatment has been reported to induce a reactivation rate of the *Hprt* and *G6pd* genes of 5×10^{-1}.[40] This is the maximum theoretical rate expected if 5AC leads to demethylation and reactivation by incorporation into DNA. Finally, the results of molecular analyses of three X-linked housekeeping genes, *Hprt*,[70,71] *G6pd*,[75] and *Pgk*,[76] show that the CpG-rich cluster at the 5′ end of these genes are hypomethylated on the active X and hypermethylated on the inactive X chromosome. Methylation analysis of 5AC-reactivated genes show the expected changes in methylation patterns in these 5′ CpG-rich regions.[148a]

These CpG-rich regions are most likely members of the HTF islands (Hpa II tiny fragments) that Bird has described and probably function as promoter regions for a great many genes.[149] He estimates that there are approximately 30,000 HTF islands which contain approximately 65% G + C content and have ten times the frequency of CpGs as bulk DNA.[150] In contrast to other CpGs in the genome, the HTFs are uniformly unmethylated except for the HTFs on the inactive X chromosome. Bird proposes that there may be bound factors at the HTF islands that sterically exclude methylase. Recently, Yang et al. have carried out a footprinting study with the 5′ *PGK* region and have identified proteins which bind to the X and protect it from digestion. It does not appear that these proteins are unique to the X chromosome.[150a] Inactivation of the gene could lead to displacement of these factors and permit access to the methylase.[150] We would postulate that the initial and critical part of this process is sex chromatin formation.

While this picture of the differentially methylated CpG-rich islands 5′ to X-linked housekeeping genes appears satisfying, it is most likely oversimplified. Thus far, methylation analysis has been carried out by restriction enzyme assays which detect only a fraction of methylatable sites, and there is already evidence that not all of the restriction sites analyzed are critical to gene expression. It will also be important to eventually work out in detail the methylation pattern of the remainder of the gene which, of course, contains many methylatable sites. Previous studies have indicated that methylation patterns in the body of genes seem to have little or no effect on gene expression; however, there is at least one site in the *Hprt* gene and one in the *G6pd* gene that are differentially methylated on the two Xs and in a pattern the opposite of the 5′ end.[70,71,151]

The methylation analysis at the *G6pd* locus appears to be quite complicated and interesting. There appears to be one CpG island at the 5′ end and two sites at the 3′ end of the gene.[75] All three sites appear to be unmethylated when the locus is expressed and reexpression of reactivants is associated with demethylation. Recently, one of the islands at the 3′ end of the *G6pd* gene was found to be 5′ to a housekeeping gene of unknown function, GdX.[151] Thus, all three islands and both genes appear to be coordinately controlled. However, since both genes are housekeeping genes and are expressed ubiquitously, it is difficult to say for certain that the CpG islands regulate different genes. It would be interesting to see if the *Gdx* gene is concurrently reactivated with the *G6pd* gene. These observations support the concept that the control of methylation is regional rather than gene by gene.

Two tissue-specific X-linked genes have been examined (Factor IX and *Otc*) for methylation patterns.[153-155] Neither of these genes has a closely linked HTF island and, in fact, Factor IX is considered to be a CpG-depleted gene.[89] Methylation studies have failed to detect any active:inactive X chromosome differences in the case of Factor IX, but in the case of *Otc* there is evidence in the mouse,[154] but not in the human,[155] for hypermethylation on the inactive X and hypomethylation on the active X. Reactivation experiments with this system have yet to be carried out. Since the Factor IX gene lacks an HTF island, it is tempting to assume that such a gene may not be immediately controlled by methylation. For example, the HTF island of a neighboring housekeeping gene may exert inactivation control over a tissue-specific gene or a completely different system of inactivation may be involved in tissue-specific genes. However, it must be remembered that in the absence of a 5′ GC-rich island, restriction analysis for methylation differences would be of limited value. It is possible that methylation at a single, properly placed CpG site might be all that is needed for expression or repression.

One further point concerns nuclease-sensitive regions in X-linked genes. Like methylation differences, nuclease-sensitive differences have been found for *Hprt*, *G6pd*, and *Pgk*, with the active genes showing nuclease sensitivity and the inactive genes, insensitivity.[72,156] The nuclease-sensitive sites and CpG-rich islands showing methylation differences are in most, if not all, cases identical. If nuclease sensitivity of a region reflects an underlying methylation pattern, then maintenance of both patterns will depend on methylation. This would appear to be the most likely sequence, though definitive evidence for methylation changes determining nuclease-sensitivity changes is lacking.

IV. SUMMARY

In this review, we have surveyed past work on the genetic biology of X inactivation and analyzed current work aimed at a molecular understanding of this system. It appears that X inactivation is initiated in extraembryonic tissues by formation of sex chromatin without any DNA modification. At present, we have no idea as to the molecular basis of sex chromatin formation. When X inactivation occurs in the embryo proper, sex chromatin formation is again the initial event, but, soon after, DNA methylation occurs, differentiating active and inactive alleles at their 5' ends. Reactivation experiments show methylation changes in keeping with a significant role for methylation in X inactivation. CpG-rich islands at the 5' end of housekeeping genes may be involved in the spreading of the methylation pattern and may also be the basis of the spreading of inactivation into ectopic sites.

REFERENCES

1. **Epstein, C. J.**, in *The Consequences of Chromosome Imbalance; Principles, Mechanisms and Models, Developmental and Cell Biology Series 18,* Barlow, P. W., Green, P. B., and Wylie, C. C., Eds., Cambridge University Press, 1986.
2. **Sandler, L. and Hecht, F.**, Genetic effects of aneuploidy, *Am. J. Hum. Genet.,* 25, 332, 1973.
3. **Muller, H. J., League, B. B., and Offermann, C. A.**, Effects of dosage changes of sex-linked genes, and compensatory effects of the gene differences between male and female, *Anat. Rec.,* 51, 110, 1931.
4. **Lyon, M. F.**, X-chromosome inactivation and development patterns in mammals, *Biol. Rev.,* 47, 1, 1972.
5. **Ohno, S.**, Evolution of sex chromosomes in mammals, *Annu. Rev. Genet.,* 3, 495, 1969.
6. **Lucchesi, J. C. and Manning, J. E.**, Gene dosage compensation in *Drosophila melanogaster, Adv. Genet.,* 24, 371, 1987.
7. **Gartler, S. M. and Riggs, A. D.**, Mammalian X-chromosome inactivation, *Annu. Rev. Genet.,* 17, 155, 1983.
8. **Lyon, M. F.**, Gene action in the X chromosome of the mouse (*Mus musculus* L.), *Nature (London),* 190, 372, 1961.
9. **Van Dyke, D. L., Flejter, W. L., Worsham, M. J., Roberson, J. G., Higgins, J. V., Herr, H. M., Knuutila, S., Wang, N., Babu, V. R., and Weiss, L.**, A practical metaphase marker of the inactive X chromosome, *Am. J. Hum. Genet.,* 39, 88, 1986.
10. **Flejter, W. L., Van Dyke, D. L., and Weiss, L.**, Location of the X inactivation center in primates and other mammals, *Hum. Genet.,* 74, 63, 1986.
11. **Russell, L. B.**, Genetics of mammalian sex chromosomes, *Science,* 133, 1795, 1961.
12. **Adler, D. A., West, J. D., and Chapman, V. M.**, Expression of alpha-galactosidase in preimplantation mouse embryos, *Nature (London),* 267, 838, 1977.
13. **Epstein, C. J., Smith, S., Travis, B., and Tucker, G.**, Both X chromosomes function before visible X chromosome inactivation in female mouse embryos, *Nature (London),* 274, 500, 1978.
14. **Kratzer, P. G. and Gartler, S. M.**, HGPRT activity changes in preimplantation mouse embryos, *Nature (London),* 274, 503, 1978.
15. **Kratzer, P. G. and Gartler, S. M.**, Hypoxanthine guanine phosphoribosyl transferase expression in early mouse development, in *Genetic Mosaics and Chimeras in Mammals,* Russell, L. B., Ed., Plenum Press, New York, 1978, 247.
16. **Monk, M.**, Biochemical studies on X-chromosome activity in preimplantation mouse embryos, in *Genetic Mosaics and Chimeras in Mammals,* Russell, L. B., Ed., Plenum Press, New York, 1978, 239.

17. **Monk, M. and Harper, M. I.,** Sequential X chromosome inactivation coupled with cellular differentiation in early mouse embryos, *Nature (London),* 281, 311, 1979.

18. **Takagi, N. and Sasaki, M.,** Preferential inactivation of the paternally derived X-chromosome in the extraembryonic membranes of the mouse, *Nature (London),* 256, 640, 1975.

19. **Gartler, S. M., Andina, R., and Gant, N.,** Ontogeny of X-chromosome inactivation in the female germ line, *Exp. Cell Res.,* 91, 454, 1975.

20. **Gartler, S. M., Rivest, M., and Cole, R. E.,** Cytological evidence for an inactive X chromosome in murine oogonia, *Cytogenet. Cell Genet.,* 28, 203, 1980.

21. **Kratzer, P. G. and Chapman, V. M.,** X chromosome reactivation in oocytes of *Mus caroli, Proc. Natl. Acad. Sci. U.S.A.,* 78, 3093, 1981.

22. **Lifschytz, E. and Lindsley, D. L.,** Sex chromosome activation during spermatogenesis, *Genetics,* 78, 323, 1974.

23. **Ropers, H. H., Wolff, G., and Hitzeroth, H. W.,** Preferential X inactivation in human placenta membranes: is the paternal X inactive in early embryonic development of female mammals?, *Hum. Genet.,* 43, 265, 1975.

23a. **Harrison, K. B.,** X chromosome inactivation in isolated human cytotrophoblast, *Am. J. Hum. Genet.,* 43, A130, 1988.

24. **Haneson, K. B. and Warburton, D.,** Preferential X-chromosome activity in human female placental tissues, *Cytogenet. Cell Genet.,* 41, 163, 1986.

25. **Migeon, B. R. and Do, T. T.,** In search of non-random X inactivation: studies of fetal membranes heterozygous for glucose-6-phosphate dehydrogenase, *Am. J. Hum. Genet.,* 31, 581, 1979.

26. **Migeon, B. R., Wolf, S. F., Axelman, J., Kaslow, D. C., and Schmidt, M.,** Incomplete X chromosome dosage compensation in chorionic villi of human placenta, *Proc. Natl. Acad. Sci. U.S.A.,* 82, 3390, 1985.

27. **Linder, D.,** Gene loss in human teratomas, *Proc. Natl. Acad. Sci. U.S.A.,* 63, 699, 1969.

28. **Rastan, S., Kaufman, M. H., Handeside, A. H., and Lyon, M. F.,** X-chromosome inactivation in extraembryonic membranes of diploid parthenogenetic mouse embryos demonstrated by differential staining, *Nature (London),* 288, 172, 1980.

29. **Therman, E. and Patau, K.,** Abnormal X chromosomes in man: origin, behavior and effects, *Hum. Genet.,* 25, 1, 1974.

30. **Takagi, N.,** Primary and secondary nonrandom X chromosome inactivation in early female mouse embryos carrying Searle's translocation T(X;16)16H, *Chromosoma,* 81, 439, 1980.

31. **Disteche, C. M., Eicher, E. M., and Latt, S. A.,** Late replication in an X-autosome translocation in the mouse: correlation with genetic inactivation and evidence for selective effects during embryogenesis, *Proc. Natl. Acad. Sci. U.S.A.,* 76, 5234, 1979.

32. **Disteche, C. M., Eicher, E. M., and Latt, S. A.,** Late replication pattern in adult and embryonic mice carrying Searle's X-autosome translocation, *Exp. Cell Res.,* 133, 357, 1981.

33. **Summitt, R. L., Tipton, R. E., Wilroy, R. S., Martens, P. R., and Phelan, J. P.,** X-autosome translocations: a review, *Birth Defects Orig. Artic. Ser.,* 14(6C), 219, 1978.

34. **Cattanach, B. M.,** Position effect variegation in the mouse, *Genet. Res.,* 23, 291, 1974.

35. **Wareham, K. A., Lyon, M. F., Glenister, P. H., and Williams, E. D.,** Age related reactivation of an X-linked gene, *Nature (London),* 327, 725, 1987.

36. **Linder, D. and Gartler, S. M.,** Glucose-6-phosphate dehydrogenase mosaicism: utilization as a cell marker in the study of leiomyomas, *Science,* 150, 67, 1965.

37. **Migeon, B. R., Wolf, S. F., Mareni, C., and Axelman, J.,** Derepression with decreased expression of the G6PD locus on the inactive X chromosome in normal human cells, *Cell,* 29, 595, 1982.

38. **Kahan, B. and DeMars, R.,** Localized derepression on the human inactive X chromosome in mouse-human cell hybrids, *Proc. Natl. Acad. Sci. U.S.A.,* 72, 1510, 1975.

38a. **Kahan, B. and DeMars, R.,** Autonomous gene expression on the human inactive X chromosome, *Som. Cell. Genet.,* 6, 309, 1980.

39. **Mohandas, T., Sparkes, R. S., and Shapiro, L. J.,** Reactivation of an inactive human X chromosome: evidence for X inactivation by DNA methylation, *Science,* 211, 393, 1981.

40. **Ellis, N., Keitges, E., Gartler, S. M., and Rocchi, M.,** High-frequency reactivation of X-linked genes in Chinese hamster human X hybrid cells, *Som. Cell Mol. Genet.,* 13, 191, 1987.

41. **Graves, J. A. M.,** 5-azacytidine induced re-expression of alleles on the inactive X chromosome in a hybrid mouse cell line, *Exp. Cell Res.,* 141, 99, 1982.

42. **Therman, E., Denniston, C., Sarto, G. E., and Ulber, M.,** X chromosome constitution and the human female phenotype, *Hum. Genet.,* 54, 133, 1980.

43. **Rouyer, F., Simmler, M.-C., Johnsson, C., Vergnaud, G., Cooke, H. J., and Weissenback, J.,** A gradient of sex linkage in the pseudoautosomal region of the human sex chromosomes, *Nature (London),* 319, 291, 1986.

44. **Cooke, H. J., Brown, W. R. A., and Rappold, G. A.,** Hypervariable telomeric sequences from the human sex chromosomes are pseudoautosomal, *Nature (London),* 317, 687, 1985.

45. **Simmler, M.-C., Rouyer, F., Vergnaud, G., Nystrom-Lahti, M., Ngo, K. Y., de la Chapelle, A., and Weissenbach, J.**, Pseudoautosomal DNA sequences in the pairing region of the human sex chromosomes, *Nature (London)*, 317, 692, 1985.

46. **Keitges, E., Rivest, M., Siniscalco, M., and Gartler, S. M.**, X-linkage of the steroid sulfatase gene in the mouse is evidence for a functional Y-linked allele, *Nature (London)*, 315, 226, 1985.

47. **Soriano, P., Keitges, E. A., Schorderet, D. G., Harbers, K., Gartler, S. M., and Jaenisch, R.**, High rate of recombination and doublecrossovers in the mouse pseudoautosomal region during male meiosis, *Proc. Natl. Acad. Sci. U.S.A.*, 84, 7218, 1987.

48. **Goodfellow, P., Pym, B., Mohandas, T., and Shapiro, L. J.**, The cell surface antigen locus, *MIC2X*, escapes X-inactivation, *Am. J. Hum. Genet.*, 36, 777, 1984.

49. **Keitges, E. and Gartler, S. M.**, Dosage of the Sts gene in the mouse, *Am. J. Hum. Genet.*, 39, 470, 1986.

50. **Muller, C. R., Migl, B., Traupe, H., and Ropers, H. H.**, X-linked steroid sulfatase: evidence for different gene-dosage in males and females, *Hum. Genet.*, 54, 197, 1980.

51. **Chance, P. F. and Gartler, S. M.**, Evidence for a dosage effect at the X-linked steroid sulfatase locus in human tissues., *Am. J. Hum. Genet.*, 35, 234, 1983.

52. **Ropers, H. H., Migl, B., and Zimmer, J.**, Activity of steroid sulfatase in fibroblasts with numerical and structural X-chromosome aberrations, *Hum. Genet.*, 57, 354, 1981.

53. **Migeon, B. R., Shapiro, L. J., Norum, R. A., Mohandas, T., Axelman, J., and Dabora, R. L.**, Differential expression of steroid sulfatase locus on active and inactive human X-chromosome, *Nature (London)*, 299, 838, 1982.

54. **Wolf, S. F. and Migeon, B. R.**, Studies of X chromosome DNA methylation in normal human cells, *Nature (London)*, 295, 667, 1982.

55. **Lindsay, S., Monk, M., Holliday, R., Huschtscha, L., Davies, K. E., Riggs, A. P., and Flavell, R. A.**, Differences in methylation on the active and inactive human X chromosomes, *Ann. Hum. Genet.*, 49, 115, 1985.

56. **Cattanach, B. M. and Isaacson, J. H.**, Controlling elements in the mouse X chromosome, *Genetics*, 57, 331, 1967.

57. **Mattei, M. G., Mattei, J. F., Vidal, I., and Giraud, F.**, Structural anomalies of the X chromosome and inactivation center, *Hum. Genet.*, 56, 401, 1981.

58. **Therman, E., Sarto, G. E., Palmer, C. G., Kallio, H., and Denniston, C.**, Position of the human X inactivation center on Xq, *Hum. Genet.*, 50, 59, 1979.

59. **Liskay, R. M. and Evans, R. J.**, Inactive X chromosome DNA does not function in DNA-mediated cell transformation for the hypoxanthine phosphoribosyltransferase gene, *Proc. Natl. Acad. Sci. U.S.A.*, 77, 4895, 1980.

60. **Chapman, V. M., Kratzer, P. G., Siracusa, L. O., Quarantillio, B. A., Evans, R., and Liskay, R. M.**, Evidence for DNA modification in the maintenance of X-chromosome inactivation of adult mouse tissues, *Proc. Natl. Acad. Sci. U.S.A.*, 79, 5357, 1982.

61. **Venolia, L. and Gartler, S. M.**, Comparison of transformation efficiency of human active and inactive X chromosome DNA, *Nature (London)*, 302, 82, 1983.

62. **Venolia, L., Gartler, S. M., Wassman, E. R., Yen, P., Mohandas, T., and Shapiro, L. J.**, Transformation with DNA from 5-azacytidine-reactivated X chromosomes, *Proc. Natl. Acad. Sci. U.S.A.*, 79, 2352, 1982.

63. **Lester, S. C., Korn, N. J., and DeMars, R.**, Derepression of genes on the human X chromosome: evidence for differences in locus-specific rates of derepression and rates of transfer of active and inactive genes after DNA-mediated transformation, *Som. Cell Mol. Genet.*, 8, 265, 1982.

64. **Kratzer, P. G., Chapman, V. M., Lambert, H., Evans, R., and Liskay, R. M.**, Differences in the DNA of the inactive X chromosome of fetal and extraembryonic tissues of mice, *Cell*, 33, 37, 1983.

65. **Lock, L. F., Takagi, N., and Martin, G. R.**, Methylation of the Hprt gene of the inactive X occurs after chromosome inactivation, *Cell*, 48, 39, 1987.

66. **Dyer, K. A., Riley, D., and Gartler, S. M.**, Analysis of inactive X chromosome structure by *in situ* nick translation, *Chromosoma*, 92, 209, 1985.

67. **Gartler, S. M., Dyer, K. A., Graves, J. A. M., and Rocchi, M.**, A two step model for mammalian X-chromosome inactivation, in *Biochemistry and Biology of DNA Methylation*, Razin, A. and Cantoni, G. L., Eds., Alan R. Liss, New York, 1985, 223.

67a. **Dyer, K. A., Canfield, T. K., and Gartler, S. M.**, Molecular cytology of active and inactive X chromatin: implications for X inactivation, *Cytogenet. Cell Genet.*, in press.

68. **Howard, G. C., Wolf, S. F., and Migeon, B. R.**, Chromosomal proteins and human X-chromosome inactivation, *J. Cell Biol.*, 82A, 1982.

69. **Hauser, L. J., Yang, T. P., and Hamkalo, B. A.**, Studies on human inactive X chromosomes, *J. Cell Biol.*, 61A, 1981.

69a. **Abe, K., Takagi, N., and Sasaki, M.,** Nonhistone nuclear proteins specific to certain mouse embryonal carcinoma clones having an inactive X chromosome, *Exp. Cell Res.,* 179, 590, 1988.

70. **Wolf, S. F., Jolly, D. J., Lunnen, K. D., Friedmann, T., and Migeon, B. R.,** Methylation of the hypoxanthine phosphoribosyltransferase locus on the human X chromosome: implications for X-chromosome inactivation, *Proc. Natl. Acad. Sci. U.S.A.,* 81, 2806, 1984.

71. **Yen, P. H., Patel, P., Chinault, A. C., Mohandas, T., and Shapiro, L. J.,** Differential methylation of hypoxanthine phosphoribosyltransferase genes on active and inactive human X chromosomes, *Proc. Natl. Acad. Sci. U.S.A.,* 81, 1759, 1984.

72. **Wolf, S. F. and Migeon, B. R.,** Clusters of CpG dinucleotides implicated by nuclease hypersensitivity as control elements of housekeeping genes, *Nature (London),* 314, 467, 1985.

73. **Wolf, S. F., Dintzis, S., Toniolo, D., Persico, G., Lunnen, K. D., Axelman, J., and Migeon, B. R.,** Complete concordance between glucose-6-phosphate dehydrogenase activity and hypomethylation of 3' CpG clusters: implications for X chromosome dosage compensation, *Nucleic Acids Res.,* 12, 9333, 1984.

74. **Toniolo, D., D'Urso, M., Martini, G., Persico, M., Tufano, V., Battistuzzi, G., and Luzzatto, L.,** Specific methylation pattern at the 3' end of the human housekeeping gene for glucose-6-phosphate de- hydrogenase, *EMBO J.,* 3, 1987, 1984.

75. **Toniolo, D., Martini, G., Migeon, B. R., and Dono, R.,** Expression of G6PD locus on the human X chromosome is associated with demethylation of three CpG islands within 100 Kb of DNA, *EMBO J.,* 7, 401, 1988.

76. **Keith, D. H., Singer-Sam, J., and Riggs, A. D.,** Active X chromosome DNA is unmethylated at eight CCGG sites clustered in a guanine-plus-cytosine-rich island at 5' end of the gene for phosphoglycerate kinase, *Mol. Cell. Biol.,* 6, 4122, 1986.

77. **Dyer, K. A. and Gartler, S. M.,** unpublished data, 1987.

78. **Riggs, A. D., Singer-Sam, J., and Keith, D. H.,** Methylation of the PGK promoter region and an enhancer way-station model of X-chromosome inactivation, in *Biochemistry and Biology of DNA Methylation,* Razin, A. and Cantoni, G. L., Eds., Alan R. Liss, New York, 1985, 211.

79. **Kunkel, L. M., Tantravahi, U., Eisenhard, M., and Latt, S. A.,** Regional localization on the human X of DNA sequences cloned from flow-sorted chromosomes, *Nucleic Acids Res.,* 10, 1557, 1982.

80. **Yang, T. P., Hansen, S. K., Oishi, K. K., Ryder, O. A., and Hamkalo, B. A.,** Characterization of a cloned repetitive DNA sequence concentrated on the human X chromosome, *Proc. Natl. Acad. Sci. U.S.A.,* 79, 6593, 1982.

81. **Camargo, M. and Cervenka, J.,** DNA replication and inactivation patterns in structural abnormality of sex chromosomes. I. X-A translocations, rings, fragments, isochromosomes, and pseudo-isodicentrics, *Hum. Genet.,* 67, 37, 1984.

82. **Keitges, E. A. and Palmer, C. G.,** Analysis of spreading of inactivation in eight X autosome translocations utilizing the high resolution RBG technique, *Hum. Genet.,* 72, 231, 1986.

83. **Mohandas, T., Geller, R. L., Yen, P. H., Rosendorff, J., Bernstein, R., Yoshida, A., and Shapiro, L. J.,** Cytogenetic and molecular studies on a recombinant human X chromosome: implications for the spreading of X chromosome inactivation, *Proc. Natl. Acad. Sci. U.S.A.,* 84, 4954, 1987.

84. **Singer-Sam, J., Keith, D. H., Tani, K., Simmer, R. L., Shively, L., Lindsay, S., Yoshida, A., and Riggs, A. D.,** Sequence of the promoter region of the gene for human X-linked 3-phosphoglycerate kinase, *Gene,* 32, 409, 1984.

85. **Melton, D. W., McEwan, C., Mckie, A. B., and Reid, A. M.,** Expression of the mouse HPRT gene: deletional analysis of the promoter region of an X chromosome linked housekeeping gene, *Cell,* 44, 319, 1986.

86. **Patel, P. I., Framson, P. E., Caskey, C. T., and Chinault, A. C.,** Fine structure of the human hypoxanthine phosphoribosyltransferase gene, *Mol. Cell. Biol.,* 6, 393, 1986.

87. **Martini, G., Toniolo, D., Vulliamy, T., Luzzatto, L., Dono, R., Vigliett, G., Paonessa, G., Durso, M., and Persico, M. G.,** Structural analysis of the X-linked gene encoding human glucose 6-phosphate dehydrogenase, *EMBO J.,* 5, 1849, 1986.

88. **Gitschier, J., Wood, W. I., Goralka, T. M., Wion, K. L., Chen, E. Y., Eaton, D. H., Vehar, G. A., Capon, D. J., and Lawn, R. M.,** Characterization of the human factor VIII gene, *Nature (London),* 312, 326, 1984.

89. **Yoshitake, S., Schach, B. G., Foster, D. C., Davie, E. W., and Kurachi, K.,** Nucleotide-sequence of the gene for human factor IX (antihemophilic factor-B), *Biochemistry,* 24, 3736, 1985.

90. **Ohno, S. and Cattanach, B. M.,** Cytological study of an X autosome translocation in *Mus musculus, Cytogenet. Cell Genet.,* 1, 129, 1962.

91. **Allderdice, P. W., Miller, O. J., Miller, D. A., and Klinger, H. P.,** Spreading of inactivation in an (X-14) translocation, *Am. J. Med. Genet.,* 2, 233, 1978.

92. **Palmer, C. P., Hubbard, T. W., Henry, G. W., and Weaver, D. D.,** Failure of X inactivation in the autosomal segment of an X/A translocation, *Am. J. Hum. Genet.,* 32, 179, 1980.

93. **Disteche, C. M., Eicher, E. M., and Latt, S. A.,** Late replication in an X autosome translocation in the mouse: correlation with genetic inactivation and evidence for selective effects during embryogenesis, *Proc. Natl. Acad. Sci. U.S.A.,* 76, 5238, 1979.

94. **Courturier, J., Dutrillaux, B., Garber, P., Raoul, O., Croquette, M., Fourlinnie, J., and Maillard, E.,** Evidence for a correlation between late replication and autosomal gene inactivation in a familial translocation t(X;21), *Hum. Genet.,* 49, 319, 1979.

95. **Russell, L. B.,** Mammalian X chromosome action: inactivation limited in spread and in region of origin, *Science,* 140, 976, 1963.

96. **Mohandas, T., Crandall, B. F., Sparkes, R. S., Passage, M. B., and Sparkes, M. C.,** Late replication studies in a human X/13 translocation: correlation with autosomal gene expression, *Cytogenet. Cell Genet.,* 29, 215, 1981.

97. **Krumlauf, R., Chapman, V. M., Hammer, R. E., Brinster, R., and Tilghman, S. M.,** Differential expression of alpha-fetoprotein genes on the inactive X-chromosome in extraembryonic and somatic tissues of a transgenic mouse line, *Nature (London),* 319, 224, 1986.

98. **Goldman, M. A., Stokes, K. R., Idzerda, R. L., McKnight, G. S., Hammer, R. E., Brinster, R. L., and Gartler, S. M.,** A chicken transferrin gene in transgenic mice escapes X-chromosome inactivation, *Science,* 236, 593, 1987.

99. **Hellkuhl, B., de la Chapelle, A., and Grezeschik, K. H.,** Different patterns of X chromosome inactivity in lymphocytes and fibroblasts of a human balanced X autosome translocation, *Hum. Genet.,* 60, 126, 1982.

100. **Keitges, E. A.,** X autosome translocations in man studied with replication banding, thesis, Indiana University, Indianapolis, 1983.

101. **Canki, N., Dutrillaux, B., and Trivadar, I.,** Dystrophie musculaire de duchenne chez une petite fille porteuse d'une translocation t(X;3)(p21; q13) de novo, *Ann. Genet.,* 22, 35, 1979.

102. **Mann, J. D. and Higgens, J.,** A case of primary amenorrhea associated with X autosomal translocation, *Am. J. Hum. Genet.,* 26, 416, 1974.

103. **Hagemeijer, A., Hoovers, J., Smit, E. M., and Bootsma, D.,** Replication pattern of the X chromosomes in three X/autosome translocations, *Cytogenet. Cell Genet.,* 18, 33, 1977.

104. **Cohen, M. M., Lin, C.-C., Sybert, V., and Oiecchio, E. J.,** Two human X autosome translocations identified by autoradiography and fluorescence, *Am. J. Hum. Genet.,* 24, 583, 1972.

105. **Mattei, M. G., Mattei, J. E., Aymes, S., Malpuech, G., and Giraud, F.,** A dynamic study in two new cases of X chromosome translocations, *Hum. Genet.,* 41, 251, 1978.

106. **Leisti, J. T., Kaback, M. M., and Rimoin, D. L.,** Human X autosome translocations: differential inactivation of the X chromosome in a kindred with an X-9 translocation, *Am. J. Hum. Genet.,* 27, 441, 1975.

107. **Pescia, G., Jotterand-Bellomo, M., de Crousaz, H., Payot, M., and Martin, D.,** Phenotype de la trisomie 9q distale chez un enfant presentant un chromosome surnumeraire remanie t(X;9), *Ann. Genet.,* 22, 158, 1979.

108. **Fraccaro, M., Maraschio, P., Pasquali, F., and Scappaticci, S.,** Women heterozygous for deficiency of the (p21→pter) region of the X chromosome are fertile, *Hum. Genet.,* 39, 283, 1977.

109. **Buhler, E. M., Jurik, L. P., Voyame, M., and Buhler, U. K.,** Presumptive evidence of two active X chromosomes in somatic cells of a human female, *Nature (London),* 265, 142, 1977.

110. **Disteche, C. M., Swisshelm, K., Forbes, S., and Pagon, R. A.,** X inactivation patterns in lymphocytes and skin fibroblasts of three cases of X-autosome translocations with abnormal phenotypes, *Hum. Genet.,* 66, 71, 1984.

111. **Zuffardi, O., Tiepolo, L., Scappaticci, S., Francesconi, D., Biunchi, C., and Di Natale, D.,** Reduced phenotypic effect of partial trisomy 1q in a X/1 translocation, *Ann. Genet.,* 20, 191, 1977.

112. **Turleau, C., Chavin-Colin, F., de Grouchy, J., Repesse, G., and Beavois, P.,** Familial t(X;2)(p223; q323) with partial trisomy 2q and male and female balanced carriers, *Hum. Genet.,* 37, 97, 1977.

113. **Morichon-Delvallez, N., Courturier, J., and Frison, B.,** Phenotype attenue de la trisomie 4p par tranlocation t(X;4)(p21.2, p13), *Ann. Genet.,* 25, 246, 1982.

114. **Gaal, M. and Laslo, J.,** X inactivation pattern in an unbalanced X autosome translocation with gonadal dysgenesis, *Hum. Hered.,* 27, 396, 1977.

115. **Tipton, R. E.,** Cytogenetic studies in subjects with mental retardation and congenital anomalies using banding techniques, thesis, University of Tennessee, Knoxville, MS, 1974.

116. **Leisti, J. T., Kaback, M. M., and Rimoin, D. L.,** Human X autosome translocations in a kindred with an X-9 translocation, *Am. J. Hum. Genet.,* 27, 441, 1975.

117. **Enjima, Y., Sasaki, M. S., Kaneka, A., Tanooka, H., Hara, Y., Hida, T., and Kinoshita, Y.,** Possible inactivation of part of chromosome 13 due to 13q Xp translocation associated with retinoblastoma, *Clin. Genet.,* 21, 357, 1982.

118. **Nichols, W. W., Miller, R. C., Sobel, M., Hoffman, E., Sparkes, R. S., Mohandas, T., Veomett, I., and Davis, J. R.,** Further observations on a 13q Xp translocation associated with retinoblastoma, *Am. J. Opthalmol.,* 89, 621, 1980.

119. **Crandall, B. F., Carrell, R. E., Howard, J., Schroeder, W., and Muller, H.,** Trisomy 13 with a 13-X translocation, *Am. J. Hum. Genet.,* 26, 385, 1974.

120. **Zabel, B. U., Baumann, W. A., Pirnthe, W., and Gerhard-Ratschow, K.,** X inactivation pattern in three cases of X/autosome translocations, *Am. J. Med. Genet.,* 1, 309, 1978.

121. **Bernstein, R., Dawson, B., Kohl, R., and Jenkins, T.,** X;15 translocation in a retarded girl: X inactivation pattern and attempt to localize the hexosaminidase A and other loci, *J. Med. Genet.,* 16, 254, 1979.

122. **Summitt, R. L., Marens, P. R., and Wilroy, R. S.,** X autosome translocations in normal mother and effectively 21-monosomic daughter, *J. Pediatr.,* 84, 539, 1974.

123. **Taysi, K., Sparkes, R. S., O'Brian, T. J., and Dengler, D. R.,** Down's syndrome phenotype and autosomal gene inactivation in a child with presumed (X;21) de novo translocation, *J. Med. Genet.,* 19, 144, 1982.

124. **Jenkins, M. B., Davis, E., Thelen, T. H., and Boyd, L.,** A familial X-22 translocation with an extra X chromosome, *Am. J. Hum. Genet.,* 26, 736, 1974.

125. **Markovic, V. D., Cox, D. W., and Wilkinson, J.,** X-14 translocation: an exception to the critical region hypothesis on the human X-chromosome, *Am. J. Med. Genet.,* 20, 87, 1985.

126. **Kajii, T., Tsukahara, M., Fukushima, Y., Hata, A., Matsuo, K., and Kuroki, Y.,** Translocation (X;13)(p11.21; q12.3) in a girl with incontinentia pigmenti and bilateral retinoblastoma, *Ann. Genet.,* 28, 219, 1985.

127. **Williams, J. and Dear, J. J.,** An unbalanced t(X;10)mat translocation in a child with congenital abnormalities, *J. Med. Genet.,* 24, 633, 1987.

128. **Rivera, H., Enziquez-Guerra, M. A., Rolon, A., Jimenez-Sainz, M. E., Nunez-Gonzalez, L., and Canta, J. M.,** Whole-arm t(X;17)(Xp17q;Xq17p) and gonadal dysgenesis, a further exception to the critical region hypothesis, *Clin. Genet.,* 29, 425, 1986.

129. **Rudak, E., Mayer, M., Jacobs, P., Sprenkel, J., Do, T., and Migeon, B.,** X/11 translocation: replication and mapping studies, *Cytogenet. Cell Genet.,* 25, 199, 1979.

130. **Sands, M. E.,** Mental retardation in association with a balanced X autosome translocation and random inactivation of the X chromosome, *Clin. Genet.,* 17, 309, 1980.

131. **Engel, W., Vogel, W., and Reinwein, H.,** Autoradiographische Untersuchungen an einer X Autosomen Translokation beim Menschen: 45 X, 15-, tan(15q;Xq+), *Cytogenet. Cell Genet.,* 10, 87, 1971.

132. **Thelen, T. H., Abrams, D. J., and Fisch, R. O.,** Multiple anomalies due to possible genetic inactivation in an X autosome translocation, *Am. J. Hum. Genet.,* 23, 410, 1971.

133. **Singh, L. and Jones, K. W.,** Sex reversal in the mouse *(Mus musculus)* is caused by a recurrent nonreciprocal crossover involving the X and an aberrant Y chromosome, *Cell,* 28, 205, 1982.

134. **Burgoyne, P. S. and McLaren, A.,** Does X-Y pairing during male meiosis protect the paired region of the X-chromosome from subsequent X-inactivation?, *Hum. Genet.,* 70, 82, 1985.

135. **Cattanach, B. M., Evans, E. P., Burtenshaw, M. D., and Barlow, J.,** Male, female and intersex development in mice of identical chromosome constitution, *Nature (London),* 300, 445, 1982.

136. **McLaren, A. and Monk, M.,** Fertile females produced by inactivation of an X chromosome of "sex reversed" mice, *Nature (London),* 300, 446, 1982.

137. **Schorderet, D. F., Keitges, E. A., Dubois, P. M., and Gartler, S. M.,** Inactivation and reactivation of the sex-linked steroid sulfatase gene in murine cell culture, *Som. Cell Mol. Genet.,* 14, 113, 1988.

138. **Ropers, H. H., Migl, B., Zimmer, J., Fraccaro, M., Maraschio, P. P., and Westerveld, A.,** Activity of steroid sulfatase in fibroblasts with numerical and structural X chromosome aberrations, *Hum. Genet.,* 57, 354, 1986.

139. **Chance, P. F. and Gartler, S. M.,** Evidence for a dosage effect at the X-linked steroid sulfatase locus in human tissues, *Am. J. Hum. Genet.,* 35, 234, 1983.

140. **Keitges, E. A., Schorderet, D. F., and Gartler, S. M.,** Linkage of steroid sulfatase gene to the sex-reversed mutation in the mouse, *Genetics,* 116, 465, 1987.

141. **Harris, M.,** Induction of thymidine kinase in enzyme-deficient Chinese-hamster cells, *Cell,* 29, 483, 1982.

142. **Harris, M.,** High frequency induction by 5-azacytidine of proline independence in CHO-K1 cells, *Som. Cell Mol. Genet.,* 10, 275, 1984.

143. **Steglich, C., Grens, A., and Scheffler, I. E.,** Chinese-hamster cells deficient in ornithine decarboxylase activity reversion by gene amplification and by azacytidine treatment, *Som. Cell Mol. Genet.,* 11, 11, 1985.

144. **Harris, M.,** Induction and reversion of asparagine auxotrophs in CHO-K1 and V79 cells, *Som. Cell Mol. Genet.,* 12, 459, 1986.

145. **Krumlauf, R., Chapman, V. M., and Tilghman, S. M.,** Differential expression of genes on the inactive X chromosome in transgenic mice, *Genet. Res.,* 47, 221, 1988.

146. **Riggs, A. D.,** X inactivation, differentiation and DNA methylation, *Cytogenet. Cell Genet.,* 14, 9, 1975.

147. **Holliday, R. and Pugh, J. E.,** DNA modification mechanisms and gene activity during development, *Science,* 187, 226, 1975.

148. **Sager, R. and Kitchin, R.,** Selective silencing of eukaryotic DNA, *Science,* 189, 426, 1975.

148a. **Hansen, R. S., Ellis, N. A., and Gartler, S. M.,** Demethylation of specific sites in the 5' region of the inactive X linked human phosphoglycerate kinase gene correlates with the appearance of nuclease sensitivity and gene expression, *Mol. Cell Biol.,* 8, 4692, 1988.

149. **Bird, A. P., Taggart, M., Fronner, M., Miller, O. J., and Macleod, D.,** A fraction of the mouse genome that is derived from islands of nonmethylated CpG rich DNA, *Cell,* 40, 91, 1985.

150. **Bird, A. P.,** CpG islands and the function of DNA methylation, *Nature (London),* 321, 209, 1986.

150a. **Yang, T. P., Singer-Sam, J., Flores, J. C., and Riggs, A. D.,** DNA binding factors for the CpG-rich island containing the promotor of the human X linked *Pgk* gene, *Som. Cell Mol. Genet.,* 14, 461, 1988.

151. **Battistuzzi, G., D'Urso, M., Toniolo, D., Persico, G. M., and Luzzatto, L.,** Tissue specific levels of human glucose-6-phosphate dehydrogenase correlate with methylation of specific sites at the 3' end of the gene, *Proc. Natl. Acad. Sci. U.S.A.,* 82, 1465, 1985.

152. **Toniolo, D., Persico, M., and Alcalay, M.,** A housekeeping gene on the X chromosome encodes a protein similar to ubiquitin, *Proc. Natl. Acad. Sci. U.S.A.,* 85, 851, 1988.

153. **Ruta Cullen, C., Hubberman, P., Kaslow, D. C., and Migeon, B. R.,** Comparison of factor IX methylation on human active and inactive X chromosomes: implications for X inactivation and transcription of tissue-specific genes, *EMBO J.,* 5, 2223, 1986.

154. **Mullins, L. J., Veres, G., Caskey, T., and Chapman, V.,** Differential methylation of the ornithine carbamoyltransferase gene of active and inactive mouse X chromosomes, *Mol. Cell. Biol.,* 7, 3916, 1987.

155. **Hannibal, M. C., Ruta Cullen, C., Kaslow, D. C., Davies, K. E., and Migeon, B. R.,** Evidence that sex differences in methylation of the ornithine transcarbamylase locus in the X chromosome are not functional, *Am. J. Hum. Genet.,* 39, A20, 1986.

156. **Riley, D. E., Goldman, M. A., and Gartler, S. M.,** Chromatin structure of active and inactive human X-linked phosphoglycerate kinase gene, *Som. Cell Mol. Genet.,* 12, 73, 1986.

Chapter 8

THE Y CHROMOSOME OF *DROSOPHILA*

Wolfgang Hennig

TABLE OF CONTENTS

I. INTRODUCTION

A. SEX CHROMOSOMES

In many animals, one of the chromosome pairs displays a heteromorphic, sex-specific combination of chromosome types. Because of this sex specificity, such chromosomes were called sex chromosomes. One of the sexes carries two identical chromosomes; in the other sex, only one of these elements is found, while the homologous chromosome is morphologically different or entirely absent. If the male sex carries two morphologically different chromosomes ("X" and "Y"; composition: "X/Y") or one chromosome without the homologous partner ("X/O") in diploid cells, it is the *heterogametic* sex since the gametes will receive only one of the two sex chromosomes (or some of them none, in an X/O combination) and therefore differ genetically. Females then carry two identical sex chromosomes and hence lack one of the sex chromosomes present in the male. They are designated the *homogametic* sex since all gametes will carry the same type of sex chromosome. Cytogenetic studies allowed the sex-linked inheritance of a particular phenotype to be directly related to specific sex chromosome content. Therefore, this difference in chromosome composition provided an important argument in support of the chromosome theory of heredity early in the 20th century.

B. THE Y CHROMOSOME

Y chromosomes have an exceptional status in the genome. Since they occur only in the hemizygous sex, they cannot carry chromosome-specific information indispensible for normal cellular functions or for functions more generally necessary to guarantee the survival of the individual. Absence of the Y chromosome in the heterogametic sex leads, however, to a mutant phenotype. The defects of this phenotype are variable, depending on the specific organism. In man, for example, individuals with a single X chromosome, but without Y (i.e., X/O), not only display a female phenotype, but also show serious deviations from normality, known as the Turner syndrome. In *Drosophila*, however, individuals of a corresponding genotype (X/O) are phenotypically normal, but sterile, males. Hence, the most simple explanation — namely, that Y chromosomes carry genetic information essential for male sex determination — only holds true for some groups of organisms and not for others. In man, the Y chromosome carries sex-determining genetic information, dominantly controlling the expression of the male phenotype (see Reference 86). This is, however, not true for other groups of organisms, including insects. In *Drosophila*, sex is determined by the ratio of X chromosomes to autosomes (reviewed by Nöthiger and Steinmann-Zwicky[1]). Nevertheless, *Drosophila* males without Y chromosomes have an incomplete spermiogenesis and are, therefore, sterile. Genes essential for male germ cell development also exist in the mammalian Y chromosome (for references, see Burgoyne[2]). Thus, even though the Y chromosome, in some organisms, is dispensable for the primary determination of the male phenotype, it still carries sex-related genetic information.

C. SEX CHROMOSOMES AND HETEROCHROMATIN

Sex chromosomes are not only heteromorphic, but are also often distinguished from the residual part of the chromosome complement by their heterochromatic character. Major parts of the sex chromosomes, or even the entire sex chromosomes, remain condensed in somatic interphase nuclei. This implies transcriptional inactivity. The Y chromosome of *Drosophila*, for example, is heterochromatic in all tissues except testes, and it does not participate in replications leading to polyteny in somatic tissues. A transcriptional inactivity in somatic cells is compatible with the dispensability of this chromosome for cellular differentiation and for functions other than those related to sperm morphogenesis. During the first meiotic prophase in males, however, the Y chromosome decondenses and displays a high level of transcriptional activity as expected because of its function in male germ cell differentiation.

A heterochromatic character of chromosomes or of chromosome regions has often been considered as reflecting the absence of genes. This idea has been supported by genetic studies inferring that the number of genes in heterochromatic chromosome regions is very small. However, other functional effects of heterochromatin have been described. The most obvious of such effects is the induction of a mosaic gene expression ("position effect variegation") in differentiated tissues if genes are translocated into heterochromatin as a consequence of a chromosome rearrangement. In addition to such *cis* effects, heterchromatin can also exert *trans* effects on position effect variegation. Variation in the amount of heterochromatin in the genome is accompanied by suppression or enhancement of position effect variegation. Also, for germ line-restricted chromosomes (B or W chromosomes), which are characteristically heterochromatic, germ line-related functions are suspected. Experimental approaches to remove germ line-restricted chromosomes entirely from the genome failed since they result in sterility or even inviability. In addition, cytological investigations have provided ample evidence that heterochromatic chromosome regions become decondensed in the germ line, usually during meiotic prophase I (see Hennig[3]). These observations document the fact that heterochromatic chromosome regions are not without function, although their sizes may be subject to large fluctuations, even within one population of a species. Heterochromatic chromosome regions are thus a universal characteristic of the eukaryotic genome and seem not to be dispensable. Their quantitative contribution to the genome and their locations are, however, not rigidly controlled and can vary within a wide range.

Molecular studies revealed a specific molecular structure of heterochromatin. One of the first findings in studying eukaryotic genome structure at the DNA level was that heterochromatin is rich in repetitive DNA. In particular, highly repetitive DNA sequences representing "satellite" or "simple sequence" DNA were found associated with heterochromatin.[4,5] *In situ* hybridization revealed that heterochromatic regions are the main genomic locations of such highly repetitive DNA sequences. This type of DNA consists of short sequences (5 to 200 nucleotides long) which occur tandemly in large clusters. Nucleotide sequence analysis excludes a protein coding character of such sequences since they have no open reading frames. Only in exceptional cases — in developing germ cells — are such sequences transcribed, but a function of these transcripts has not yet been defined.[6-8]

Highly repetitive DNA sequences display — besides their general transcriptional inactivity — other characteristics typical of heterochromatic parts of the genome. In particular, their amount and genomic localization are extremely variable. Usually, little sequence similarity exists between the highly repetitive DNAs of closely related species. This implies that blocks of highly repetitive DNA can be rapidly exchanged with blocks of a new sequence type. We believe that occasional amplification cycles of DNA sequences originally present in only minor copy numbers or even as unique DNA sequences are responsible for the creation of a new, highly repetitive sequence type in the genome.

Highly repetitive DNA is, however, not the only constituent of heterochromatic chromosome regions. It has been repeatedly shown that other repetitive DNA sequences of a different sequence character occur interspersed between satellite DNA sequence blocks. There are good reasons to assume that such sequences often belong to the class of mobile DNA elements. The analysis of such DNA has, nevertheless, been extremely tedious and laborious, and unfortunately, has hardly contributed to our understanding of the actual function of heterochromatin as yet.

II. THE Y CHROMOSOME OF *DROSOPHILA*

A. GENETIC ASPECTS

Genetic studies on the Y chromosome of *Drosophila melanogaster* have been carried out since the early days of *Drosophila* genetics. Bridges[9] recognized the importance of this chromosome for spermatogenesis. Since then, many investigators have tried to resolve its actual biological function, but even today it is still not definitely understood. Several reasons account for the difficulties of understanding the role of Y chromosomal genes. Major problems for an experimental analysis arise from the unusual genetic and — as now evident — unusual molecular structure of this chromosome. The prominent problem for genetic investigations is the absence of suitable phenotypic markers which could help to precisely map this chromosome with a resolution comparable to the resolution obtained elsewhere in the *Drosophila* genome. In addition, the possibility of mapping by conventional crossing-over techniques does not exist since crossing-over does not take place during the male meiosis of *Drosophila*. Molecular studies are severely hampered by the (probably exclusive) composition of repetitive DNA of the Y chromosome. There may, however, be another, more fundamental reason for our failure to understand the biological role of the Y chromosomal genes. As will be shown below, the Y chromosomal fertility genes differ in their structure from conventional protein- or structural RNA-coding genes. This means that different experimental approaches are necessary to reveal the actual biological function of such genes.

In conventional genetic experiments on the fine structure of the Y chromosomal fertility genes, two methods were used for mapping the Y chromosome. One method is the combination of different X-Y or autosome-Y translocation chromosomes. Alternatively, mutated Y chromosomes were studied in complementation experiments with (partial) duplications of the Y chromosome (for details, see Reference 10). The classic work, in particular of Brosseau,[11] established the presence of seven male fertility genes in the Y chromosome of *D. melangaster*. By similar approaches in combination with cytogenetic studies, the existence of six of these genes has recently been confirmed by Kemphues et al.[12] and Gatti et al.[13]

Another site allelic to an X chromosomal site, the Stellate locus, exists close to fertility locus K12 of *D. melanogaster*.[14] The Y chromosomal Stellate locus is, however, not essential for male fertility. It represents the only genetic site on the Y chromosome of *D. melanogaster*, besides the nucleolus organizer region, which is not directly involved in, and is possibly even dispensible for, sperm development.

In another *Drosophila* species, *D. hydei*, the biology of the Y chromosome has been extensively studied because of many cytological advantages, compared with *D. melanogaster*. But genetic experimentation in this species is much more difficult because of the lack of genetic markers and of balancer chromosomes, which are available for *D. melanogaster*. However, using cytological markers on the Y chromosome, which are not available in *D. melanogaster*, it became possible to genetically dissect the Y chromosome of *D. hydei*.[15,16] This study revealed that up to 16 fertility genes may be present (for a critical evaluation, see References 10, 17). Five of the Y chromosomal male fertility genes of *D. hydei* develop giant lampbrush loops during their transcription in primary spermatocytes.[15,18] They can be used as cytological markers in genetic studies of the Y chromosome. The site-specific morphology of the lampbrush loops permitted a regional dissection of the Y chromosome by cytological studies of primary spermatocyte nuclei (for a recent review, see Reference 19). This cytological property of the fertility genes has essentially contributed to the investigation of the Y chromosomal fertility genes in *D. hydei* not only by cytogenetic,[20-23] cytological,[23,24] and genetic means,[10,16,25] but also by molecular methods.[17] Although some evidence indicates that the Y chromosome of *D. melanogaster* in its molecular properties is comparable to that of *D. hydei,* this will have to be ascertained in future investigations.

Since the Y chromosomal lampbrush loops in primary spermatocytes of *D. hydei* play a central role in all studies of this chromosome due to their favorable cytology, a short summary of their basic cytological features will be given. Details are contained in several recent articles.[10,17,19,26,27]

B. CYTOLOGICAL ASPECTS

Lampbrush loops in primary spermatocytes of *D. hydei* are only formed by Y chromosomal loci. The time of their appearance coincides with the only period of activity of the Y chromosomal fertility genes, which is the first meiotic prophase. Just as with the formation of lampbrush loops in oocyte chromosomes of many organisms,[29] the Y chromosomal lampbrush loops must be considered to be a consequence and cytological manifestation of the transcriptional activity of the respective genes.[18] Repeatedly, the argument has been made that it is by no means clear whether the actual genetic information resides inside or outside the lampbrush loop-forming chromosome regions. However, this argument is artificial and irrelevant since it has been amply demonstrated that the actual presence of (cytologically normal) lampbrush loops in primary spermatocytes is a prerequisite for the development of fertile sperm.[15,16,20,25,30]

The uncoiling of the Y chromosomal DNA shortly after the onset of the primary spermatocyte stage is accompanied by an increasing intensity of transcription of the Y chromosomal loci and of other genomic sites.[23,26] This period of male germ cell development seems to be more generally important for the onset of gene activity essential for the morphogenesis of sperm. It has recently been demonstrated that, just as for the transcription of the Y chromosomal genes (see Figure 9c, Reference 26), the testis-specific β2-tubulin gene activity also begins early in the primary spermatocyte stage.[12,31]

A characteristic feature of meiotic prophase chromosomes is the formation of a chromomere pattern. This type of chromosomal morphology is replaced by the Y chromosome in favor of an open structure of the chromatids.[23,24] But the X chromosome and the autosomes also display an unusual behavior since they do not pass through the characteristic meiotic prophase chromosome structures, but remain decondensed until shortly before the onset of the first meiotic metaphase.[24] Then all chromosomes condense rapidly and pass through the first meiotic division. This unconventional behavior of the chromosomes may be related to the absence of synaptonemal complexes and the absence of crossing-over during the male meiosis of *Drosophila*. Extended periods of pairing of the homologs is consequently not required. The transcriptional activity of the male genome ceases at the first meiotic metaphase. Even with the resolution provided by autoradiography, postmeiotic transcriptional activity has never been observed in *Drosophila*.[23,32] Postmeiotic genomic activity — if present at all — must occur at a low level not relevant for the production of major structural components of sperm. If it occurs at all, then primarily regulatory effects would be expected from such low levels of postmeiotic transcription.

C. MOLECULAR COMPOSITION OF LAMPBRUSH LOOPS

All five Y chromosomal lampbrush loops display a highly loop-specific morphology (Figure 1). Histochemical studies established that the major components of the loops are proteins and RNA, while DNA is hardly detectable.[23,24,33-35] Since early studies, it has been suggested that the particular morphology of the loops is governed by the interaction of the transcripts with associated proteins.[23] This idea has been substantiated by the results of Miller spreading experiments and by molecular studies on the lampbrush loop DNA, its transcripts, and loop proteins (see below).[27,36-39]

The first step toward an understanding of the molecular structure of the Y chromosomal fertility genes of *D. hydei* were DNA/RNA hybridization experiments demonstrating the transcription of middle-repetitive Y chromosomal DNA sequences in testes (Figure 2).[40,41] Further insight into the character and structure of the transcripts had, however, to await the development of specialized techniques such as recombinant DNA technology to make Y chromosomal DNA open to detailed studies. Identification of Y chromosomal DNA sequences by conventional centrifugation techniques turned out to be unsuccessful.[42,43] But even molecular cloning techniques were initially not fully successful since discrimination

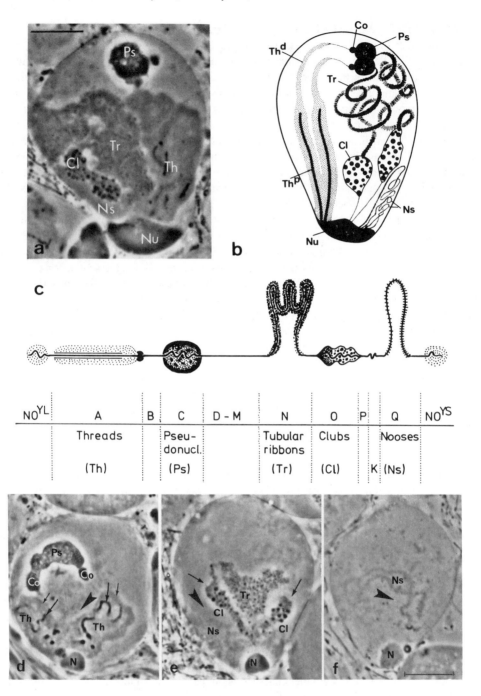

FIGURE 1 a to f.

of cloned Y chromosomal DNA fragments simply by comparing their capacity to hybridize with the DNA of males or females proved to be of limited potential to recover Y chromosomal DNA.[44-46] Only more specialized techniques, in particular cloning at a microscale level after microdissection of lampbrush loops (Figure 3),[39,47-49] allowed progress to be made toward understanding the complex molecular DNA structure of the Y chromosomal fertility genes.[27,49-53]

Studies of recombinant DNA fragments recovered by a semiquantitative screening

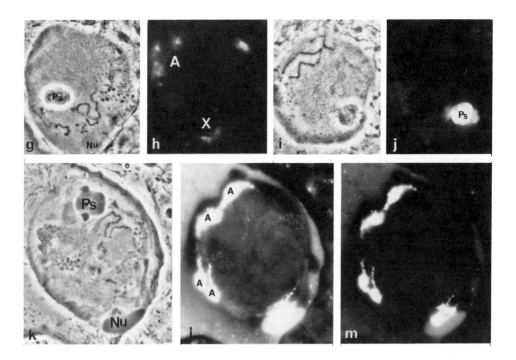

FIGURE 1 g to m.

FIGURE 1.   Primary spermatocyte nuclei of *Drosophila hydei* displaying the Y chromosome lampbrush loops. (a) Phase contrast; (b) diagram; (c) schematic drawing of a single chromatid with the lampbrush loop positions, the designations of the loops, the genetic loci (A to Q), and the positions of the nucleolus organizer regions indicated; (d) to (f) spermatocyte nuclei of males with partial deletions of the Y chromosome. In (d), only the "threads" and the "pseudonucleolus" are present; in (e) only the "tubular ribbons", "clubs", and "nooses"; and in (f), only the "nooses". Note the specific morphology of each of the loops. Spermatocyte nucleus (g) displaying the binding sites of histone H2A antiserum by immunofluorescence (h). Compare the distribution of this core histone fraction with the distribution of DNA, as seen by DAPI staining in (k to m). Spermatocyte nucleus (i) displaying the binding sites of histone H1 antiserum by immunofluorescence (j). In contrast to core histones, histone H1 is not found in the chromatin, but resides exclusively in one of the lampbrush loops (pseudonucleolus). Since the DNA concentration in this loop is low (see m), it must have accumulated within this loop for reasons other than by association with nucleosomes. Nucleus (k) after DAPI staining indicate the location of DNA within the nucleus (l, m). Different exposures are shown to demonstrate the low signal derived from the lampbrush loop DNA (l) and the strong fluorescence derived from the autosomes and the X chromosome (m). In (l), the axis of some Y chromosomal lampbrush loops are faintly visible. Note the peripheral location of the autosomes. Nu, nucleolus; Ns, nooses; Co, cones; Cl, clubs; Tr, tubular ribbons; Ps, pseudonucleolus; Th, threads. Bar represents 10 μm. (From Kremer, H., Hennig, W., and Dijkhof, R., *Chromosoma*, 94, 147, 1986; Hennig, W., *Adv. Genet.*, 23, 179, 1985. With permission.)

technique[45] are the basis of our present understanding of the molecular structure of the Y chromosome. The investigation led to two fundamental conclusions. The first conclusion was that at least the majority, if not all, Y chromosomal DNA is repetitive; the second, that the fertility genes accommodate two main classes of repetitive DNA sequences (Figure 4). One of these classes ("Y-specific") is a gene-specific (i.e., lampbrush loop-specific) DNA type with much similarity to simple sequence DNA (highly repetitive DNA) because it is composed of short (up to 400 bp long) tandemly arranged repeat units. In contrast to "highly-repetitive" DNA fractions, it occurs in copy numbers of only several hundred to 2000. The second class of DNA sequences includes repetitive DNA sequences of even lower copy numbers (10 to 60), but with a higher sequence complexity (several kb in length). These DNA sequences are not tandemly arranged within the Y chromosome and they occur, in

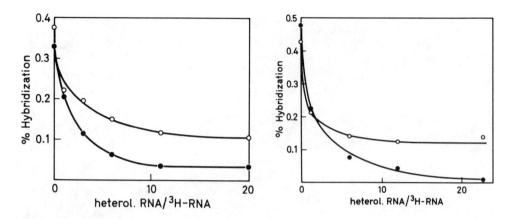

FIGURE 2.   Competition hybridization experiment with uniformly labeled testis RNA of *Drosophila hydei* (a) or testes RNA labeled for 48 h and subsequently chased for 5 d (b) against unlabeled competitor RNA from males without testes. DNA of X/Y/Y males (O) or females (X/X) (●) was bound to membrane filters and incubated with constant amounts of labeled testis RNA and an increasing excess of unlabeled competitor RNA. The competition experiment shows the presence of RNA species reacting with Y chromosomal DNA. The pulse-chase experiment shows that such RNA is still present after meiosis. (From Hennig, W., *J. Mol. Biol.*, 38, 227, 1968. With permission.)

addition to their Y chromosomal location, with the majority of their sequence family members in other sites spread all over the genome. This DNA class was designated as "Y-associated".

D. MOLECULAR STRUCTURE OF ONE OF THE FERTILITY GENES

The most detailed investigations at the DNA level have so far been carried out with the lampbrush loop "nooses". This locus has the advantage of being the only fertility gene on the short arm, and all other Y chromosomal loci can be deleted (except the terminal Y chromosome nucleolus organizer region[54,55]). The location of the lampbrush loop nooses on the short arm of the Y chromosome allowed us to unequivocally identify this active gene in Miller spreading experiments[36] and facilitated the visualization of the complete lampbrush loop (Figure 5). The Miller spreading technique has provided important information about the molecular structure of genes during their transcription. However, only in exceptional cases could distinct eukaryotic genes be identified and analyzed.[56] The fertility gene expressed in the lampbrush loop nooses is one of these genes. The length of DNA involved in transcription can be estimated to be 260 kb, a size closely in agreement with the earlier estimates based on cytology.[41] The transcripts display a high degree of secondary structure and cover the entire length of the loop DNA. Their enormous sizes imply that each loop is transcribed over a major part or even over its entire length. If some "traveling" of the completed transcripts along the DNA axis of the loop should occur without further transcription, the distance of such a movement can only be small. Active secondary initiation sites, which would be recognized by the presence of smaller transcripts, have not been discovered in this loop (but were occasionally found in others[37]).

DNA sequences of this lampbrush loop pair were identified by the following methods.[45,50,51,57,58] In a first approach, cloned DNA fragments were identified as Y chromosomal in origin by semiquantitative hybridization with labeled genomic DNA of male or female flies. Clones with hybridization signals more prominent with the DNA of males were collected and analyzed in Southern blots of restriction digests of genomic DNA from male or female flies. The presence of male-specific bands was considered as evidence for a Y chromosomal location of the male-specific restriction fragments. This location was further established by *in situ* hybridization to metaphase chromosomes of male neuroblasts. Hybridization of a cloned DNA fragment to the Y chromosome not only gave confirmation of the Y chro-

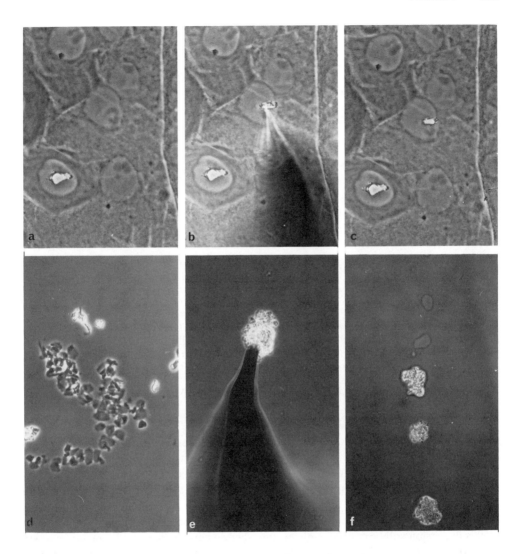

FIGURE 3.   Cloning of lampbrush loop DNA at the microscale level. Dissection of a lampbrush loop (here, pseudonucleolus, cf. Figure 1) is carried out with a glass needle (a to c). The dissected loops are collected (d), taken up with the tip of a glass needle (e), and transferred into a small droplet of buffer (1 nl) (d) to allow DNA extraction, restriction digest of the purified DNA, and ligation to a lamba vector. The procedure is carried out under the microscope in an oil chamber. Photographs by W. Hennig.

mosomal location of at least some copies of this DNA, but also provided evidence concerning its approximate location within the chromosome. This allowed more precise mapping by studying the hybridization to partially deleted Y chromosomes in Southern blots.

The next question of major importance was whether the recovered DNA sequences are transcribed in primary spermatocytes. This question could be answered by application of the transcript *in situ* hybridization technique to testes squashes. Hybridization to spermatocyte nuclei reveals whether DNA sequences similar or identical to the cloned fragments are transcribed in one (or several) of the lampbrush loops (Figure 6). Hence, not only the transcription of a DNA fragment can be recognized in this way, but it can also be assigned to transcripts of specific lampbrush loop pairs. Moreover, the use of strand-specific labeled probes allows the identification of the transcribed strand.

In this way, we identified a series of clones originating from the short arm of the Y chromosome which are represented within the transcripts of the lampbrush loop nooses and

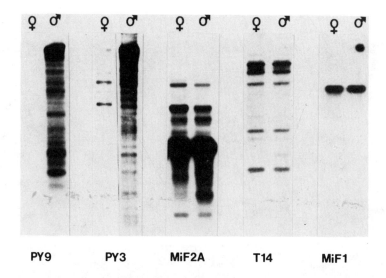

FIGURE 4.   Classes of Y chromosomal DNA. Y-specific DNA reacts only with the DNA of males (PY9, to the left), Y-associated DNA reacts with the DNA of both males and females and shows usually only few male-specific bands (MiF1, to the right). Labeled DNA fragments containing both classes of DNA hybridize preferentially with the DNA of males, but also, to a variable extent, with the DNA of females, depending on the relative contribution of both DNA classes within the cloned DNA fragment (the three clones PY3, dhMiF2A, and cDhT14). Blots of genomic DNA of males (♂) or females (♀) after restriction digests were hybridized with labeled, cloned DNA fragments. (Data from Vogt, P., Huijser, P., and Brand, R. C., Nijmegen.)

studied them by DNA sequence analysis. These DNA sequences belong to the two classes of DNA described before, i.e., to Y-specific or to Y-associated DNA. The first, Y-specific class is represented by a DNA sequence family characteristic of this lampbrush loop pair. This sequence family, called the ay1 sequence family,[50] can be related to a basic repeat unit "ay1" with a length of 400 bp. Such repeat units occur preferentially in small tandem clusters of repeats which are spread all along the lampbrush loop nooses (Figure 7). Evidence for this dispersion will be considered later. Interspersed between these ay1 sequence clusters are DNA sequences of the second DNA class, the Y-associated sequences. They are represented, for example, by the DNA fragment of the clone dhMY3[51] or dhNo19 (Figure 8)[28] not displaying sequence similarity to ay1.

The lampbrush loop specificity of Y-specific ay1 DNA sequences permitted the collection of a set of additional cloned fragments hybridizing with this Y chromosome-specific DNA probe. In principle, this approach should permit the recovery of all DNA segments of a particular lampbrush loop pair, provided that no large sections of the lampbrush loop DNA are devoid of the respective Y-specific DNA.

We have completed the first step in such an extensive analysis of the DNA of the lampbrush loop pair nooses. Clones from a *Bam*H1 clone bank in EMBL3 lambda phages were screened with labeled ay1 sequences and yielded clones with sequence similarities. In total, these clones cover approximately 210 kb of the lampbrush loop DNA. This represents 80% of the 260 kb estimated on basis of Miller spreads to be the actual amount of DNA in the loop.

The majority of these clones with sequence similarity to ay1 have been analyzed with respect to their restriction patterns and their regions of ay1 similarity. In general, ay1 sequences are restricted to a minority of the clones, except for a few clones which are entirely or almost entirely composed of ay1 sequences. Adding up the DNA fragments similar to ay1 gives approximately 90 kb of ay1 sequences. This value agrees rather well

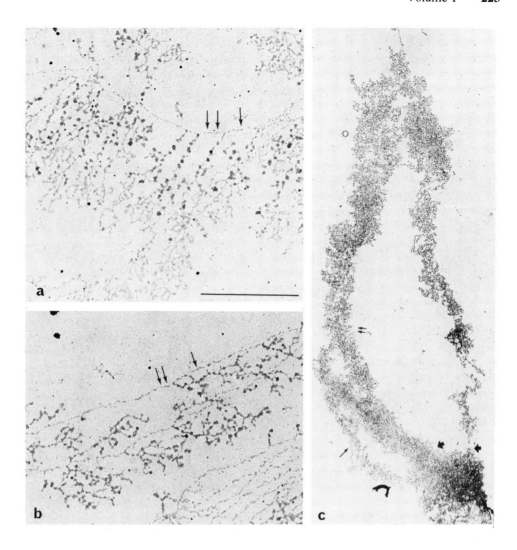

FIGURE 5.   Transcripts of the lampbrush loop pair nooses visualized by Miller spreading. (a) and (b) show sections of the lampbrush loop with giant transcripts displaying a high degree of secondary structure and covered with protein granules. Nucleosomes along the DNA axis are indicated by arrows. (c) The complete lampbrush loop nooses after Miller spreading. It represents a single transcription unit extending over 260 kb of DNA. Compare with Figure 1e! Bars represent 10 μm. (From Grond, C. J., Slegmund, I., and Hennig, W., *Chromosoma*, 88, 50, 1983. With permission.)

with earlier, more indirect estimates that approximately one third of the loop DNA represents sequences related to ay1.[50] The ay1 distribution pattern in Figure 8 demonstrates that ay1 clusters occur interspersed with DNA of the Y-associated class.

The size of ay1 clusters is, in general, relatively small. Obviously, each cluster contains only a few repeats. Several such ay1 clusters have now been sequenced and the results allow some general conclusions.[28] Each cluster represents a unique sequence arrangement of the repeat units which differs from the basic ay1 repeat unit (Figure 9). Duplications and intrachromosomal amplifications of this ancestral ay1 sequence have probably occurred in several subsequent steps since the different clusters have an internally more homogeneous structure than found in different clusters.[50,51,87] Such modifications probably occurred simultaneously with, or subsequent to, the insertion of a Y-associated DNA sequence into the loop (see below). If the exact order of ay1 clusters within the lampbrush loop is known,

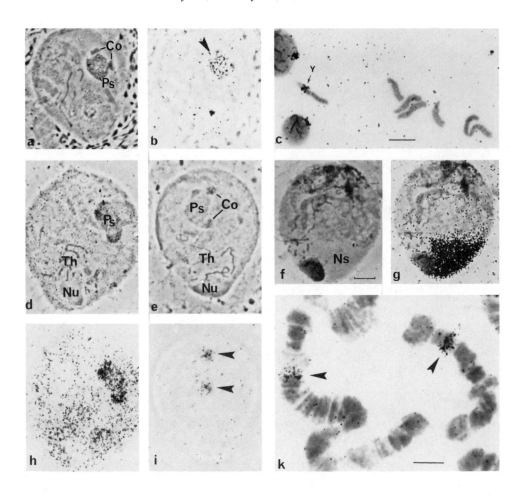

FIGURE 6.   *In situ* hybridization of various cloned Y chromosome DNA fragments. (a, b, d to i) Transcript *in situ* hybridization to primary spermatocyte nuclei. Hybridization occurs to growing transcripts (cf. Figure 5) of the lampbrush loops. Hybridization is found on the "cones" (e, i), the "pseudonucleolus" (a, b), the "threads" and the "pseudonucleolus" (d, h), or the "nooses" (f, g). (c) Hybridization to metaphase chromosomes of neuroblasts to localize the fragments within the Y chromosome (arrow). The Y chromosome is also condensed in the interphase as the localized label indicates. (k) Hybridization to autosomal loci in polytene chromosomes of salivary glands. Bars represent 10 μm.

it will be of interest to compare the actual sequences in adjacent ay1 clusters. This will give insight into the gradual evolutionary development of the molecular structure of this lampbrush loop pair (for a discussion, see Reference 59).

The next question to be addressed is that of the origin of the DNA sequences interspersed between ay1 clusters. The first sequencing data did not allow direct conclusions about the nature of such sequences, except that they are unlikely to code for proteins.[51] However, *in situ* hybridization of several Y-associated DNA sequences to polytene chromosomes of *D. hydei* and two closely related species with an almost identical banding pattern,[60] *D. neohydei* and *D. eohydei*, showed that, in contrast to the stability of the chromosome banding patterns, the locations of the Y-associated DNA sequences in sites other than the Y chromosome varied drastically between the three species. This suggested that Y-associated DNA represents families of transposable elements.[27] The low copy numbers and the average lengths of the Y-associated DNA fragments also agree with their nature as transposable elements. This concept of transpositions contributing DNA elements to the fertility genes is also attractive

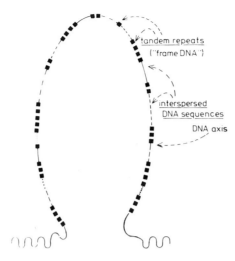

FIGURE 7. Scheme for the molecular organization of the lampbrush loop nooses. Y-specific DNA is scattered in small blocks of tandemly arranged repeats all along the gene. Interspersed are Y-associated DNA sequences which occur in only a few copies within the gene, with the majority of their copies elsewhere in the genome. The small blocks in the nooses are represented by the ay1 sequence family, which most likely is the only Y-specific sequence type of this lampbrush loop pair. (From Hennig, W., in *Results and Problems in Cell Differentiation*, Vol. 14, *Structure and Function of Eukaryotic Chromosomes*, Hennig, W., Ed., Springer-Verlag, Berlin, 1987, 137. With permission.)

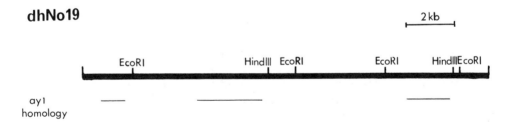

FIGURE 8. Restriction map of the recombinant DNA fragment dhNo19 from the lampbrush loop nooses. Regions with homology to the ay1 DNA sequence family are indicated by lines underneath the restriction map. This DNA fragment displays the characteristic interspersion pattern within the loop DNA. The ay1 cluster shown in Figure 9B is the left-hand cluster. Sequence data indicate that the region between the left-hand ay1 cluster and the cluster in the middle of the clone is a transposable sequence. (Unpublished data from Pötgens, A., Dijkhof, R., de Graf, R., and Hennig, W.)

if it is considered that transposable elements may be specifically active in the germ line (see Reference 61) and that the open chromatin structure in transcribed chromosome regions, such as the Y chromosomal fertility genes, would be advantageous for insertion events.

This concept was recently validated.[49,62] Y-associated DNA sequences isolated by microcloning DNA from lampbrush loop-pair threads and pseudonucleolus were identified as fragments of a new family of retrotransposons, which was called the micropia family.[49,62] This family of retrotransposons shares all basic features with other copia-like elements in *Drosophila* and other retrotransposons, but also displays some unique features.[62] While elements of this sequence family located in Y chromosome lampbrush loops are defective,

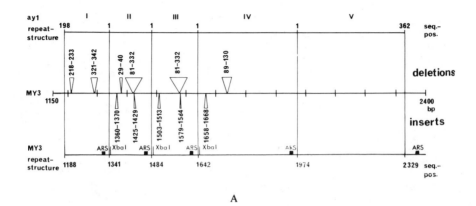

A

FIGURE 9. Molecular structure of two clusters of the lampbrush loop-specific DNA sequence family ay1. Both clusters, localized in different regions of the loop, display a unique and highly region-specific molecular structure which indicates a separate evolutionary history of the clusters after an initial amplification of the ancestral sequence. (a) Cluster of the cloned fragment MY3. (From Vogt, P. and Hennig, W., *Chromosoma*, 94, 459, 1986. With permission.) Five incomplete tandem repeats display a deletion-insertion pattern which argues for a successive duplication and divergence of the repeats. (b) Cluster of the cloned fragment dhNo19BE (left ay1 cluster in Figure 8). (Unpublished data from Pötgens, A., de Graf, R., and Hennig, W.) Two repeats of ay1 flanked by Y-associated DNA show a construction different from that of the MY3 cluster of the ay1 repeats.

other genomic copies are likely to be functional. These observations are full compatible with the conclusions of Hareven et al.[63] and Vogt et al.[59] on the conservation of these sequences between different *Drosophila* species and their distribution within the genome. Both groups found that Y-associated DNA sequences of the lampbrush loops of *D. hydei* are exclusively autosomal in other, more distantly related species.

In summarizing these data on Y chromosome DNA, the following picture emerges which at present can most typically be assigned to the lampbrush loop-pair nooses, but which most likely holds true for the other Y chromosomal lampbrush loops as well, since their basic features are closely comparable to those of the nooses.

Lampbrush loop DNA is constructed from small clusters of a short, tandemly repeated sequence type (up to 400 bp as basic repeat units) which are separated by other repeated DNA elements, probably without question of a transposable nature. One loop may carry several complete or incomplete members of this transposable sequence family, but the majority of the transposable elements reside outside the Y chromosome. Since this type of DNA represents a substantial portion of the lampbrush loop DNA and since it is highly interspersed between Y-specific DNA, it has in the past been difficult to identify and isolate Y chromosomal DNA by conventional techniques.

E. THE BIOLOGICAL FUNCTION OF THE FERTILITY GENES

This basic picture of the molecular structure of the lampbrush loops raises the question of their biological role. What is the function of their transcripts? Do they code for proteins? What is the function of the different sequence elements, Y-associated and Y-specific? To answer these questions, we must reconsider knowledge concerning lampbrush loops from this perspective.

1. Transcription of the Lampbrush Loops

Earlier in this article, it was pointed out that the major constituents of the lampbrush loops are proteins and RNA. The relationship between RNA and lampbrush loops has long

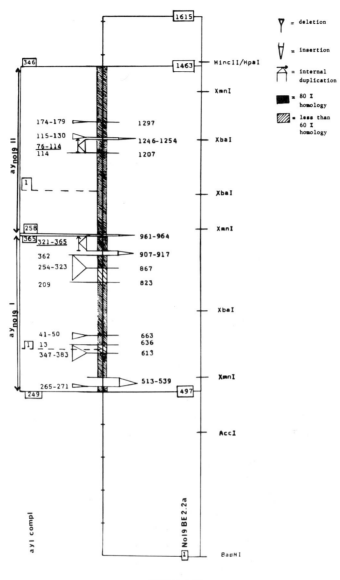

FIGURE 9B

been established by the classic experiments of Gall and Callan, who demonstrated that lampbrush loops in oocytes are chromosomal regions of intense transcription. Meyer[18] concluded that the spermatocyte structures of *Drosophila* are functional equivalents of oocyte lampbrush loops. Consistent with this, autoradiographic studies of spermatocyte nuclei confirmed that the Y chromosome lampbrush loops are also extensively transcribed[23] and DNA/RNA hybridization experiments demonstrated that Y chromosome repetitive DNA sequences are represented in testes transcripts.[40,41] The structural properties of transcription in Y chromosome lampbrush loops were documented by Miller spreading.[36,37,41,64,65] These Miller spreading micrographs consistently demonstrated that giant transcripts are synthesized along the lampbrush loops, while only in exceptional cases were small transcripts found interspersed between large ones. These small transcripts were interpreted as indicating the presence of secondary initiation sites within the loops.[37] This interpretation was later substantiated by nucleotide sequence analysis of the DNA of the respective lampbrush loops, revealing the presence of potential initiation sites for transcription within the loop DNA. An additional important feature emerging from Miller spreading experiments is the loop specificity of the transcription on patterns, both with respect to the length of growing transcripts and their high degree of secondary structure.[37]

2. Lampbrush Loop Proteins

Data concerning the lampbrush loop proteins are scarce, but still allow some fundamental conclusions. Glätzer and Kloetzel[66] showed that proteins typical of transcriptionally active loci (ribonucleoproteins) are associated with the lampbrush loops. This reflects the presence of proteins involved in the metabolism of the transcripts in RNA packaging and transport, as they are characteristically associated with other active genes. The association of comparable proteins with lampbrush loops in amphibian oocytes was established by Sommerville et al.[67,68] and has recently been confirmed by Roth and Gall[69] and Ragghianti et al.[70] With the use of antisera obtained against testis proteins, the presence of loop-specific proteins in *Drosophila* spermatocyte lampbrush loops has also been discovered.[24,38] A comparable site-specific assignment of proteins to single active loci has, for other active genes, only been demonstrated previously in exceptional cases. However, recently, studies on amphibian lampbrush chromosomes have revealed a similar site specificity of chromosomal proteins.[69,70]

3. Selective Protein Binding to Transcripts

What are the mechanisms responsible for determining the site-specific accumulation of particular proteins? What is the nature and function of such proteins?

Is is reasonable to assume that the specific character of transcripts within an active lampbrush loop is responsible for the selective binding of a protein to this chromosomal site. Before we consider the evidence for this and a potential biological role for such site-specific chromosomal proteins, let us first complete our picture of lampbrush loop structure by evaluating our knowledge of the molecular composition of lampbrush loops at the DNA level since this is essential for a discussion of their function.

4. Lampbrush Loops Formed by the Fertility Genes Have Functions Other than Protein Coding

If we evaluate all known nucleotide sequences of Y chromosome lampbrush loops, we must conclude that there is no evidence for functional protein-coding DNA sequences.[28,39,48-53] The micropia retrotransposon has the potential to code for proteins, e.g., for reverse transcriptase, but the micropia elements found in the Y chromosome make it unlikely that functional proteins are encoded because of the large amount of stop codons within all reading frames.[49]

One might still argue that it is impossible to exclude the possibility that single protein-coding elements exist between the overwhelming amount of noncoding repetitive DNA elements which would escape detection. This argument can only be considered invalid after completing our knowledge of the lampbrush loop nucleotide sequences. However, there exist good reasons to consider that the Y chromosome lampbrush loops have functions different from protein coding.

One of these reasons is the formation itself of conspicuous lampbrush loop pairs. We know of no other transcriptionally active loci with such a high degree of site-specific structural properties. The functional significance of this cytological structure is emphasized by the observation that structurally modified lampbrush loops usually result in sterility of the males due to incomplete sperm development.[25] Moreover, their fertility genes display an unusual mutation pattern with mutation frequencies at least $10\times$ above the frequencies of conventional protein-coding genes. This high mutation rate is compatible with the large size of the genes and implies that most of the mutations within a fertility gene interfere with its normal function. Obviously, this cannot be understood on the basis of an intron-exon structure with small exons embedded within large introns. The lampbrush loops must, therefore, form large functional units with highly specific structural requirements.[25]

Because of the unique cytological properties of the Y chromosome lampbrush loops, one is reminded of the situation of the heat-shock locus 93D of *D. melanogaster* or 48B of *D. hydei* which in polytene salivary gland chromosomes display an unusual and particular arrangement of RNP granules. Remarkably enough, these loci do not code for proteins and have a molecular structure not unlike that of Y chromosome lampbrush loops (see Reference 71). They are constructed from short, tandemly repeated DNA sequences and a section of unique DNA, both of which are transcribed, but do not code for proteins, and serve unknown purposes. Hence, two types of eukaryotic genes, the heat-shock loci and the lampbrush loops, have attracted interest because of their particular cytological and (ultra)structural features. Both types of genes may represent eukaryotic genes of functional relevance different from that assigned to conventional gene functions, i.e., that of coding for structural RNA or for proteins.

Other reasons to assume that the lampbrush loop-forming genes serve functions other than protein coding are derived from studies of the protein composition of testes. Various investigators have tried to relate testis-specific proteins to the activity of Y chromosome genes.[72,73] Remarkably enough, in none of these studies did the investigators succeed in demonstrating a direct relationship between the activity of particular Y chromosomal loci and the presence of testis proteins. On the contrary, several lines of evidence suggest a regulatory role of Y chromosomal loci in the testis-specific expression of proteins. Livak[74] investigated the expression of the Stellate locus on the X chromosome. This site is responsible for the production of a protein forming large protein crystals in the testis of *D. melanogaster*,[14] particularly in X/O males. Livak demonstrated that the Stellate locus carries multiple copies of the protein-coding region. Additional copies were found on the Y chromosome close to the fertility gene kl2.[74] These Y chromosomal copies do not, however, produce mRNA of the Stellate protein, but rather regulate the amount of protein formed in a so far unknown fashion.[75] The presence of the Y chromosomal copies of the protein-coding sequence prevents the formation of excess Stellate protein, as occurs in X/O cells. Comparable observations were made by Hulsebos et al.[73] These authors demonstrated that the Y chromosome is required in testes to assure the production of sufficient amounts of three testis-specific proteins, the sp155 and sp35 proteins and tubulin. Additional studies in our laboratory have since shown that the production of tubulin is regulated at the translational level since mRNA for tubulin in X/O testes is synthesized in amounts similar to wild-type testis.[88]

These various data suggest a role of Y chromosome fertility genes other than a coding function for sperm constituents. This conclusion is not only supported by the data discussed above, but also by cytological observations of the effects of Y chromosome deficiencies in X/O cells on spermatogenesis. In *D. melanogaster*, absence of the Y chromosome still allows considerable progression in postmeiotic spermiogenesis.[76] It seems that primarily chromatin condensation and the final shaping of sperm heads is affected by the absence of the Y. In *D. hydei*, more severe defects are found, depending on the nature of the Y chromosome deficiency. It has even been claimed that the absence of the Y chromosome prevents meiosis, but this is incorrect.[77] Kremer[89] found that the stage of interruption of spermiogenesis depends strongly on temperature. Sperm differentiation proceeds farther at lower temperatures (e.g., 18°C) than at higher temperatures (e.g., 24°C). Irrespective of these temperature effects, for *D. hydei* no major structural components of sperm are missing in the absence of the Y chromosome. All data, therefore, agree with the conclusion that Y chromosome genes in *Drosophila* have functions other than coding for structural components of sperm.

The observation that shaping of the sperm head in *D. melanogaster* is incomplete in the absence of the Y chromosome has been substantiated by equivalent and more detailed observations for *D. hydei*, which even more strongly suggest a connection between Y chromosome function and chromatin composition in postmeiotic development. Kremer et al.[24] described a complex cycle of condensation and decondensation of the chromatin of *D. hydei* after meiosis, before the final condensation of chromatin takes place. In young spermatid nuclei, the chromosomes first condense in a small area of the nucleus close to the attachment site of the flagellum. Then they decondense and the chromatin spreads out uniformly over all of the elongating spermatid nucleus. Only late during sperm formation does the final condensation of chromatin occur (see Grond[78]). This postmeiotic chromatin condensation/decondensation cycle is characteristically changed in males of an X/O genotype. In the absence of the Y chromosome, the initial condensation does not take place and a final condensation of the chromatin is not observed. In the latter case, it is not clear whether this is a consequence of the incomplete development of the spermatids or whether it represents a primary defect.

Cytogenetic experiments revealed that the entire Y chromosome is not responsible for these effects on chromatin condensation. Two regions are required to allow a normal chromatin condensation/decondensation cycle.[89] The most important region in the Y chromosome is the distal end of the long arm accommodating threads and pseudonucleolus (loci A to C). But the sole presence of the region close to the kinetochore can, also at least in part, promote chromatin condensation and decondensation. These observations indicate either additive or qualitatively identical effects of different regions of the Y chromosome. Earlier genetic studies indicated only qualitative differences between the different fertility genes.[79] However, the limited possiblity for genetic manipulation in *D. hydei* did not allow testing for all possible combinations of Y chromosome fragments in those experiments. Moreover, since the results of DNA investigations gave evidence for the presence of similar or identical sequences in different regions of the Y chromosome,[47,52,53] purely quantitative effects cannot with certainty be excluded. Therefore, identical functions residing in different regions of the Y chromosome are still feasible.

These cytogenetic data can now be related to observations on lampbrush loop proteins. A relationship between the Y chromosome lampbrush loops and chromosomal proteins was suggested by histochemical studies which demonstrated the strong basic nature of the lampbrush loop-associated proteins.[18,78] The correlation with chromosomal proteins is further supported by proteins recognized by various antibodies which react specifically with particular lampbrush loops in immunofluorescence studies (Figure 10). Kremer et al.[24] demonstrated that a polyclonal histone H1 antiserum strongly and specifically interacts with the lampbrush loop pseudonucleolus (see also Hennig[26]). A monoclonal antiserum obtained against a DNA-associated protein of *D. melanogaster*, Bv96,[80] reacts in testes specifically with the lampbrush loop pair nooses. Postmeiotically, it also reacts with sperm heads.[87] In salivary gland polytene chromosomes, this antiserum decorates a restricted number of chromosomal sites (approximately 30 condensed bands). On blots of testis protein after gel electrophoresis, the antiserum strongly reacts with a protein of M_r 35,000. In *Drosophila*, histone H1 characteristically migrates at this apparent molecular weight. The binding to particular bands in polytene chromosomes, to the nooses, and to sperm heads suggests that the protein identified might correspond to a histone H1 variant or a related chromosomal protein.

The immunological data point, although indirectly, toward the possibility that histones or histone-like proteins are, apart from their usual presence in nucleosomes, associated with the transcripts in certain lampbrush loops. It must be pointed out that antisera against core histones do not detectably react with the lampbrush loops, although the presence of nucleosomes was documented by Miller spreading experiments. This indicates that the actual amount of histones in nucleosomes on the strongly despiralized Y chromosomal DNA is too small to be detected by conventional immunofluorescence methods. Also, histone H1 antisera do not react detectably with lampbrush loops other than the pseudonucleolus. This indicates an early depletion of histone H1 during the early primary spermatocyte stage.[24] Somatic

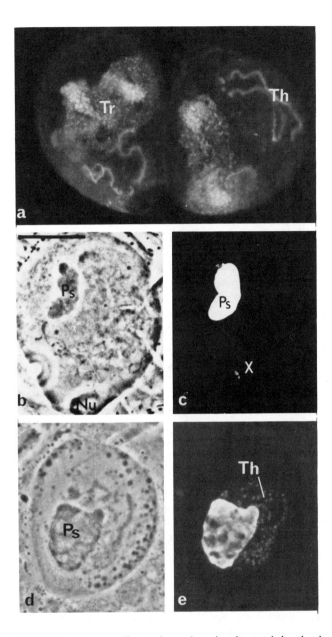

FIGURE 10.   Loop-specific proteins can be assigned to certain lampbrush
loops by immunofluorescence. (a) Antiserum K7 (Hennig, W. et al., unpublished) is specific for the loop tubular ribbons (Tr) and threads (Th);
(b,c) antiserum sph155 decorates the pseudonucleolus (From Hulsebos,
T. J. M., Hackstein, J. H. P., and Hennig, W., *Proc. Natl. Acad. Sci.
U.S.A.*, 81, 3404, 1984. With permission.); (d,e) in a mutant, the same
antiserum displays the presence of antigenic sites in the neighboring lampbrush loop-pair threads as small dots. (From Hackstein, J. H. P., Hulsebos,
T. J. M., and Hennig, W., submitted.) Bars represent 10 μm.

histone H1 cannot be detected in any stage of spermatogenesis beyond early primary spermatocytes. This implies that it is functionally replaced by other chromosomal proteins. The
substitution of histones during spermiogenesis by more basic chromosomal proteins is a
well-known phenomenon, but it has not been studied in *Drosophila* by methods other than

histochemical approaches.[81] The observations on incomplete chromatin condensation in the absence of the Y chromosome suggest that one of the biological functions of Y chromosome lampbrush loops is related to the process of chromosomal protein substitution during sperm development.

5. The Y Chromosomal Fertility Genes Function by Protein Binding

These observations on the chromatin composition of the developing male germ cell, together with the arguments against a protein-coding function of these genes, induced us to propose that the role of the Y chromosomal fertility genes is to bind chromosomal proteins by means of their growing transcripts. This working concept is, in more detail, based on the following considerations. The main features distinguishing Y chromosome lampbrush loops from all other genes active in the male germ line of *Drosophila* are their size and their peculiar morphology. As shown before, size and morphology are determined by the large units of transcription. However, the size of a transcription unit is not sufficient to account for morphology since other transcription units of large size are known — for example, the bithorax locus, the Antennapedia locus, or the Notch locus of *D. melanogaster* — which form no structures with such a conspicuous morphology as the Y chromosome lampbrush loops. The morphology of the lampbrush loops is, therefore, a special property of these loci and is likely to represent an expression of their specific function. All evidence indicates that such a function is protein assembly or storage — or even a sink function for proteins to be removed from chromatin — during the first meiotic prophase. The enormous sizes of the transcription units and the proportionate sizes of the transcripts may serve to supply a sufficiently large number of target sites for such proteins which are recognized and bound with the aid of the transcripts. We believe that the sequence elements of primary importance within a loop are the Y-specific sequences, while Y-associated sequences may simply serve to increase the size of the transcription units.[17] Since this results in the availability of larger numbers of transcripts traveling simultaneously along the loop axis, the number of target sites for protein binding becomes considerably enlarged in this way. In addition, RNA segments may be required to provide an appropriate spacing of the binding sites to prevent close contact between the bound protein molecules, which could lead to undesirable interactions. More recent sequencing data have indicated the presence of DNA sequences within Y-associated DNA segments with the potential to act as replication origins or as enhancers of transcription.[90] Also, the presence of $(CA/GT)_n$ blocks within lampbrush loop DNA,[48,82] which characteristically provide the potential for DNA to form a Z-configuration, suggests that sequences which support the maintenance of an open chromatin conformation may be essential for the synthesis of transcripts of such an enormous size. Transposable elements typically contain DNA sequence elements important for the initiation and enhancement of transcription. In this sense, the presence of transposable elements as insertions within the genes may be an additional essential feature of the fertility gene structure.

In summarizing these various aspects, we arrive at a working model which considers that Y chromosomal fertility genes are required to assemble proteins through binding to their site-specific transcripts. Such proteins are either required in the process of chromatin substitution during germ cell development or they may be involved in regulatory processes during postmeiotic development. We also cannot exclude the possibility that parts of the transcripts are important in regulatory processes since postmeiotically they are found in the cytoplasm[24,39,41] and remain stable for extended periods (see Figure 2).[20]

The question arises whether similar functions can explain the formation and special properties of lampbrush chromosomes in oogenesis. This question cannot be answered. But in particular those lampbrush loops with a highly specialized morphology known as landmark loops may be candidates for such special, possibly germ line-specific, functions. Our knowledge of the processes accompanying germ cell development is very limited in both sexes and new conceptual approaches may be necessary to obtain insight into the specific requirements of germ cell development.

F. EVOLUTIONARY ASPECTS OF Y CHROMOSOME STRUCTURE

The complex structure of the Y chromosomal fertility genes raises the question of their evolutionary history. A comparison of the evolutionary conservation of both types of DNA can provide us with the key to their evolutionary relationship since their presence in other *Drosophila* species differs remarkably. Y-specific DNA is, in general, species specific or conserved only between very closely related *Drosophila* species.[39,59,63] Y-associated sequences, on the other hand, were also found in evolutionarily distant *Drosophila* species.[49,59,62,63] However, in more distant species the "Y-associated" DNA sequences are usually not found on the Y chromosome.[59,63] This observation argues for a relatively recent evolutionary invasion of Y-associated DNA sequences into the Y chromosome. This implication is compatible with their character as transposable elements.

This situation makes it unlikely that the quality of the genetic information of Y-associated DNA sequences is of primary relevance for the biological function of the Y chromosomal fertility genes. I have argued before that the molecular information residing within the transposable elements may, nevertheless, become important in the context of the molecular construction of these genes.

These considerations imply that the DNA sequences primarily important for the biological function of the fertility genes are of the Y-specific type. How, then, is the high degree of species specificity of such DNA sequences compatible with the supposedly more general functions of such genes during sperm development? This question, however, is easily answered in the context of our model of gene function. It has been amply demonstrated that specific protein binding to nucleic acids can be mediated by different nucleotide sequences. The same lampbrush loop protein may therefore become bound by different sequence types in different species. Supporting experimental evidence has been provided by the immunofluorescence studies of Hulsebos et al.[38] (see also Reference 25). Antigenic sites reacting with the sph155 serum were found in the primary spermatocyte nuclei of most of the *Drosophila* species studied, although lampbrush loop morphology is subject to drastic alterations in different, even closely related species. These observations at once emphasize that transcriptionally active genes behaving similarly to the lampbrush loops of *D. hydei* probably occur in all *Drosophila* spermatocyte nuclei, even though they may not be detectable by cytology.

These arguments still do not seem to explain why Y-specific DNA sequences are evolutionarily restricted to single species or very closely related species. However, in this context the specific properties of tandemly arranged repetitive DNA sequences should be considered. It is well known that tandemly arranged repetitive DNA sequences are either under strict control mechanisms to assure their sequence identity — like that keeping ribosomal DNA genes alike — or that they rapidly diverge from each other — as has been shown for "satellite" DNAs. In addition, the (perhaps functionally favorable) insertion of Y-associated DNA elements may interfere with correction mechanisms as they act on rDNA or other repeated gene families, or may even induce local amplifications. A rapid mutational divergence of distinct Y-specific DNA fractions may be the consequence, as has been experimentally demonstrated to occur in two such sequence families, the ay1 family [28,50] and the rally family.[39] Protein binding functions may not necessarily require strict control of the maintenance of the nucleic acid sequences adjacent to the direct binding sites and, therefore, a considerable degree of divergence in some loop regions may be accepted. However, at some point, the divergence between the different repeat units may achieve a degree which increasingly interferes with their proper function. At this point, new intrachromosomally amplified DNA sections may provide a selective advantage by overtaking the

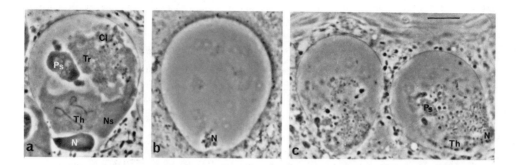

FIGURE 11.   Spermatocyte nuclei of males of *D. hydei*. (a) Wild type; (b) ms(3)5/ms(3)5 male with normal sex chromosome composition (X/Y). The homozygous composition of the recessive male-sterility mutation prevents the formation of Y chromosome lampbrush loops at high temperature. (c) The same composition as in (b) allows a reduced degree of transcription of the Y chromosome genes at low temperatures, i.e., the mutation is conditional. But even at this temperature, males remain sterile. This is most likely correlated to the abnormally low expression of the Y chromosome genes. This mutation demonstrates that the size and shape of the primary spermatocyte nuclei are not determined by the expression of the Y chromosome lampbrush loops, but are independently established. Bar represents 10 μm. (From Hackstein, J. H. P., Hennig, W., and Steinman-Zwicky, M., *Wilhelm Roux Arch. Dev. Biol.*, 196, 119, 1987. With permission.)

particular protein binding function. The occurrence of such events is demonstrated by the evolutionarily young rally sequence family in the lampbrush loop pair threads, which is not found in very closely related species.[39]

A final point for discussion concerns the extreme variability in size and structure of Y chromosome lampbrush loops within the genus *Drosophila*. Our knowledge allows only speculation, but a few observations may be intriguing enough to justify this. If the primary spermatocytes of different *Drosophila* species are compared, it becomes clear that the prominence of lampbrush loop morphology is correlated with nuclear size. This seems trivial if one accepts that lampbrush loop size determines the size of nuclei. However, we have shown that the size and shape of primary spermatocyte nuclei in *D. hydei* are independently regulated.[83] The recessive autosomal male fertility mutant ms(3)5 displays no Y chromosomal lampbrush loops in the primary spermatocytes if homozygous, although a Y chromosome is present. The nuclei, however, are still as large as wild-type nuclei (Figure 11). It is, on the other hand, well known that cell volume and the sizes of nuclei are related to one another. Spermatocyte sizes are widely different between *Drosophila* species. This may explain why the nuclear sizes in spermatocytes of different species show considerable differences. The variability of spermatocyte size, in turn, may be related to the sizes of the spermatozoa. In *D. melanogaster,* sperm length reaches about 1.8 mm, while in *D. hydei* it exceeds 12 mm. Other species with large sperm sizes also have large primary spermatocytes. This correlation is functionally plausible since in the primary spermatocytes material required for postmeiotic sperm morphogenesis must be produced and stored. Considering our model for the function of the Y chromosome lampbrush loops as targets for chromosomal proteins, one might consider that it is necessary to fill the primary spermatocyte nuclei as much as possible with lampbrush loop material to allow close contact with the residual chromosomes. Such a close proximity between the lampbrush loops and the rest of the chromatin may be required to promote the proper substitution of chromosomal proteins. In this sense, an indirect correlation may exist between sperm length and the size of lampbrush loops in primary spermatocytes.

III. Y CHROMOSOMAL DNA OF OTHER ORGANISMS

I would like to complete this evaluation of our knowledge of Y chromosomal DNA sequences and their function with a comparison to data available from other organisms.

However, besides *Drosophila,* Y chromosomal DNA has only been studied in some detail in man.

A summary of knowledge concerning human Y chromosomal DNA has recently been given by Smith et al.[84] In contrast to *Drosophila,* the human Y chromosome carries regions which are considered homologous to the X chromosome. Moreover, it is involved in the primary sex determination mechanism by determining the male sex,[85,86] but it also carries genes required for sperm development.[2] Part of the long arm of the human Y chromosome is heterochromatic and can vary in length. In comparing the heterochromatic character of the Y chromosomes of *Drosophila* and man, it is most remarkable that in both organisms DNA sequence types corresponding to those known from *Drosophila* are found, i.e., Y-specific and Y-associated DNA classes. One wonders whether Y chromosomal DNA in particular, and heterochromatin in general, are constructed according to similar principles and have similar biological functions. An answer is not yet available. The difficulties in analyzing repetitive DNA sequences from structural and functional points of view will still require considerable efforts to understand the precise organization of genomic regions rich in heterochromatin. Recent progress in the molecular analysis of such regions and the observation of the expression of repetitive DNA in the germ line may permit us to obtain insight into new biological principles of eukaryotic genome structure and function.

ACKNOWLEDGMENTS

The ideas and models presented in this article are mainly based on the conceptual and practical work of my research group. In particular, Drs. R. C. Brand, J. Hackstein, R. Hochstenbach, P. Huijser, H. Kremer, D.-L. Lankenau, and P. Vogt have contributed data and discussions. I am indebted to Hannie Kremer for critically reading the manuscript.

REFERENCES

1. **Nöthiger, R. and Steinmann-Zwicky, M.,** Genetics of sex determination, in *Results and Problems in Cell Differentiation,* Vol. 14, Hennig, W., Ed., Springer-Verlag, Berlin, 1984, 271.
2. **Burgoyne, P. S.,** The role of the mammalian Y chromosome in spermatogenesis, *Dev. Suppl.,* 101, 133, 1987.
3. **Hennig, W.,** Heterochromatin and germ line-restricted DNA, in *Results and Problems in Cell Differentiation,* Vol. 13, *Germ Line — Soma Differentiation,* Hennig, W., Ed., Springer-Verlag, Berlin, 1986, 175.
4. **Hennig, W.,** Molecular hybridization of DNA and RNA in situ, *Int. Rev. Cytol.,* 36, 1, 1973.
5. **John, B. and Micklos, G. L. G.,** Functional aspects of satellite DNA and heterochromatin, *Int. Rev. Cytol.,* 58, 1, 1979.
6. **Macgregor, H. C. and Andrews, C.,** The arrangement and transcription of "middle repetitive" DNA sequences on lampbrush chromosomes of *Triturus, Chromosoma,* 63, 109, 1977.
7. **Jamrich, M., Warrior, R., Steele, R., and Gall, J. G.,** Transcription of repetitive sequences on *Xenopus* lampbrush chromosomes, *Proc. Natl. Acad. Sci. U.S.A.,* 80, 3364, 1983.
8. **Epstein, L. M., Mahon, K. A., and Gall, J. G.,** Transcription of a satellite DNA in the newt, *J. Cell. Biol.,* 103, 1137, 1986.
9. **Bridges, C. B.,** Non-disjunction as proof of the chromosome theory of heredity, *Genetics,* 1, 1, 107, 1916.
10. **Hackstein, J. H. P.,** Spermatogenesis in *Drosophila hydei,* in *Results and Problems in Cell Differentiation,* Vol. 15, *Spermatogenesis: Genetic Aspects,* Hennig, W., Ed., Springer-Verlag, Berlin, 1987, 63.
11. **Brosseau, G. E.,** Genetic analysis of the male fertility factors on the Y chromosome of *Drosophila melanogaster, Genetics,* 45, 257, 1960.
12. **Kemphues, K. J., Kaufman, T. C., Raff, R. A., and Raff, E. C.,** The testis-specific β-tubulin subunit in *Drosophila melanogaster* has multiple functions in spermatogenesis, *Cell,* 31, 655, 1982.
13. **Gatti, M. and Pimpinelli, S.,** Cytological and genetic analysis of the Y chromosome of *Drosophila melanogaster.* I. Organization of the fertility factors, *Chromosoma,* 88, 349, 1983.

14. **Hardy, R. W., Lindsley, D. L., Livak, K. J., Lewis, B., Silverstein, A. L., Joslyn, G. L., Edwards, J., and Bonaccorsi, S.,** Cytogenetic analysis of a segment of the Y chromosome of *Drosophila melanogaster, Genetics,* 107, 591, 1984.

15. **Hess, O.,** Strukturdifferenzierungen im Y-Chromosom von *Drosophila hydei* und ihre Beziehungen zu Gen-Aktivitäten. III. Sequenz und Lokalisation der Schleifenbildungsorte, *Chromosoma,* 16, 222, 1965.

16. **Hackstein, J. H. P., Leoncini, O., Beck, H., Peelen, G., and Hennig, W.,** Genetic fine structure of the Y chromosome of *Drosophila hydei, Genetics,* 101, 257, 1982.

17. **Hennig, W.,** Spermatogenesis in *Drosophila,* in *Primers in Developmental Biology,* Vol. 3, Malacinski, G., Ed., Macmillan, New York, 1988.

18. **Meyer, G. F.,** Die Funktionsstrukturen des Y-Chromosoms in den Spermatocytenkernen von *Drosophila hydei, D. neohydei, D. repleta* und einigen anderen *Drosophila*-Arten, *Chromosoma,* 14, 207, 1963.

19. **Hennig, W.,** The Y chromosomal lampbrush loops of *Drosophila, in Results and Problems in Cell Differentiation,* Vol. 14, *Structure and Function of Eukaryotic Chromosomes,* Hennig, W., Ed., Springer-Verlag, Berlin, 1987, 137.

20. **Hess, O.,** The function of the lampbrush loops formed by the Y chromosome of *Drosophila hydei* in spermatocyte nuclei, *Mol. Gen. Genet.,* 103, 58, 1968.

21. **Hess, O. and Meyer, G. F.,** Genetic activities of the Y chromosome in *Drosophila* during spermatogenesis, *Adv. Genet.,* 14, 171, 1968.

22. **Meyer, G. F.,** Spermiogenesis in normalen und Y-defizienten Männchen von *Drosophila melanogaster* und *D. hydei, Z. Zellforsch. Mikrosk. Anat.,* 84, 141, 1968.

23. **Hennig, W.,** Untersuchungen zur Struktur und Funktion des Lampenbürsten-Y-Chromosoms in der Spermatogenese von *Drosophila, Chromosoma,* 22, 294, 1967.

24. **Kremer, H., Hennig, W., and Dijkhof, R.,** Chromatin organization in the male germ line of *Drosophila hydei, Chromosoma,* 94, 147, 1986.

25. **Hackstein, J. H. P., Hulsebos, T. J. M., and Hennig, W.,** Genetic and cytogenetic analysis of the ''Th'' — ''Ps'' region of the Y chromosome of *Drosophila hydei,* submitted.

26. **Hennig, W.,** Y chromosome function and spermatogenesis in *Drosophila hydei, Adv. Genet.,* 23, 179, 1985.

27. **Hennig, W., Brand, R. C., Hackstein, J., Hochstenbach, R., Kremer, H., Lankenau, D.-H., Lankenau, S., and Miedema, K.,** Y chromosomal fertility genes of *Drosophila:* a new type of eukaryotic gene, *Genome,* xxx, xxx, 1988.

28. **Hennig, W., Dijkhof, R., Pötgens, A., de Graaf, R., and Bremers, R.,** Arrangement of clusters of ay1-like repetitive sequences in the lampbrush loop nooses of the Y chromosomes of *Drosophila hydei,* in preparation.

29. **Callan, H. G.,** *Lampbrush Chromosomes,* Springer-Verlag, Berlin, 1986.

30. **Leoncini, O.,** Temperature sensitive Mutanten im Y-Chromosom von *Drosophila hydei, Chromosoma,* 63, 329, 1977.

31. **Michiels, F., Gasch, A., Kaltschmidt, B., and Renkawitz-Pohl, R.,** A 14 bp sequence motif necessary for testis-specific β2 tubulin gene expression is conserved across species borders in *Drosophila, EMBO J.,* in press.

32. **Olivieri, G. and Olivieri, A.,** Autoradiographic study of nucleic acid synthesis during spermatogenesis in *Drosophila melanogaster, Mutat. Res.,* 2, 366, 1965.

33. **Yamasaki, N.,** Differential staining of Y chromosomal loops in *Drosophila hydei, Chromosoma,* 83, 679, 1981.

34. **Yamasaki, N.,** Selective staining of Y chromosomal loops in *Drosophila hydei, D. Neohydei* and *D. eohydei, Chromosoma,* 60, 27, 1977.

35. **Grond, C. J., Rutten, R. G. J., and Hennig, W.,** Ultrastructure of the Y chromosome lampbrush loops in primary spermatocytes of *Drosophila hydei, Chromosoma,* 89, 85, 1984.

36. **Grond, C. J., Siegmund, I., and Hennig, W.,** Visualization of a lampbrush loop-forming fertility gene in *Drosophila hydei, Chromosoma,* 88, 50, 1983.

37. **deLoos, F., Dijkhof, R., Grond, C. J., and Hennig, W.,** Lampbrush loop-specificity of transcript morphology in spermatocyte nuclei of *Drosophila hydei, EMBO J.,* 3, 2845, 1984.

38. **Hulsebos, T. J. M., Hackstein, J. H. P., and Hennig, W.,** Lampbrush loop-specific protein of *Drosophila hydei, Proc. Natl. Acad. Sci. U.S.A.,* 81, 3404, 1984.

39. **Huijser, P. and Hennig, W.,** Ribosomal DNA-related sequences in a Y chromosomal lampbrush loop of *Drosophila hydei, Mol. Gen. Genet.,* 260, 441, 1987.

40. **Hennig, W.,** Ribonucleic acid synthesis of the Y chromosome of *Drosophila hydei, J. Mol. Biol.,* 38, 227, 1968.

41. **Hennig, W., Meyer, G. F., Hennig, I., and Leoncini, O.,** Structure and function of the Y chromosome of *Drosophila hydei, Cold Spring Harbor Symp. Quant. Biol.,* 38, 673, 1974.

42. **Hennig, W.,** Highly repetitive DNA sequences in the genome of *Drosophila hydei.* I. Preferential localization in the X chromosome heterochromatin, *J. Mol. Biol.,* 71, 407, 1972.

43. **Renkawitz, R.,** Isolation of twelve satellite DNAs from *Drosophila hydei, Int. J. Biol. Macromol.,* 1, 133, 1979.

44. **Lifschytz, E.,** A procedure for the cloning and identification of Y-specific middle repetitive sequences in *Drosophila hydei, J. Mol. Biol.,* 133, 267, 1979.

45. **Vogt, P. and Hennig, W.,** Y chromosomal DNA of *Drosophila hydei, J. Mol. Biol.,* 167, 37, 1983.

46. **Awgulewitsch, A. and Bünemann, H.,** Isolation of Y-chromosomal repetitive DNA sequences of *Drosophila hydei* via enrichment of chromosome-specific sequences by heterogeneous hybridization between female and male DNA, *J. Biochem. Biophys. Methods,* 12, 37, 1986.

47. **Hennig, W., Huijser, P., Vogt, P., Jäckle, H., and Edström, J.-E.,** Molecular cloning of microdissected lampbrush loop DNA sequences of *Drosophila hydei, EMBO J.,* 2, 1741, 1983.

48. **Huijser, P., Hennig, W., and Dijkhof, R.,** Poly(dC-dA/dG-dT) repeats in the *Drosophila* genome: a key function for dosage compensation and position effects? *Chromosoma,* 95, 209, 1987.

49. **Huijser, P. Kirchhoff, C., Lankenau, D.-H., and Hennig, W.,** Retrotransposon-like sequences are expressed in Y chromosomal lampbrush loops of *Drosophila hydei, J. Mol. Biol.,* 203, 689, 1988.

50. **Vogt, P. and Hennig, W.,** Molecular structure of the lampbrush loop nooses of the Y chromosome of *Drosophila hydei.* I. The Y chromosome-specific repetitive DNA sequence family ay1 is dispersed in the loop DNA, *Chromosoma,* 94, 449, 1986.

51. **Vogt, P. and Hennig, W.,** Molecular structure of the lampbrush loop nooses of the Y chromosome of *Drosophila hydei.* II. DNA sequences with homologies to multiple genomic locations are a major constituent of the loop, *Chromsoma,* 94, 459, 1986.

52. **Wlaschek, M., Awgulewitsch, A., and Bünemann, H.,** Structure and function of Y chromosomal DNA. I. Sequence organization and localization of four families of repetitive DNA on the Y chromosome of *Drosophila hydei, Chromosoma,* 96, 145, 1988.

53. **Trapitz, P., Wlaschek, M., and Bünemann, H.,** Structure and function of Y chromosomal DNA. II. Analysis of lampbrush loop associated transcripts in nuclei of primary spermatocytes of *Drosophila hydei* by in situ hybridization, *Chromosoma,* 96, 159, 1988.

54. **Meyer, G. F. and Hennig, W.,** The nucleolus in primary spermatocytes of *Drosophila hydei, Chromosoma,* 46, 121, 1974.

55. **Hennig, W., Vogt, P., Jacob, G., and Siegmund, I.,** Nucleolus organizer regions in *Drosophila* species of the repleta group, *Chromosoma,* 87, 279, 1982.

56. **Miller, O. L., Jr.,** The nucleolus, chromosomes and visualization of genetic activity, *J. Cell Biol.,* 91, 155, 1981.

57. **Vogt, P., Hennig, W., and Siegmund, I.,** Identification of transcribed cloned Y chromosomal DNA sequences from a lampbrush loop of *Drosophila hydei, Proc. Natl. Acad. Sci. U.S.A.,* 79, 5132, 1982.

58. **Lifschytz, E., Hareven, D., Azriel, A., and Brodsly, H.,** DNA clones and RNA transcripts of four lampbrush loops from the Y chromosome of *Drosophila hydei, Cell,* 32, 191, 1983.

59. **Vogt, P., Hennig, W., tenHacken, D., and Verbost, P.,** Evolution of Y chromosomal lampbrush loop DNA sequences of *Drosophila, Chromosoma,* 94, 458, 1986.

60. **Hennig, I.,** Vergleichend-cytologische und -genetische Untersuchungen am Genom der Fruchtfliegen-Arten *Drosophila hydei, neohydei* und *eohydei* (Diptera: Drosophilidae), *Entomol. Ber. (Berlin),* 4, 211, 1978.

61. **Fawcett, D. H., Lister, C. K., Kellet, E., and Finnegan, D. J.,** Transposable elements in eukaryotes, *Int. Rev. Cytol.,* 93, 281, 1985.

62. **Lankenau, D.-H., Huijser, P., Miedema, K., Janssen, E., and Hennig, W.,** Micropia: a retrotransposon of *Drosophila* combines structural features of DNA viruses, retroviruses, and non-viral transposable elements, *J. Mol. Biol.,* 204, 233, 1989.

63. **Hareven, D., Zuckerman, M., and Lifschytz, E.,** Origin and evolution of the transcribed repeated sequences of the Y chromosome lampbrush loops of *Drosophila hydei, Proc. Natl. Acad. Sci. U.S.A.,* 83, 125, 1986.

64. **Meyer, G. F. and Hennig, W.,** Molecular aspects of the fertility factors in *Drosophila,* in *The Functional Anatomy of the Spermatozoon,* Afzelius, B. A., Ed., Pergamon Press, Oxford, 1974, 69.

65. **Glätzer, K. H. and Meyer, G. F.,** Morphological aspects of the genetic activity in primary spermatocyte nuclei of *Drosophila hydei, Biol. Cell,* 41, 165, 1981.

66. **Glätzer, K. H. and Kloetzel, P. M.,** Preservation of nuclear RNP antigens in male germ cell development of *Drosophila hydei, Mol. Gen. Genet.,* 196, 236, 1984.

67. **Sommerville, J.,** Immunolocalization and structural organization of nascent RNP, in *The Cell Nucleus,* Vol. 8, Busch, H., Ed., Academic Press, New York, 1988, 1.

68. **Sommerville, J., Crichton, C., and Malcolm, D.,** Immunofluorescent localization of transcriptional activity on lampbrush chromosomes, *Chromosoma,* 66, 99, 1978.

69. **Roth, M. B. and Gall, J. G.,** Monoclonal antibodies that recognize transcription unit proteins on newt lampbrush chromosomes, *J. Cell. Biol.,* 105, 1047, 1987.

70. **Ragghianti, M., Bucci, S., Mancino, G., Lacroix, J. C., Boucher, D., and Charlemagne, J. A.,** Novel approach to cytotaxonomic studies in the genus *Triturus* by monoclonal antibodies to lampbrush chromosome antigens, *Chromosoma,* 97, 134, 1988.

71. **Pardue, M. L., Bendena, W. G., and Garbe, J. C.,** Heat shock: puffs and response to environmental stress, in *Results and Problems in Cell Differentiation,* Vol. 14, *Structure and Function of Eukaryotic Chromosomes,* Hennig, W., Ed., Springer-Verlag, Berlin, 1984, 121.

72. **Ingman-Baker, J. and Candido, E. P. M.,** Proteins of the *Drosophila melanogaster* reproductive system: two-dimensional gel patterns of proteins synthesized in XO, XY, and XYY testis and paragonial gland and evidence that the Y chromosome does not code for structural sperm proteins, *Biochem. Genet.,* 18, 809, 1980.

73. **Hulsebos, T. J. M., Hackstein, J. H. P., and Hennig, W.,** Involvement of Y chromosomal loci in the synthesis of *Drosophila hydei* sperm proteins, *Dev. Biol.,* 100, 238, 1983.

74. **Livak, K. J.,** Organization and mapping of a sequence on the *Drosophila melanogaster* X and Y chromosomes that is transcribed during spermatogenesis, *Genetics,* 107, 611, 1984.

75. **Lovett, J., Kaufman, T. C., and Mahowald, A. P.,** A locus on the X chromosome apparently controlled by the Y chromosome during spermatogenesis in *Drosophila melanogaster, Eur. J. Cell. Biol.,* 22, 49, 1980.

76. **Kiefer, B. I.,** Genetics of sperm development in *Drosophila,* in *Genetic Mechanisms in Development,* Ruddle, E. H., Ed., Academic Press, London, 1973, 47.

77. **Hennig, W., Hennig, I., and Leoncini, O.,** Some observations on spermatogenesis of *Drosophila hydei, Drosophila Inf. Serv.,* 51, 127, 1974.

78. **Grond, C.,** Spermatogenesis in *Drosophila hydei,* Ph.D. thesis, Universiteit Nijmegen, The Netherlands, 1984.

79. **Hess, O.,** Complementation of genetic activity in translocated fragments of the Y chromosome in *Drosophila hydei, Genetics,* 56, 283, 1967.

80. **Frasch, M.,** Charakterisierung chromatinassoziierter Kernproteine von *Drosophila melanogaster* mit Hilfe monoklonaler Anikörper, Ph.D. thesis, University of Tübingen, 1985.

81. **Hauschteck-Jungen, E. and Hartl, D. L.,** Defective histone transition during spermiogenesis in heterozygous segregation distorter males of *Drosophila melanogaster, Genetics,* 101, 57, 1982.

82. **Huijser, P., Beckers, L., Top, B., Hermans, M., Sinke, R., and Hennig, W.,** Poly[d(C-A)]· poly[d(G-T)] is highly transcribed in testes of *Drosophila hydei,* submitted.

83. **Hackstein, J. H. P., Hennig, W., and Steinmann-Zwicky, M.,** Autosomal control of lampbrush-loop formation during spermatogenesis in *Drosophila hydei* by a gene also affecting somatic sex determination, *Wilhelm Roux Arch. Dev. Biol.,* 196, 119, 1987.

84. **Smith, K. D., Young, K. E., Talbot, C. C., Jr., and Schmeckpeper, B.,** Repeated DNA of the human Y chromosome, *Dev. Suppl.,* 101, 77, 1987.

85. **Page, D. C.,** Sex reversal: deletion mapping the male-determining function of the human Y chromosome, *Cold Spring Harbor Symp. Quant. Biol.,* 51, 229, 1986.

86. **Page, D. C., Mosher, R., Simpson, E. M., Fisher, E. M. C., Mardon, G., Pollack, J., McGillivray, B., de la Chapelle, A., and Brown, L. G.,** The sex-determining region of the human Y chromosome encodes a finger protein, *Cell,* 51, 1091, 1987.

87. **Hennig, W., Dijkhof, R., and Pötgens, A.,** unpublished data.

88. **Brand, R. C. et al.,** in preparation.

89. **Kremer, H.,** unpublished observations.

90. **Hochstenbach, R.,** unpublished data.

Index

INDEX